T0206012

An Elementary Course on Partial Differential Equations

Differential equations are very important due to the extensive applications in various fields such as pure and applied mathematics, physics, engineering, biology, and economics. At present differential equations form the foundation of mathematical modelling applied to solving real-life problems that may not be solved directly. This subject is a part of higher mathematics curriculum and taught at undergraduate and postgraduate levels. Differential equations are taught as two courses, namely ordinary differential equations and partial differential equations.

This book is designed to serve as a textbook for the first course on partial differential equations, which is often taught after a first course on ordinary differential equations. There are numerous books on this subject but most such standard books run very quickly through the elementary partial differential equations, which is not sufficient for undergraduate students. The authors have utilized their teaching experience of several years to fill this gap wherein special attention is paid to elementary partial differential equations. This book covers important techniques such as Lagrange's method, Charpit's method, Jacobi's method, Monge's method, Monge–Ampere type non-linear equations, Fourier method, reduction of first and second order linear equations in canonical forms, derivation of first and second order equations, higher-order equations with constants coefficients, and their reduction to variable coefficients, and so on.

Aftab Alam is Assistant Professor in the Department of Mathematics, Aligarh Muslim University, India. His domain of research interests includes non-linear analysis, general topology, operator theory, and differential equations. He has published a number of research papers in SCI listed journals of international repute.

Mohammad Imdad is Professor (retired) and the former Chairman of the Department of Mathematics, Aligarh Muslim University, India. His research interests are ordered metric spaces, non-linear functional analysis (fixed point theory), best approximation, fuzzy spaces, and complex-valued metric spaces.

AN ELEMENTARY COURSE ON PARTIAL DIFFERENTIAL EQUATIONS

Aftab Alam
Mohammad Imdad

CAMBRIDGE
UNIVERSITY PRESS

CAMBRIDGE
UNIVERSITY PRESS

Shaftesbury Road, Cambridge CB2 8EA, United Kingdom

One Liberty Plaza, 20th Floor, New York, NY 10006, USA

477 Williamstown Road, Port Melbourne, VIC 3207, Australia

314–321, 3rd Floor, Plot 3, Splendor Forum, Jasola District Centre, New Delhi – 110025, India

103 Penang Road, #05–06/07, Visioncrest Commercial, Singapore 238467

Cambridge University Press is part of Cambridge University Press & Assessment,
a department of the University of Cambridge.

We share the University's mission to contribute to society through the pursuit of
education, learning and research at the highest international levels of excellence.

www.cambridge.org
Information on this title: www.cambridge.org/9781009201445

First published 2022

A catalogue record for this publication is available from the British Library

ISBN   978-1-009-20144-5   Paperback

To our parents and teachers

Contents

Preface

This book is designed to serve as a textbook for the first course on partial differential equations. After the introduction of the semester system in various universities of India, normally there are two courses of differential equations at undergraduate levels, one is on ordinary differential equations (ODEs), whereas the other is exclusively on partial differential equations (PDEs). Generally, most of the books used at the undergraduate level include ODEs as well as PDEs, and all such books emphasise more on ODEs, which, consequently, affect the coverage of the PDEs. On the other hand, all the standard textbooks on PDEs cover the contents of graduate as well as postgraduate levels, wherein often undergraduate materials are discussed in haste. Overall, it is necessary to write a separate book exclusively for undergraduate students of Indian as well as foreign universities. With a view to have a detailed discussion on elementary topics of PDEs, we endeavour to write this book.

In fact, this book is based on our lecture notes, which we have given to our students at Aligarh Muslim University (AMU) over the last several years. We hope that this book will meet the requirement as well as expectations of undergraduate students. While preparing this book, we have examined deeply the syllabi of all such courses of B.Sc. (Mathematics) of all Indian universities. The course contents covered by this book can be described as the union of all syllabi prescribed by various Indian universities as well as UGC curriculum up to undergraduate level. The course contents are presented in such a manner that they can be equally useful in various competitive examinations such as UPSC, UGC-CSIR NET, and GATE. On the other hand, from the application point of view, the book also contains the relevant contents of Mathematical Physics/Engineering Mathematics/Applied Mathematics, which are the parts of course contents in B.Sc. (Physics) as well as B.E./B.Tech. We attempt to strike a balance between theory and problems. Consequently, our book remains equally useful to both pure as well as applied mathematicians. We have made every attempt to have a simpler and lucid presentation without scarifying theoretical rigour. The book provides different types of examples, updated references, and applications in diverse fields. All methods of solution and necessary concepts are arranged in the form of theorems. The proofs of theorems may be omitted for an undergraduate course or a course in other disciplines.

This book consists of six chapters organised in a natural order. The topics are discussed in a systematic way. The aim of the first chapter is an attempt to make this book as self-contained as possible, wherein we have collected the necessary background material from ODEs, multivariable calculus, and geometry required for the subsequent chapters. Chapter 2 deals with basic concepts and necessary notions regarding PDEs and their originations. The object of Chapter 3 is to study some basic techniques to solve certain types of PDEs. Chapters 4, 5, and 6 are devoted, respectively, to first-, second-, and higher order PDEs.

All suggestions and constrictive criticisms for further improvement of the book will be received thankfully by us (at emails aafu.amu@gmail.com and mhimdad@gmail.com) to serve the cause of imparting correct and valuable information to learners.

Aftab Alam
Mohammad Imdad

Acknowledgements

The first author owes a special debt to Professor M. S. Jamil Asghar, who advised him to write this book and constantly encouraged him at every stage of the project until it was completed. The second author is very grateful to all his former teachers, but more particularly to Professor M. Saeed Khan, Professor Neyamat Zaheer, (late) Professor M. Muhsin, (late) Professor Syed Izhar Hussein, (late) Professor M. Zubair Khan, and similar others for their academic inspirations and encouragements in the beginning of his academic career. We avail this opportunity to express our thanks to our colleagues, namely, Professor Shahid Ali, Professor Viqar Azam Khan, Professor M. Kalimuddin Ahmad, Dr Rafiquddin, Dr Qamrul Haq Khan, Dr Javid Ali, Dr Abdur Raheem Khan, and Dr M. Arif, who read carefully different parts of the book and gave valuable suggestions and comments. We are also thankful to our several of our students, namely, Dr M. Asim, Mr M. Hasanuzzaman, Mr Faruk SK, Mr M. Yasir Arafat, Mr M. Jubair, Mr Based Ali, Mr Asik Hossain, and Mr Inzamamul Haq, who spared their valuable time to carry out critical readings of the entire manuscript and pointed out many typos and egregious errors. We are indebted to our friend Mr Adil Paddar, who helped us to draw all the figures in this book. We are also thankful to eight anonymous learned referees for their fruitful and constructive comments on the earlier version of the book.

At this moment of accomplishment, we would like to offer thanks to a large number of authors who have contributed to this subject in the past. We have never attempted to attribute various problems and their solutions to their original discoverers. However, we avail this opportunity to express our heartfelt gratitude to all of them.

It is our pleasure to thank the publisher, Cambridge University Press, for showing keen interest in getting this book published for the benefit of students and workers in the field. We express our extreme gratitude and deepest appreciation to especially Dr Vaishali Thapliyal, Senior Commissioning Editor, Cambridge University Press, for helping us in formatting and editing the entire book. Her constant persuasion enabled us to complete the book within the stipulated time. It is our pleasure to acknowledge the sincere efforts of Ms Qudsiya Ahmad, Head of Publishing-Academic, Cambridge University Press, India, for her excellent cooperation in bringing out this book both at the editorial and production stages.

Last but not least, we are sincerely thankful to our family members, who stood with us until the completion of this project.

Aftab Alam
Mohammad Imdad

Symbols

NOTATIONS

$\neq$	non-equality
$:=$	equal by definition
$\equiv$	equivalence
$\varnothing$	empty set
$\mathbb{N}$	the set of natural numbers
$\mathbb{R}$	the set of real numbers
$\mathbb{R}^n$	n-dimensional Euclidean space
$\in$	belongs to
$\notin$	does not belong to
$\exists$	there exists
$\Rightarrow$	implies
$\Leftrightarrow$	logical equivalence

ACRONYMS

iff	if and only if
w.r.t.	with respect to
Eq., Eqs	equation, equations
PDE, PDEs	partial differential equation/equations
ODE, ODEs	ordinary differential equation/equations
BC	boundary condition
IC	initial condition
BVP	boundary value problem
IVP	initial value problem
IBVP	initial-boundary value problem
LHS	left-hand side
RHS	right-hand side
AE	auxiliary equation
CF	complementary function
PI	particular integral
IF	integrating factor

Relevant Pre-requisites and Terminologies

Before embarking on the formal study of partial differential equations (PDEs), it is essential to equip ourselves with background materials from multivariable calculus, geometry of curves and surfaces, and ordinary differential equations (ODEs; including total and simultaneous total differential equations). Selected ideas from earlier mentioned courses play an important role in the study of PDEs. For the sake of completeness and to make our text self-contained, we briefly discuss the preliminaries as indicated earlier, which are needed in the subsequent chapters. The approach adopted in this chapter is somewhat different from the one used in the rest of the chapters. This chapter is descriptive in nature, wherein the utilised arguments are intended towards plausibility and understanding rather than adopting traditional rigorous ways. The preliminaries are divided into five sections.

1.1 Partial Derivatives and Allied Topics

PDEs involve at least two independent variables. Consequently, the tools of the calculus of several variables are instrumentals in PDEs. In the following lines, we recall some relevant notions and terminologies from differential calculus of two and three variables.

Vectors: An element of Euclidean space $\mathbb{R}^n$ is an n-tuple of the form $\mathbf{x} = (x_1, x_2, \ldots, x_n)$, which is called an n-vector or simply, a vector. The real numbers $x_1, x_2, \ldots, x_n$ are called components of this element. The sum and scalar multiplication in $\mathbb{R}^n$ are defined as component-wise sum and component-wise scalar multiplication. The dot product of two elements $\mathbf{x} = (x_1, x_2, \ldots, x_n)$ and $\mathbf{y} = (y_1, y_2, \ldots, y_n)$ is defined as

$$\mathbf{x} \cdot \mathbf{y} = x_1 y_1 + x_2 y_2 + \cdots + x_n y_n.$$

The *norm* or *length* of a vector $\mathbf{x} = (x_1, x_2, \ldots, x_n)$ is defined as

$$\left\| \mathbf{x} \right\| = \sqrt{x_1^2 + x_2^2 + \cdots + x_n^2}.$$

Thus, $\mathbf{x} \cdot \mathbf{x} = \left\| \mathbf{x} \right\|^2$. A vector of length 1 is called a *unit vector*.

Standard Basis Vectors: The two vectors $e_1 = (1, 0)$ and $e_2 = (0, 1)$ are called *standard basis vectors of* $\mathbb{R}^2$. Any vector $u = (a, b) \in \mathbb{R}^2$ can be written as a linear combination of e_1 and e_2, so that

$$u = ae_1 + be_2.$$

Similarly, $e_1 = (1, 0, 0)$, $e_2 = (0, 1, 0)$, and $e_3 = (0, 0, 1)$ are called *standard basis vectors of* $\mathbb{R}^3$.

Neighbourhoods: The δ-neighbourhood of a point $(x_0, y_0) \in \mathbb{R}^2$ is an open sphere centred at (x_0, y_0) with radius $\delta > 0$, which can be represented by

$$N_\delta(x_0, y_0) = \{(x, y) \in \mathbb{R}^2 : \sqrt{(x - x_0)^2 + (y - y_0)^2} < \delta\}.$$

Regions: Let R be a region (subset) of $\mathbb{R}^2$ and (x_0, y_0) a point in R. We say that (x_0, y_0) is an *interior point* of R if there exists a δ-neighbourhood of (x_0, y_0), which lies entirely in R. If every point in R is an interior point, then R is called an *open region*. We say that (x_0, y_0) is a *boundary point* of R if every δ-neighbourhood of (x_0, y_0) contains points that lie outside of R as well as points that lie in R. If R contains all its boundary points, then R is called a *closed region*. The set of all interior points of R is called the *interior* of R and usually it is denoted by R°. Similarly, the set of all boundary points of R is called the *boundary* of R and often it is denoted by ∂R. The *closure* of R, often denoted by $\overline{R}$, remains the union of all boundary points of R with R, that is, $\overline{R} = R \cup \partial R$.

Partial Derivatives: Let $z = f(x, y)$ be a real-valued function of two variables. The partial derivative of z w.r.t. x, denoted by $\dfrac{\partial z}{\partial x}$ or $\dfrac{\partial f}{\partial x}$ or f_x, is defined as

$$\frac{\partial f}{\partial x} = \lim_{h \to 0} \frac{f(x + h, y) - f(x, y)}{h}$$

provided the limit exists. Similarly, the partial derivative of z w.r.t. y, denoted by $\dfrac{\partial z}{\partial y}$ or $\dfrac{\partial f}{\partial y}$ or f_y, is defined as

$$\frac{\partial f}{\partial y} = \lim_{k \to 0} \frac{f(x, y + k) - f(x, y)}{k}$$

provided the limit exists. The partial derivatives at a particular point (x_0, y_0) are often denoted by

$$\frac{\partial f}{\partial x}(x_0, y_0) \quad \text{or} \quad \left. \frac{\partial f}{\partial x} \right|_{(x_0, y_0)} \quad \text{or} \quad f_x(x_0, y_0)$$

and

$$\frac{\partial f}{\partial y}(x_0, y_0) \quad \text{or} \quad \left. \frac{\partial f}{\partial y} \right|_{(x_0, y_0)} \quad \text{or} \quad f_y(x_0, y_0).$$

Thus, we have

$$\frac{\partial f}{\partial x}(x_0, y_0) = \lim_{h \to 0} \frac{f(x_0 + h, y_0) - f(x_0, y_0)}{h}$$

and

$$\frac{\partial f}{\partial y}(x_0, y_0) = \lim_{k \to 0} \frac{f(x_0, y_0 + k) - f(x_0, y_0)}{k}.$$

In practice, f_x can be computed as ordinary derivative of f w.r.t. x by treating y as a constant. Similarly, f_y is determined as an ordinary derivative of f w.r.t. y by treating x as a constant.

Notice that f_x and f_y are called *first-order partial derivatives*. In the similar manner, we can define partial derivatives for a function of more than two variables. Thus far, a function $u = f(x, y, z)$ possesses three first-order partial derivatives, namely, u_x, u_y, and u_z.

Higher Order Derivatives: As the partial derivatives are themselves functions, we can take their partial derivatives to obtain higher order derivatives. Higher order derivatives are denoted by the order in which the differentiation occurs. For instance, the function $z = f(x, y)$ admits the following four different second-order partial derivatives.

(i) Differentiate twice w.r.t. x:

$$f_{xx} = \frac{\partial^2 f}{\partial x^2} = \frac{\partial}{\partial x} \left(\frac{\partial f}{\partial x} \right).$$

(ii) Differentiate twice w.r.t. y:

$$f_{yy} = \frac{\partial^2 f}{\partial y^2} = \frac{\partial}{\partial y} \left(\frac{\partial f}{\partial y} \right).$$

(iii) Differentiate first w.r.t. x and then w.r.t. y:

$$f_{xy} = (f_x)_y = \frac{\partial}{\partial y}(f_x) = \frac{\partial}{\partial y} \left(\frac{\partial f}{\partial x} \right) = \frac{\partial^2 f}{\partial y \partial x}.$$

(iv) Differentiate first w.r.t. y and then w.r.t. x:

$$f_{yx} = (f_y)_x = \frac{\partial}{\partial x}(f_y) = \frac{\partial}{\partial x} \left(\frac{\partial f}{\partial y} \right) = \frac{\partial^2 f}{\partial x \partial y}.$$

The last two derivatives f_{xy} and f_{yx} are called *mixed partial derivatives*. Notice that the mixed partial derivatives are not necessarily equal.

Homogeneous Functions: A function $z = f(x, y)$ is said to be a *homogenous function* of degree n if for every positive real number λ, it satisfies

$$f(\lambda x, \lambda y) = \lambda^n f(x, y).$$

Euler's Theorem: Let $f(x, y)$ be a continuously differentiable and homogenous function of degree n. Then

$$x\frac{\partial f}{\partial x} + y\frac{\partial f}{\partial y} = nf.$$

Moreover, if $f(x, y)$ has continuous second-order partial derivatives, then

$$x^2\frac{\partial^2 f}{\partial x^2} + 2xy\frac{\partial^2 f}{\partial x \partial y} + y^2\frac{\partial^2 f}{\partial y^2} = n(n - 1)f.$$

Directional Derivatives: Let $z = f(x, y)$ be a function and $u = (\alpha, \beta)$ a unit vector. Then, the *directional derivative* of f at a point (x_0, y_0) in the direction of u, denoted by $D_u f(x_0, y_0)$, is defined as

$$D_u f(x_0, y_0) = \lim_{t \to 0} \frac{f(x_0 + t\alpha, y_0 + t\beta) - f(x_0, y_0)}{t}$$

provided the limit exists. Here, u is called the *direction vector.*

Clearly, $D_{e_1} f(x_0, y_0) = f_x(x_0, y_0)$ and $D_{e_2} f(x_0, y_0) = f_y(x_0, y_0)$, that is, the directional derivatives in the directions of standard basis vectors coincide with partial derivatives. In other words, the partial derivatives are directional derivatives along coordinate axes.

Similarly, we can define the directional derivatives of $u = f(x, y, z)$ at a point (x_0, y_0, z_0) in the direction of unit vector $u = (\alpha, \beta, \gamma)$ as

$$D_u f(x_0, y_0, z_0) = \lim_{t \to 0} \frac{f(x_0 + t\alpha, y_0 + t\beta, z_0 + t\gamma) - f(x_0, y_0, z_0)}{t}.$$

Vector Fields: A *vector field in plane* is a function $\mathbf{F} : \mathbb{R}^2 \to \mathbb{R}^2$. Thus, we write

$$\mathbf{F}(x, y) = (f_1(x, y), f_2(x, y))$$

wherein f_1 and f_2 are real-valued functions of two variables and are called the components of $\mathbf{F}$.

Similarly, a *vector field in space* is a function $\mathbf{F} : \mathbb{R}^3 \to \mathbb{R}^3$ so that

$$\mathbf{F}(x, y, z) = (f_1(x, y, z), f_2(x, y, z), f_3(x, y, z)).$$

Gradient: Let $z = f(x, y)$ be a function such that f_x and f_y exist. Then the *gradient* of f, denoted $\nabla f(x, y)$, is a vector field defined as

$$\nabla f(x, y) = (f_x(x, y), f_y(x, y)).$$

Thus far, the gradient of f at a particular point (x_0, y_0) is

$$\nabla f(x_0, y_0) = (f_x(x_0, y_0), f_y(x_0, y_0)).$$

Similarly, the gradient of $u = f(x, y, z)$ is defined as

$$\nabla f(x, y, z) = (f_x(x, y, z), f_y(x, y, z), f_z(x, y, z)).$$

Fact: The directional derivative is the dot product of the gradient and direction vector, that is,

$$D_u f = \nabla f \cdot u.$$

Total Differentials: Let $z = f(x, y)$ be a function such that f_x and f_y exist. If dx and dy define the differentials of independent variables, then the *total differential dz* is defined as

$$dz = f_x(x, y)dx + f_y(x, y)dy.$$

Chain Rules: Let $z = f(x, y)$ be a differentiable function of two variables. If x and y are differentiable functions of one independent variable 't', that is, $x = \phi(t)$ and $y = \psi(t)$, then $z = f(\phi(t), \psi(t))$ is a (composite) function of t. Thus, the chain rule states that

$$\frac{dz}{dt} = \frac{\partial z}{\partial x}\frac{dx}{dt} + \frac{\partial z}{\partial y}\frac{dy}{dt}.$$

We now consider the situation where the function $z = f(x, y)$ is differentiable but each of x and y is a differentiable function of two variables 's' and 't', that is, $x = \phi(s, t)$ and $y = \psi(s, t)$. Then, z is indirectly a function of s and t and hence the chain rule states that

$$\frac{\partial z}{\partial s} = \frac{\partial z}{\partial x}\frac{\partial x}{\partial s} + \frac{\partial z}{\partial y}\frac{\partial y}{\partial s}$$

and

$$\frac{\partial z}{\partial t} = \frac{\partial z}{\partial x}\frac{\partial x}{\partial t} + \frac{\partial z}{\partial y}\frac{\partial y}{\partial t}.$$

Jacobian: Let $u = u(x, y)$ and $v = v(x, y)$ be two differentiable functions of two independent variables x and y. The *Jacobian* of u and v w.r.t. x and y is the second-order functional determinant defined by

$$\frac{\partial(u, v)}{\partial(x, y)} = \begin{vmatrix} \dfrac{\partial u}{\partial x} & \dfrac{\partial u}{\partial y} \\ \dfrac{\partial v}{\partial x} & \dfrac{\partial v}{\partial y} \end{vmatrix}.$$

Functional Dependence: Two functions $u = u(x, y)$ and $v = v(x, y)$ defined on a set $D \subset \mathbb{R}^2$ are said to be *functionally dependent* or simply, *dependent* if there exists a real-valued continuously differentiable function $F(u, v)$ such that for all $(x, y) \in D$, we have

$$F(u(x, y), v(x, y)) = 0 \quad \text{and} \quad \nabla F(u(x, y), v(x, y)) \neq 0.$$

Two functions are called *functionally independent* or *independent* if they are not dependent.

Fact: Two functions $u = u(x, y)$ and $v = v(x, y)$ are independent iff

$$\frac{\partial(u, v)}{\partial(x, y)} \neq 0.$$

Implicit Functions: Let x and y be two variables, which are related by a functional equation of the form $F(x, y) = 0$. A function $y = f(x)$ is called *implicit function* defined by $F(x, y) = 0$ if $F(x, f(x)) = 0$ is satisfied for all x in the domain of f. The equation $F(x, y) = 0$ is called implicit form of $y = f(x)$. The well-known *Implicit Function Theorem* provides the conditions under which $F(x, y) = 0$ defines y implicitly as a function of x. The derivative of $y = f(x)$ can be determined as follows:

$$\frac{dy}{dx} = -\frac{F_x}{F_y}$$

whenever $F_y \neq 0$.

Inverse Function Theorem: Let $\mathbf{T} : \mathbb{R}^2 \to \mathbb{R}^2$ be a continuously differentiable function, which transforms the variables (x, y) to the variables (u, v) so that $u = u(x, y)$ and $v = v(x, y)$. If (x_0, y_0) is a point in the domain of $\mathbf{T}$ such that

$$J\mathbf{T}(x_0, y_0) := \left. \frac{\partial(u, v)}{\partial(x, y)} \right|_{(x_0, y_0)} \neq 0$$

and $(u_0, v_0) = \mathbf{T}(x_0, y_0)$, then there exist neighbourhoods N_0 of (x_0, y_0) and N_0' of (u_0, v_0) such that the restriction $\mathbf{T}|_{N_0}$ has an inverse $\mathbf{T}^{-1} : N_0' \to N_0$, which is differentiable at (u_0, v_0) and $J\mathbf{T}^{-1}(u_0, v_0) = [J\mathbf{T}(x_0, y_0)]^{-1}$.

Invertible Transformation: In lieu of inverse function theorem, a coordinate transformation for which the Jacobian $J \neq 0$ is called an *invertible transformation* or *non-singular transformation*.

1.2 Curves and Surfaces

The integrals of PDEs are surfaces, while the particular solution of an initial value problem remains a surface passing through a given curve. For the appreciation of the methods of solutions and for the interpretation of the solutions of PDEs, a study of curves and surfaces is essential. Henceforth, in the following lines, we briefly discuss the relevant aspects of theory of curves in $\mathbb{R}^2$ and $\mathbb{R}^3$ and surfaces in $\mathbb{R}^3$.

Direction Ratios: If a line in space makes the angles α, β, γ with the axes of x, y, z, respectively, then the quantities $l = \cos \alpha, m = \cos \beta, n = \cos \gamma$ are called the *direction cosines* of the line. Further, the numbers a, b, c proportional to directions cosines, so that $a = \lambda l, b = \lambda m, c = \lambda n$ for any arbitrary real number λ, are called the *direction ratios* of the line.

Plane Curves: A functional equation involving two variables x and y of the form

$$F(x, y) = 0$$

represents a *plane curve*. The explicit representation of the plane curve is defined by a function of single variable of the form

$$y = f(x).$$

Straight line, circle, ellipse, parabola, and hyperbola are natural examples of plane curves.

Surfaces: A functional equation involving three variables x, y, and z of the form

$$F(x, y, z) = 0 \tag{1.1}$$

represents a *surface*. The explicit representation of the surface is defined by a function of two variables of the form

$$z = f(x, y).$$

Plane: A surface is called a *plane* if it can be represented by the general equation of first degree in x, y, z, that is, the equation of the form

$$ax + by + cz + d = 0. \tag{1.2}$$

Every plane determines a (unique) perpendicular line through the origin to the plane, which is called the *normal* of the plane. It can be pointed out that the coefficients a, b, c in Eq. (1.2) represent the direction ratios of the normal $\mathbf{n}$ of the plane.

If a plane represented by (1.2) passes through a point $P_0(x_0, y_0, z_0)$, then $d = -ax_0 - by_0 - cz_0$ so that (1.2) reduces to

$$a(x - x_0) + b(y - y_0) + c(z - z_0) = 0$$

which is the equation of the plane passing through P_0 and with the normal having a, b, c as direction ratios.

Conicoid: A surface is called *conicoid* or *quadric* if it can be represented by the general equation of second degree in x, y, z, that is, the equation of the form

$$ax^2 + by^2 + cz^2 + 2fyz + 2gzx + 2hxy + 2ux + 2vy + 2wz + d = 0.$$

Conical Surface: A surface $F(x, y, z) = 0$ is said to be *cone* with vertex $A(x_0, y_0, z_0)$ if for each point $P(x, y, z)$ of the surface and for all $\lambda \in \mathbb{R}$, the point $\lambda P + (1 - \lambda)A$ lies on the surface, that is,

$$F\left(\lambda x + (1 - \lambda)x_0, \lambda y + (1 - \lambda)y_0, \lambda z + (1 - \lambda)z_0\right) = 0.$$

Tangent Plane and Normal to a Surface: Equation of tangent plane to the surface (1.1) at a point $P_0(x_0, y_0, z_0)$ is

$$\frac{\partial F}{\partial x}\bigg|_{P_0} (x - x_0) + \frac{\partial F}{\partial y}\bigg|_{P_0} (y - y_0) + \frac{\partial F}{\partial z}\bigg|_{P_0} (z - z_0) = 0.$$

It follows that the components of the gradient $\nabla F = (F_x, F_y, F_y)$ evaluated at a point represent the direction ratios of the normal of the tangent plane and, hence, to the surface at that point. Thus, the equation of the normal to the surface (1.1) at $P_0(x_0, y_0, z_0)$ is

$$\frac{x - x_0}{\dfrac{\partial F}{\partial x}\bigg|_{P_0}} = \frac{y - y_0}{\dfrac{\partial F}{\partial y}\bigg|_{P_0}} = \frac{z - z_0}{\dfrac{\partial F}{\partial z}\bigg|_{P_0}}.$$

Let us consider the equation of a surface in explicit form:

$$z = f(x, y).$$

Denote

$$p := \frac{\partial z}{\partial x} = \frac{\partial f}{\partial x} \quad \text{and} \quad q := \frac{\partial z}{\partial y} = \frac{\partial f}{\partial y}.$$

If we write $F(x, y, z) \equiv f(x, y) - z = 0$, then we have

$$\frac{\partial F}{\partial x} = \frac{\partial f}{\partial x} = p$$

$$\frac{\partial F}{\partial y} = \frac{\partial f}{\partial y} = q$$

$$\frac{\partial F}{\partial z} = -1$$

which implies that at any point (x, y, z), the normal to the surface $z = f(x, y)$ has direction ratios $p, q, -1$ and the equation of tangent plane to this surface is

$$p(x - x_0) + q(y - y_0) = (z - z_0).$$

Family of Surfaces: A family of surfaces usually means an infinite set of surfaces. In most of the cases, the surfaces are all of the same type, for example, all spheres differ only in size or position. If each member of a family of surfaces is attached with an arbitrary constant a, we may represent the whole family by the single equation

$$F(x, y, z, a) = 0. \tag{1.3}$$

Substituting a particular value of a in (1.3), the equation represents a specific surface that is assigned with this value of a. Equation (1.3) is called *one-parameter family of surfaces* and a is called *parameter* of this family. In the similar manner, we can define *two-parameter family of surfaces*, which depends on two arbitrary constants (parameters) a and b, as follows:

$$F(x, y, z, a, b) = 0. \tag{1.4}$$

Envelopes: A surface is called the *envelope* of a family of surfaces if it is touched by all members of the family. The equation of envelope of the one-parameter family (1.3) (if exists) is obtained by eliminating 'a' between the following two equations:

$$\left. \begin{array}{l} F = 0 \\ \dfrac{\partial F}{\partial a} = 0. \end{array} \right\}$$

Similarly, the equation of envelope of two-parameter family (1.4), if exists, is obtained by eliminating 'a' and 'b' between the following three equations:

$$\left. \begin{array}{l} F = 0 \\ \dfrac{\partial F}{\partial a} = 0 \\ \dfrac{\partial F}{\partial b} = 0. \end{array} \right\}$$

In a manner similar to the family of surfaces, we can define family of curves and their envelope.

Space Curves: A system of two functional equations involving three variables x, y, and z of the form

$$\left. \begin{array}{l} F(x, y, z) = 0 \\ G(x, y, z) = 0 \end{array} \right\}$$

represents a *space curve*. Thus, a space curve can be determined as an intersection of two surfaces.

The following result is used to determine the equation of the surface, which passes through a given space curve.

Theorem 1.1. *Let Γ be a space curve represented by*

$$\left. \begin{array}{l} u(x, y, z) = c_1 \\ v(x, y, z) = c_2. \end{array} \right\}$$

Then the functional relation

$$F(u, v) = 0$$

such that $F(c_1, c_2) = 0$ represents a surface passing through Γ.

Parameterised Curves: A plane curve can be viewed as a mapping $\gamma : I \subset \mathbb{R} \to \mathbb{R}^2$ so that

$$\gamma(t) = (x(t), y(t))$$

where I is an interval and the components $x(t)$ and $y(t)$ of γ are continuous functions of t. The domain I is called the parameter interval and the variable $t \in I$ is called the parameter of the curve. Hence, the equations

$$x = x(t), y = y(t)$$

are called the parametric equations of the curve.

Similarly, a parameterised space curve can be represented by the function $\gamma : I \subset \mathbb{R} \to \mathbb{R}^3$ so that

$$\gamma(t) = (x(t), y(t), z(t))$$

whereas the equations

$$x = x(t), y = y(t), z = z(t)$$

are called the parametric equations of the curve. Here, it can be highlighted that any curve can be represented by different sets of parametric equations. Now, we present examples of several well-known curves in their Cartesian as well as parametric forms.

Example 1.1. *Parametric equations of the **circle** $x^2 + y^2 = a^2$ are*

$$x = a\cos t, \; y = a\sin t, \quad 0 \le t \le 2\pi.$$

Another possible parametric equations of the circle are

$$x = a\sin t, \; y = a\cos t, \quad \frac{\pi}{2} \le t \le \frac{5\pi}{2}.$$

Example 1.2. *Parametric equations of the **ellipse** $\dfrac{x^2}{a^2} + \dfrac{y^2}{b^2} = 1$ are*

$$x = a\cos t, \; y = b\sin t, \quad 0 \le t \le 2\pi$$

or

$$x = a\sin t, \; y = b\cos t, \quad \frac{\pi}{2} \le t \le \frac{5\pi}{2}.$$

Example 1.3. *Parametric equations of the **hyperbola** $\dfrac{x^2}{a^2} - \dfrac{y^2}{b^2} = 1$ are*

$$x = a\sec t, \; y = b\tan t, \quad -\frac{\pi}{2} \le t \le \frac{\pi}{2}$$

or

$$x = a\cosh t, \; y = b\sinh t, \quad -\infty < t < \infty.$$

Example 1.4. *Parametric equations of the* **parabola** $y^2 = 4ax$ *are*

$$x = at^2, \; y = 2at, \quad -\infty < t < \infty.$$

Example 1.5. *The simplest example of a space curve is* **straight line**. *Equation of a straight line passing through a fixed point* (x_0, y_0, z_0) *having the direction ratios* a, b, c *is*

$$\frac{x - x_0}{a} = \frac{y - y_0}{b} = \frac{z - z_0}{c}. \tag{1.5}$$

Equation (1.5) is a system of linear equations, which shows that a straight line in space remains the intersection of two planes. These equations are called symmetric equations of a line. Parametric equations are obtained by equating each ratio of Eq. (1.5) to t, which are given as under

$$x = x_0 + at, \; y = y_0 + bt, \; z = z_0 + ct.$$

Example 1.6. *A* **right circular helix** *is a space curve lying on a circular cylinder. The parametric equations of helix are*

$$x = a\cos \omega t, \; y = a\sin \omega t, \; z = kt$$

where a, ω, and k are constants.

Remark 1.1. *Two lines having direction ratios* a_1, b_1, c_1 *and* a_2, b_2, c_2 *are orthogonal iff*

$$(a_1, b_1, c_1).(a_2, b_2, c_2) = a_1 a_2 + b_1 b_2 + c_1 c_2 = 0.$$

Remark 1.2. *Two planes are orthogonal iff their normals are orthogonal.*

Tangent Vector to a Curve: At each point of space curve $x = x(t), y = y(t), z = z(t)$, the component of the vector $\left(\dfrac{dx}{dt}, \dfrac{dy}{dt}, \dfrac{dz}{dt} \right)$ represents the direction ratios of the tangent to the curve at that point. Due to this fact, this vector is called *tangent vector* of the curve. In the similar manner, we can define the tangent vector to a plane curve.

Parameterised Surfaces: The parametric representation of a surface can be expressed as the function $\sigma : \mathbb{R}^2 \to \mathbb{R}^3$ so that

$$\sigma(s, t) = (x(s, t), y(s, t), z(s, t)).$$

Thus, the equations

$$x = x(s,t), y = y(s,t), z = z(s,t)$$

are called the parametric equations of the surface.

Likewise, in the case of curves, the parametric equations of a surface need not be unique. Now, we give examples of several well-known quadric surfaces in their Cartesian as well as parametric forms.

Example 1.7. *The surface $x^2 + y^2 + z^2 = a^2$ represents a **sphere**. Its parametric equations are*

$$x = a\cos s\cos t, \ y = a\sin s\cos t, \ z = a\sin t.$$

Another parametric equations of sphere are

$$x = \frac{a(1-t^2)}{1+t^2}\cos s, \ y = \frac{a(1-t^2)}{1+t^2}\sin s, \ z = \frac{2at}{1+t^2}.$$

Example 1.8. *The surface $x^2 + y^2 = a^2$ represents a **cylinder**. Its parametric equations are*

$$x = a\cos s, \ y = a\sin s, \ z = t.$$

Example 1.9. *The surface $\dfrac{x^2}{a^2} + \dfrac{y^2}{b^2} + \dfrac{z^2}{c^2} = 1$ represents an **ellipsoid**. Its parametric equations are*

$$x = a\cos s\cos t, \ y = b\sin s\cos t, \ z = c\sin t.$$

Example 1.10. *The surface $\dfrac{x^2}{a^2} + \dfrac{y^2}{b^2} - \dfrac{z^2}{c^2} = 1$ represents a **hyperboloid of one sheet**. Its parametric equations are*

$$x = a\cosh s\cos t, \ y = b\cosh s\sin t, \ z = c\sinh t.$$

Example 1.11. *The surface $\dfrac{x^2}{a^2} - \dfrac{y^2}{b^2} - \dfrac{z^2}{c^2} = 1$ represents a **hyperboloid of two sheets**. Its parametric equations are*

$$x = a\cosh s, \ y = b\sinh s\cos t, \ z = c\sinh s\sin t.$$

Example 1.12. *The surface $\dfrac{x^2}{a^2} + \dfrac{y^2}{b^2} = \dfrac{2z}{c}$ represents an **elliptic paraboloid**. Its parametric equations are*

$$x = as\cos t, \ y = bs\sin t, \ z = \frac{1}{2}cs^2.$$

Example 1.13. *The surface* $\dfrac{x^2}{a^2} - \dfrac{y^2}{b^2} = \dfrac{2z}{c}$ *represents a* **hyperbolic paraboloid**. *Its parametric equations are*

$$x = as \cosh t, \ y = bs \sinh t, \ z = \frac{1}{2}cs^2.$$

Example 1.14. *The surface* $\dfrac{x^2}{a^2} + \dfrac{y^2}{b^2} - \dfrac{z^2}{c^2} = 0$ *represents a* **cone**. *Its parametric equations are*

$$x = as \cos t, \ y = bs \sin t, \ z = cs.$$

1.3 Ordinary Differential Equations

Several notions and methods from ODEs yield natural extensions to the case of PDEs. In this section, we include such instances briefly.

Differential Equations: An equation containing the derivatives or differentials of one or more dependent variables w.r.t. one or more independent variables is called a *differential equation*.

Ordinary Differential Equations: A differential equation in which each involved dependent variable is a function of a single independent variable is known as an *ordinary differential equation* (abbreviated as ODE).

Order and Degree of ODE: The order of an ODE is the order of the highest order derivative appearing in the equation. The degree of an ODE is the power of highest order derivative occurring in the equation, when differential coefficients are made free from radicals and fractions, for example, the ODE $\dfrac{d^3y}{dx^3} - 6x\left(\dfrac{dy}{dx}\right)^2 - 4y = 0$ is of order 3 and degree 1.

Classification of ODE: ODEs are classified into linear and non-linear ODEs. An ODE of order n is called **linear** if it can be expressed as

$$P_0\frac{d^ny}{dx^n} + P_1\frac{d^{n-1}y}{dx^{n-1}} + P_2\frac{d^{n-2}y}{dx^{n-2}} + \ldots + P_{n-1}\frac{dy}{dx} + P_ny = Q \tag{1.6}$$

where $P_0, P_1, \ldots, P_n$ and Q are either constants or functions of the variable x and $P_0 \neq 0$. An ODE that is not linear is called a **non-linear** ODE.

Linear ODEs are further classified into **homogeneous** and **non-homogeneous** equations according to $Q \equiv 0$ and $Q \not\equiv 0$, respectively.

Solution of ODE: A solution (or integral/primitive) of an ODE is an explicit or implicit relation between the involved variables that do not contain derivatives and satisfy the given ODE. Geometrically speaking, the solution of an ODE represents the family of plane curves (containing arbitrary constants). Due to this, the solutions of an ODE are also referred as 'integral curves'.

General Solution: A solution of an ODE that contains a number of arbitrary constants equal to the order of the equation is called the *general solution* or the *complete solution*; for example, $y = a\cos x + b\sin x$ is the general solution of the equation $y'' + y = 0$. Although $y = a\cos x$ and $y = b\sin x$ both satisfy the given equation, yet they are not general solutions as each of them contains only one arbitrary constant.

Particular Solution: In lieu of the concept of general solution, it can be highlighted that a single ODE can possess an infinite number of solutions corresponding to the unlimited number of choices for the arbitrary constants. A solution obtained by giving particular values to arbitrary constants in the general solution is called a *particular solution*.

First-Order Linear ODE: A linear ODE of order one takes the form

$$\frac{dy}{dx} + P(x)y = Q(x).$$

The integrating factor (abbreviated as IF) of this equation is defined as

$$\text{IF} = e^{\int P dx}.$$

The general solution of the given ODE is

$$y \cdot \text{IF} = \int (Q \cdot \text{IF}) dx + c$$

where c is an arbitrary constant.

Higher Order Linear ODE with Constant Coefficients: Consider the homogeneous linear ODE of order n of the form

$$a_n \frac{d^n y}{dx^n} + a_{n-1} \frac{d^{n-1} y}{dx^{n-1}} + \cdots + a_1 \frac{dy}{dx} + a_0 y = 0 \tag{1.7}$$

where the coefficients $a_n, a_{n-1}, \cdots, a_0$ are all constants provided $a_n \neq 0$. Rewrite Eq. (1.7) in symbolic form as

$$P(D)y = 0$$

where $P(D) \equiv \left(a_n D^n + a_{n-1} D^{n-1} + \cdots + a_1 D + a_0 \right)$ is a polynomial in differential operator $D := \frac{d}{dx}$.

Definition 1.1. *An equation of order n in m of the form*

$$a_n m^n + a_{n-1} m^{n-1} + \cdots + a_1 m + a_0 = 0, \qquad a_n \neq 0 \tag{1.8}$$

is called auxiliary equation (abbreviated as AE) of Eq. (1.7).

Clearly, AE is obtained by $P(m) = 0$. Notice that Eq. (1.8) being an algebraic equation in m of degree n has n roots. There are the following three cases:

I: All the roots are real and distinct.

II: All the roots are real but some of them are repeated.

III: Roots are complex numbers.

Case-I: If the roots, say, $m_1, m_2, \ldots, m_n$ of AE, are real and distinct, then the general solution of (1.7) is

$$y = c_1 e^{m_1 x} + c_2 e^{m_2 x} + \cdots + c_n e^{m_n x}$$

where $c_1, c_2, \ldots, c_n$ are constants.

Case-II: If $m_1, m_2, \ldots, m_n$ are roots of AE such that $m_1 = m_2 = \cdots = m_r = m$ (say) and the remaining $(n - r)$ roots are distinct, then the general solution of (1.7) is

$$y = (c_1 + c_2 x + c_3 x^2 + \cdots + c_r x^{r-1}) e^{mx} + c_{r+1} e^{m_{r+1} x} + \cdots + c_n e^{m_n x}.$$

Thus far, in particular, if all n roots are equal, that is, $m_1 = m_2 = \cdots = m_n = m$ (say), then the general solution is

$$y = (c_1 + c_2 x + c_3 x^2 + \cdots + c_n x^{n-1}) e^{mx}.$$

Case-III: If $\alpha \pm i\beta$ are two complex conjugate roots[1] of an AE, then the terms of general solution of (1.7) corresponding to these roots are

$$e^{\alpha x}[c_1 \cos \beta x + c_2 \sin \beta x].$$

If $\alpha + i\beta$ and $\alpha - i\beta$ each occurs twice as a root, then the terms of general solution corresponding to these roots are

$$e^{\alpha x}[(c_1 + c_2 x) \cos \beta x + (c_3 + c_4 x) \sin \beta x].$$

The general solution of a non-homogeneous linear ODE of the form

$$a_n \frac{d^n y}{dx^n} + a_{n-1} \frac{d^{n-1} y}{dx^{n-1}} + \cdots + a_1 \frac{dy}{dx} + a_0 y = Q(x) \tag{1.9}$$

where $a_1, a_2, \ldots, a_n$ are constants, $a_n \neq 0$ and $Q(x) \neq 0$, is

$$y = y_c + y_p$$

---

[1] The well-known **Complex Conjugate Root Theorem** states that if $\alpha + i\beta$ is a root of a polynomial in one variable with real coefficients, then its complex conjugate $\alpha - i\beta$ is also a root of this polynomial.

where y_c is called *complementary function* (CF) and y_p is called *particular integral* (PI). The complementary function of Eq. (1.9) is the general solution of homogeneous equation corresponding to (1.9). Particular integral of $P(D)y = Q(x)$ is defined as

$$y_p = \frac{1}{P(D)} Q(x)$$

where $\frac{1}{P(D)}$, a rational expression in D, remains the inverse of the polynomial $P(D)$. A particular integral never contains any arbitrary constant. The operator $\frac{1}{P(D)}$ is linear, that is,

$$\frac{1}{P(D)}[Q_1(x) + Q_2(x)] = \frac{1}{P(D)} Q_1(x) + \frac{1}{P(D)} Q_2(x)$$

and $\qquad \dfrac{1}{P(D)}[cQ(x)] = c\dfrac{1}{P(D)} Q(x), \qquad$ where c is a constant.

In the following lines, we write certain formulae for finding the particular integrals.

(i) If $P(D) = D^k$, then PI is

$$y_p = \frac{1}{D^k} Q(x) = \underbrace{\int \int \cdots \int Q(x)(dx)^k}_{k \ times}.$$

(ii) If $P(D) = D + a$, then PI is

$$y_p = \frac{1}{D + a} Q(x) = e^{-ax} \int e^{ax} Q(x)dx.$$

(iii) If $Q(x) = x^k$, then PI is

$$y_p = [P(D)]^{-1} x^k$$

where $[P(D)]^{-1}$ can be expanded by the binomial theorem in ascending power of D as far as the result of operation on x^k is zero.

(iv) If $Q(x) = e^{ax}$ with $P(a) \neq 0$, then PI is

$$y_p = \frac{1}{P(D)} e^{ax} = \frac{e^{ax}}{P(a)}.$$

(v) If $Q(x) = e^{ax}$ with $P(a) = 0$, then there exists a natural number k such that $P(D) = (D - a)^k f(D)$, where $f(a) \neq 0$. In this case, PI is

$$y_p = \frac{1}{(D - a)^k f(D)} e^{ax} = \frac{x^k}{k!} \frac{e^{ax}}{f(a)}.$$

(vi) If $Q(x) = \sin ax$ or $\cos ax$, then PI are

$$y_p = \frac{1}{P(D^2)} \sin ax = \frac{\sin ax}{P(-a^2)}, \quad P(-a^2) \neq 0$$

and

$$y_p = \frac{1}{P(D^2)} \cos ax = \frac{\cos ax}{P(-a^2)}, \quad P(-a^2) \neq 0.$$

(vii) If $Q(x) = \sin ax$ or $\cos ax$ and $P(-a^2) = 0$, then $D^2 + a^2$ is a factor of $P(D^2)$. In this case, we use the following:

$$\frac{1}{D^2 + a^2} \sin ax = -\frac{x}{2a} \cos ax$$

and

$$\frac{1}{D^2 + a^2} \cos ax = \frac{x}{2a} \sin ax.$$

(viii) If $Q(x) = e^{ax} V(x)$, then PI is

$$y_p = \frac{1}{P(D)} e^{ax} V(x) = e^{ax} \frac{1}{P(D+a)} V(x).$$

(ix) If $Q(x) = xV(x)$, then PI is

$$y_p = \frac{1}{P(D)} xV(x) = x \frac{1}{P(D)} V(x) - \frac{P'(D)}{[P(D)]^2} V(x)$$

where $P'(D)$ is the derivative of $P(D)$ w.r.t. D.

1.4 Total Differential Equations

Total differential equations are frequently used in the theory of PDEs. An equation of the form

$$P dx + Q dy + R dz = 0 \tag{1.10}$$

where P, Q, R are continuous functions of x, y, z, is called a **total differential equation** or **Pfaffian differential equation** named after the German mathematician *Johann Friedrich Pfaff* (1765−1825). The expression on the LHS of Eq. (1.10) is called the **Pfaffian differential form**. An integral (solution) of Eq. (1.10), if exists, is of the form

$$u(x, y, z) = c.$$

Geometrical Interpretation: The solution of Eq. (1.10) forms a one-parameter family of surfaces such that the direction ratios of the normal to each surface of the family are P, Q, R at any point (x, y, z) of the surface.

A necessary and sufficient condition for integrability of a total differential equation can be represented by the following theorem.

Theorem 1.2. *Equation (1.10) is integrable iff*

$$P\left[\frac{\partial Q}{\partial z} - \frac{\partial R}{\partial y}\right] + Q\left[\frac{\partial R}{\partial x} - \frac{\partial P}{\partial z}\right] + R\left[\frac{\partial P}{\partial y} - \frac{\partial Q}{\partial x}\right] = 0.$$

Example 1.15. *Show that the Pfaffian equation $(3xz + 2y)dx + xdy + x^2dz = 0$ is integrable.*

Solution: Here, $P = 3xz + 2y$, $Q = x$, and $R = x^2$. Now, we have

$$P\left[\frac{\partial Q}{\partial z} - \frac{\partial R}{\partial y}\right] + Q\left[\frac{\partial R}{\partial x} - \frac{\partial P}{\partial z}\right] + R\left[\frac{\partial P}{\partial y} - \frac{\partial Q}{\partial x}\right]$$
$$= (3xz + 2y)(0 - 0) + x(2x - 3x) + x^2(2 - 1)$$
$$= 0.$$

Hence, the equation is integrable.

When the condition of integrability is satisfied, we can apply a suitable method to find the required integral of a given total differential equation. Here, we present different methods to find the solution of a total differential equation.

Method I (For Exact Total Differential Equations): A total differential equation

$$P\,dx + Q\,dy + R\,dz = 0$$

is exact if the following conditions are satisfied:

$$\frac{\partial P}{\partial y} = \frac{\partial Q}{\partial x}, \quad \frac{\partial Q}{\partial z} = \frac{\partial R}{\partial y}, \quad \frac{\partial R}{\partial x} = \frac{\partial P}{\partial z}.$$

If a total differential equation is exact, then its solution is obtained by the regrouping of terms.

Example 1.16. *Solve $(x - y)dx - xdy + zdz = 0$.*

Solution: Here, $P = x - y$, $Q = -x$, $R = z$. We have

$$\frac{\partial P}{\partial y} = \frac{\partial Q}{\partial x} = -1$$

$$\frac{\partial Q}{\partial z} = \frac{\partial R}{\partial y} = 0$$

$$\frac{\partial R}{\partial x} = \frac{\partial P}{\partial z} = 0.$$

Therefore, the given total equation is exact. On regrouping the terms, we get

$$xdx - (xdy + ydx) + zdz = 0.$$

On integrating, we get

$$\frac{x^2}{2} - xy + \frac{z^2}{2} = c.$$

Hence, the required solution is

$$x^2 - 2xy + z^2 = a$$

where $a = 2c$ is an arbitrary constant.

Example 1.17. *Solve $xdx + zdy + (y + 2z)dz = 0$.*

Solution: Here, $P = x$, $Q = z$, $R = y + 2z$. We have

$$\frac{\partial P}{\partial y} = \frac{\partial Q}{\partial x} = 0$$

$$\frac{\partial Q}{\partial z} = \frac{\partial R}{\partial y} = 1$$

$$\frac{\partial R}{\partial x} = \frac{\partial P}{\partial z} = 0.$$

Thus, the given equation is exact. On regrouping the terms, we get

$$xdx + (ydz + zdy) + 2zdz = 0.$$

On integrating, the above equation gives rise to

$$\frac{1}{2}x^2 + yz + z^2 = c$$

which is the required solution.

Method II (Solution by Inspection): If a total differential equation is not exact, sometimes it may be possible to find an integrating factor that makes the equation as exact.

Example 1.18. *Solve $y^2dx - zdy + ydz = 0$.*

Solution: Here, $P = y^2$, $Q = -z$, $R = y$. It can be verified that the given equation is not exact. But, the equation is integrable as

$$P\left[\frac{\partial Q}{\partial z} - \frac{\partial R}{\partial y}\right] + Q\left[\frac{\partial R}{\partial x} - \frac{\partial P}{\partial z}\right] + R\left[\frac{\partial P}{\partial y} - \frac{\partial Q}{\partial x}\right]$$
$$= y^2[-1 - 1] - z[0 - 0] + y[2y - 0]$$
$$= 0.$$

Dividing the given equation by y^2, we get

$$\frac{y^2 dx - z dy + y dz}{y^2} = 0$$

or

$$dx + \frac{y dz - z dy}{y^2} = 0.$$

On integrating, we get

$$x + \frac{z}{y} = c$$

which is the required solution.

Method III (For Homogeneous Equations): A total differential equation

$$P\,dx + Q\,dy + R\,dz = 0$$

is called homogeneous if P, Q, R are homogeneous functions of x, y, z. In such equations, one of the variables can be separated from the other two by using the transformation

$$x = uz \ \text{ and } \ y = vz$$

so that

$$dx = z\,du + u\,dz \ \text{ and } \ dy = z\,dv + v\,dz.$$

Making use of the above relation, the given equation reduces to the form in which either the coefficient of dz is zero or not zero. In either case, the new equation may easily be integrated.

Example 1.19. *Solve $yz dx - z^2 dy - xy dz = 0$.*

Solution: Here, $P = yz$, $Q = -z^2$, $R = -xy$. Given equation is homogeneous as each of P, Q, R is a homogeneous function of degree 2. Consider the transformation

$$x = uz \ \text{ and } \ y = vz$$

so that

$$dx = z\,du + u\,dz \ \text{ and } \ dy = z\,dv + v\,dz.$$

Using this transformation, the given equation reduces to

$$(vz)z(z du + u dz) - z^2(z dv + v dz) - (uz)(vz)dz = 0$$

or

$$z^2[v(zdu + udz) - (zdv + vdz) - uvdz] = 0$$

or

$$z^2(vzdu - zdv - vdz) = 0.$$

On dividing by z^2, we get

$$vzdu - zdv - vdz = 0$$

which again dividing by vz, becomes

$$du = \frac{dv}{v} - \frac{dz}{z}.$$

On integrating, we get

$$u + \log c = \log v + \log z$$

implying thereby

$$vz = ce^u.$$

Putting the values of u and v in the above equation, we get

$$y = ce^{x/z}$$

which is the required solution.

Method IV (One Variable Regarded as Constant): If two terms of $P\,dx + Q\,dy + R\,dz = 0$, say, $P\,dx + Q\,dy$ can be easily integrable, then the third variable z is taken as constant so that $dz = 0$. We integrate $P\,dx + Q\,dy = 0$ and use $\phi(z)$ instead of constant of integration. Finally, we take total differential of obtained integral and compare its coefficients with the given total differential equation, to determine $\phi(z)$.

Example 1.20. *Solve $yzdx + (xz - yz^3)dy - 2xydz = 0$.*

Solution: It can be easily verified that the given equation is integrable. Consider y as a constant, so that $dy = 0$. Then, the given equation reduces to

$$yzdx - 2xydz = 0$$

which dividing by y becomes

$$zdx - 2xdz = 0$$

or

$$\frac{dx}{x} = 2\frac{dz}{z}.$$

On integrating, we get

$$\log x = 2\log z + \log \phi(y)$$

implying thereby

$$x = z^2\phi(y). \tag{1.11}$$

Taking the total derivative of (1.11), we get

$$dx = z^2 d\phi + \phi(y) \cdot 2zdz$$

or

$$dx - 2\phi(y)zdz - z^2 d\phi = 0.$$

From (1.11), we have $\phi(y) = \dfrac{x}{z^2}$. *Hence, the above equation becomes*

$$dx - 2\left[\frac{x}{z^2}\right]zdz - z^2 d\phi = 0$$

or

$$zdx - 2xdz - z^3 d\phi = 0.$$

Multiplying the above equation by y, we obtain

$$yzdx - yz^3 d\phi - 2xydz = 0. \tag{1.12}$$

Comparing (1.12) with the given total differential equation, we get

$$-yz^3 d\phi = (xz - yz^3)dy$$

which by using (1.11) reduces to

$$-yz^3 d\phi = [z^3 \phi(y) - yz^3]dy$$

yielding thereby

$$\phi(y)dy + yd\phi - ydy = 0$$

or

$$d[y \cdot \phi] - d\left(\frac{1}{2}y^2\right) = 0.$$

On integrating, we get

$$y\phi(y) - \frac{1}{2}y^2 = \frac{c}{2}$$

or

$$\phi(y) = \frac{1}{2}y + \frac{c}{2y}.$$

Using this value of $\phi(y)$, (1.11) becomes

$$x = z^2\left[\frac{1}{2}y + \frac{c}{y}\right]$$

or

$$2xy = y^2z^2 + cz^2$$

which is the required solution.

1.5 Simultaneous Total Differential Equations

Simultaneous total differential equations are frequently used in the study of first-order PDEs, about which we will study in Chapter 4. The equations

$$\left.\begin{array}{l} P_1 dx + Q_1 dy + R_1 dz = 0 \\ P_2 dx + Q_2 dy + R_2 dz = 0 \end{array}\right\} \tag{1.13}$$

where $P_1, Q_1, R_1, P_2, Q_2, R_2$ are functions of the variables x, y and z, are called **simultaneous total differential equations**. Using Cramer's rule, the system (1.13) can also be written as

$$\frac{dx}{Q_1 R_2 - Q_2 R_1} = \frac{dy}{R_1 P_2 - R_2 P_1} = \frac{dz}{P_1 Q_2 - P_2 Q_1}$$

or

$$\frac{dx}{P} = \frac{dy}{Q} = \frac{dz}{R} \tag{1.14}$$

where

$$P = Q_1 R_2 - Q_2 R_1, \ Q = R_1 P_2 - R_2 P_1, \ R = P_1 Q_2 - P_2 Q_1.$$

Equation (1.14) is the standard form of simultaneous total differential equations. A solution of (1.13) (or (1.14)), if exists, is of the form

$$\left.\begin{array}{l} u(x, y, z) = c_1 \\ v(x, y, z) = c_2 \end{array}\right\}$$

where c_1 and c_2 are arbitrary constants.

Geometrical Interpretation: A solution of Eq. (1.14) forms a two-parameter family of space curves such that the direction ratios of the tangent to each curve of the family are P, Q, R at any point (x, y, z) of the curve. Furthermore, these curves are orthogonal to the surfaces represented by the total differential equation $Pdx + Qdy + Rdz = 0$.

There are two basic methods for solving simultaneous total differential equations, namely, *grouping method* and *multipliers method*.

1. Grouping Method: If it is possible to take two fractions, say, $\dfrac{dx}{P} = \dfrac{dy}{Q}$ of Eq. (1.14), wherein z can be cancelled or is absent, then the equation $\dfrac{dx}{P} = \dfrac{dy}{Q}$ involves x and y only and hence by integrating gives rise to

$$u(x, y) = c_1. \tag{1.15}$$

In the similar manner, if the equation $\dfrac{dx}{P} = \dfrac{dz}{R}$ contains x and z only (not y), then by integrating it, we get

$$v(x, z) = c_2. \tag{1.16}$$

Equations (1.15) and (1.16) taken together constitute the solution of Eq. (1.14).

Example 1.21. *Solve* $\dfrac{dx}{z^2 y} = \dfrac{dy}{z^2 x} = \dfrac{dz}{y^2 x}.$

Solution: *Taking the first two fractions, we get*

$$\frac{dx}{z^2 y} = \frac{dy}{z^2 x}$$

or

$$xdx - ydy = 0.$$

On integrating, we get

$$x^2 - y^2 = c_1 \tag{1.17}$$

where c_1 is an arbitrary constant. Now, taking the last two fractions, we get

$$\frac{dy}{z^2 x} = \frac{dz}{y^2 x}$$

or

$$y^2 dy - z^2 dx = 0.$$

Integrating it, we get

$$y^3 - z^3 = c_2 \tag{1.18}$$

where c_2 is an arbitrary constant. Equations (1.17) and (1.18) taken together form the required solution.

Special Case of Grouping Method: Suppose that one integral of Eq. (1.14) is determined by the method of grouping. Nevertheless, another pair of fractions involves all the three variables x, y, z such that none of them can be cancelled. Consequently, the second integral cannot be obtained directly from this pair. In such a case, we eliminate one suitable variable by using the first integral and then apply the method of grouping to obtain the second integral. In the second integral, the constant of integration of the first integral must be removed later on.

Example 1.22. *Solve* $\dfrac{dx}{xy} = \dfrac{dy}{y^2} = \dfrac{dz}{xyz - 2x^2}.$

Solution: *Taking the first two fractions, we have*

$$\frac{dx}{xy} = \frac{dy}{y^2}$$

or

$$\frac{dx}{x} = \frac{dy}{y}.$$

On integrating, we get

$$\log x = \log(c_1 y)$$

or

$$x = c_1 y. \tag{1.19}$$

Now, considering the last two fractions, we have

$$\frac{dy}{y^2} = \frac{dz}{xyz - 2x^2}$$

which by using (1.19) reduces to

$$\frac{dy}{y^2} = \frac{dz}{c_1 y^2 z - 2c_1^2 y^2}.$$

By separating the variables, the above equation becomes

$$c_1 dy = \frac{dz}{z - 2c_1}.$$

On integrating, we get

$$c_1 y = \log(z - 2c_1) + c_2.$$

Putting the value of c_1 from (1.19), the above equation becomes

$$x = \log\left(z - \frac{2x}{y}\right) + c_2. \tag{1.20}$$

Thus, (1.19) and (1.20) constitute the required solution of the given equations.

2. Multipliers Method: Let l, m, n be functions of x, y, z, then by well-known principle of algebra, each fraction of Eq. (1.14) is equal to

$$\frac{ldx + mdy + ndz}{lP + mQ + nR}. \tag{1.21}$$

Here l, m, n are called multipliers. In this method, if possible, we choose multipliers such that the numerator of Eq. (1.21) is the exact differential of its denominator. Consequently, Eq. (1.21) can be combined with a suitable fraction in (1.14) to give an integral. However, in some problems, another set of multipliers l', m', n' are so chosen that the fraction

$$\frac{l'dx + m'dy + n'dz}{l'P + m'Q + n'R} \tag{1.22}$$

is integrable and hence fractions (1.21) and (1.22) are combined to get an integral. To get another independent solution of Eq. (1.14), either the above method may be repeated or sometimes grouping method can be applied.

Example 1.23. *Solve* $\dfrac{dx}{x^2 - y^2 - z^2} = \dfrac{dy}{2xy} = \dfrac{dz}{2xz}.$

Solution: *Taking the last two fractions, we get*

$$\frac{dy}{2xy} = \frac{dz}{2xz}$$

or

$$\frac{dy}{y} = \frac{dz}{z}.$$

Integrating it, we get

$$\log y = \log z + \log c_1$$

or

$$y = c_1 z. \tag{1.23}$$

Using multipliers x, y, z in the given equation, we get

$$each\ ratio = \frac{xdx + ydy + zdz}{x(x^2 - y^2 - z^2) + 2xy^2 + 2xz^2}$$

$$= \frac{xdx + ydy + zdz}{x(x^2 - y^2 - z^2 + 2y^2 + 2z^2)}$$

$$= \frac{xdx + ydy + zdz}{x(x^2 + y^2 + z^2)}.$$

Taking it with second fraction, we get

$$\frac{xdx + ydy + zdz}{x(x^2 + y^2 + z^2)} = \frac{dy}{2xy}$$

implying thereby

$$\frac{2xdx + 2ydy + 2zdz}{x^2 + y^2 + z^2} = \frac{dy}{y}$$

or

$$\frac{d(x^2 + y^2 + z^2)}{x^2 + y^2 + z^2} = \frac{dy}{y}.$$

On integrating it, we get

$$\log(x^2 + y^2 + z^2) = \log y + \log c_2$$

or

$$x^2 + y^2 + z^2 = c_2 y. \tag{1.24}$$

Equations (1.23) and (1.24) together form the required solution.

Special Case of Multipliers Method: If we are able to choose the multipliers l, m, n in (1.21) such that $ldx + mdy + ndz$ is integrable and $lP + mQ + nR = 0$, then we have

$$\text{each fraction} = \frac{ldx + mdy + ndz}{0},$$

which implies that

$$ldx + mdy + ndz = 0.$$

This gives us one independent solution of (1.14).

Example 1.24. *Solve* $\dfrac{dx}{x(y^2 - z^2)} = \dfrac{dy}{-y(z^2 + x^2)} = \dfrac{dz}{z(x^2 + y^2)}.$

Solution: *Using multipliers x, y, z, we get*

$$\begin{aligned}
\text{each ratio} &= \frac{xdx + ydy + zdz}{x^2(y^2 - z^2) - y^2(z^2 + x^2) + z^2(x^2 + y^2)} \\
&= \frac{xdx + ydy + zdz}{x^2y^2 - x^2z^2 - y^2z^2 - x^2y^2 + x^2z^2 + y^2z^2} \\
&= \frac{xdx + ydy + zdz}{0}
\end{aligned}$$

which implies that

$$xdx + ydy + zdz = 0.$$

On integrating, the above equation gives rise to

$$x^2 + y^2 + z^2 = c_1. \tag{1.25}$$

Now, using the multipliers $\dfrac{1}{x}, -\dfrac{1}{y}, -\dfrac{1}{z}$, *we get*

$$\begin{aligned}
\text{each ratio} &= \frac{\dfrac{1}{x}dx - \dfrac{1}{y}dy - \dfrac{1}{z}dz}{(y^2 - z^2) + (z^2 + x^2) - (x^2 + y^2)} \\
&= \frac{\dfrac{dx}{x} - \dfrac{dy}{y} - \dfrac{dz}{z}}{0}
\end{aligned}$$

which implies that

$$\frac{dx}{x} = \frac{dy}{y} + \frac{dz}{z}.$$

Integrating it, we get

$$\log x + \log c_2 = \log y + \log z$$

or

$$yz = c_2 x. \tag{1.26}$$

Equations (1.25) and (1.26) together form the required solution.

Direction Fields and Integral Curves: Let $\mathbf{E} : \big(P(x, y, z), Q(x, y, z), R(x, y, z)\big)$ be a vector field. A space curve that is tangential to the direction of $\mathbf{E}$ at each of its points is called an *integral curve* or *trajectory* of the vector field $\mathbf{E}$. Thus, the tangent to the integral curve has direction ratios P, Q, R at each point (x, y, z) on the curve. Consequently, the slope of the integral curve is represented by the simultaneous total differential equations:

$$\frac{dx}{P} = \frac{dy}{Q} = \frac{dz}{R}. \tag{1.27}$$

Here, the vector field $\mathbf{E}$ is called the *direction field* or *slope field* for Eq. (1.27).

Integral Curves in Plane: An integral curve of a vector field $\mathbf{E} : (P(x, y), Q(x, y)))$ is a plane curve such that $\mathbf{E}$ assigns a tangent vector at any point of the curve. The slope of the integral curve thus can be represented by the following ODE:

$$\frac{dy}{dx} = \frac{Q}{P}$$

so that $\mathbf{E}$ is called the *direction field* associated with the above equation.

▶ Exercises

1. If $z(x + y) = x^2 + y^2$, then show that

$$\left[\frac{\partial z}{\partial x} - \frac{\partial z}{\partial y}\right]^2 = 4\left[1 - \frac{\partial z}{\partial x} - \frac{\partial z}{\partial y}\right].$$

2. If $u = \frac{y}{z} + \frac{z}{x}$, then show that $x\frac{\partial u}{\partial x} + y\frac{\partial u}{\partial y} + z\frac{\partial u}{\partial z} = 0$.

3. If $z = \sin^{-1}\left(\frac{y}{x}\right)$, then prove that $\frac{\partial^2 z}{\partial x \partial y} = \frac{\partial^2 z}{\partial y \partial x}$.

4. If $z = x^y$, then show that $\frac{\partial^3 z}{\partial x^2 \partial y} = \frac{\partial^3 z}{\partial x \partial y \partial x}$.

5. If $z = \tan^{-1}\left(\frac{2xy}{x^2 - y^2}\right)$, then prove that $\frac{\partial^2 z}{\partial x^2} + \frac{\partial^2 z}{\partial y^2} = 0$.

6. If $z = e^x(x \cos y - y \sin y)$, then prove that $\frac{\partial^2 z}{\partial x^2} + \frac{\partial^2 z}{\partial y^2} = 0$.

7. If $\dfrac{x^2}{a^2+u} + \dfrac{y^2}{b^2+u} + \dfrac{z^2}{c^2+u} = 1$, then prove that

$$\left(\dfrac{\partial u}{\partial x}\right)^2 + \left(\dfrac{\partial u}{\partial y}\right)^2 + \left(\dfrac{\partial u}{\partial z}\right)^2 = 2\left[x\dfrac{\partial u}{\partial x} + y\dfrac{\partial u}{\partial y} + z\dfrac{\partial u}{\partial z}\right].$$

8. If $u = \log(x^2 + y^2 + z^2)$, then prove that

$$(x^2 + y^2 + z^2)\left[\dfrac{\partial^2 u}{\partial x^2} + \dfrac{\partial^2 u}{\partial y^2} + \dfrac{\partial^2 u}{\partial z^2}\right] = 1.$$

9. Find the directional derivative of $z = \dfrac{x^2 - y^2}{x^2 + y^2 + 1}$ at the point $(2, 1)$ in the direction of $(3, 2)$.

   Ans. $-\dfrac{1}{6\sqrt{2}}$.

10. Find the gradient of $f(x, y, z) = x^2 y^2 + xy^2 - z^2$ at the point $(3, 1, 1)$.
    Ans. $(7, 24, -2)$.

11. If $z = f(x, y)$, $x = e^{2s} + e^{-2t}$, $y = e^{-2s} + e^{2t}$, then using chain rule, show that

$$\dfrac{\partial z}{\partial s} - \dfrac{\partial z}{\partial t} = 2\left[x\dfrac{\partial z}{\partial x} - y\dfrac{\partial z}{\partial y}\right].$$

12. If $z = x^3 - xy + y^3$, $x = r\cos\theta$, $y = r\sin\theta$, then find $\dfrac{\partial z}{\partial r}$ and $\dfrac{\partial z}{\partial \theta}$.

    Ans. $\dfrac{\partial z}{\partial r} = (3x^2 - y)\cos\theta + (3y^2 - x)\sin\theta$, $\dfrac{\partial z}{\partial \theta} = (3x^2 - y)(-r\sin\theta) + (3y^2 - x)r\cos\theta$.

13. If $u = x^2 - 2y$ and $v = x + y$, then prove that $\dfrac{\partial(u, v)}{\partial(x, y)} = 2x + 2$.

14. Show that the functions $u = \dfrac{x+y}{1-xy}$ and $v = \tan^{-1}x + \tan^{-1}y$ are functionally dependent. Find also a functional relationship between u and v.   Ans. $u = \tan v$.

15. Determine the inverse mapping of the transformation: $u = x - 2y$, $v = 2x + y$. Also, evaluate the Jacobian of the mapping and that of the inverse mapping.

    Ans. $x = \dfrac{1}{5}(u + 2v)$, $y = \dfrac{1}{5}(v - 2u)$; $5, \dfrac{1}{5}$.

16. Find the equation of the plane determined by the points $P(2, -1, 3)$, $Q(-1, 4, 2)$, and $R(3, 0, -2)$.
    Ans. $3x + 2y + z - 7 = 0$.

17. Find the equation to the tangent plane and the normal line to the surface $x^2 + 2y^2 + 3z^2 = 12$ at the point $(1, 2, -1)$.

    Ans. $x + 4y - 3z = 12$; $\dfrac{x-1}{1} = \dfrac{y-2}{4} = \dfrac{z+3}{-1}$.

Find the envelopes of the families of surfaces described in problems 18–21.

18. $(x - a)^2 + y^2 + z^2 = 1$ Ans. $y^2 + z^2 = 1$.
19. $a^3 - 3a^2x + 3ay - z = 0$ Ans.

$$(xy - z)^2 - 4(x^2 - y)(y^2 - xz) = 0.$$

20. $z = ax + by + a^2 + b^2$

Ans. $x^2 + y^2$.

21. $(x - a)^2 + (y - b)^2 + z^2 = 1$

Ans. $z = \pm 1$.

Solve the ordinary differential equations described in problems 22–34.

22. $(x^2 + 1)\dfrac{dy}{dx} + 2xy = 4x^2$.

Ans. $3(x^2 + 1)y = 4x^3 + c$.

23. $(1 + y^2)dx + (x - e^{-\tan^{-1}y})dy = 0$.

Ans. $xe^{-\tan^{-1}y} = -\tan^{-1}y + c$.

24. $\dfrac{d^3y}{dx^3} + 6\dfrac{d^2y}{dx^2} + 11\dfrac{dy}{dx} + 6y = 0$.

Ans. $y = c_1 e^{-x} + c_2 e^{-2x} + c_3 e^{-3x}$.

25. $\dfrac{d^3y}{dx^3} - 3\dfrac{dy}{dx} + 2y = 0$.

Ans. $y = (c_1 x + c_2)e^x + c_3 e^{-2x}$.

26. $\dfrac{d^4y}{dx^4} + 8\dfrac{d^2y}{dx^2} + 16y = 0$.

Ans. $y = (c_1 + c_2 x)\cos 2x + (c_3 + c_4 x)\sin 2x$.

27. $(D^3 - 2D + 4)y = x^4 + 3x^2 - 5x + 2$.

Ans.
$y = c_1 e^{-2x} + e^x(c_2 \cos x + c_3 \sin x) + \dfrac{1}{4}\left(x^4 + 2x^3 + 6x^2 - 5x - \dfrac{7}{2}\right)$.

28. $(D^3 - D^2 - 4D + 4)y = e^{3x}$.

Ans.
$y = c_1 e^x + c_2 e^{2x} + c_3 e^{-2x} + \dfrac{1}{10}e^{3x}$.

29. $(D^2 + 4D + 4)y = e^{2x} - e^{-2x}$.

Ans. $y = (c_1 + c_2 x)e^{-2x} + \dfrac{1}{16}e^{2x} - \dfrac{1}{2}x^2 e^{-2x}$.

30. $(D^3 + D^2 - D - 1)y = \cos 2x$.

Ans. $y = c_1 e^x + (c_2 + c_3 x)e^{-x} - \dfrac{1}{25}(2\sin 2x + \cos 2x)$.

31. $(D^3 - D^2 + 4D - 4)y = \sin 3x$.

Ans. $y = c_1 e^x + c_2 \cos 2x + c_3 \sin 2x + \dfrac{1}{50}(3\cos 3x + \sin 3x)$.

32. $(D^2 - 3D + 2)y = 3\sin 2x$.

Ans. $y = c_1 e^x + c_2 e^{2x} + \dfrac{3}{20}(3\cos 2x - \sin 2x)$.

33. $(D^2 - 2D + 5)y = e^{2x}\sin x$.

Ans. $y = e^x(c_1 \cos 2x + c_2 \sin 2x) - \dfrac{1}{10}e^{2x}(\cos x - 2\sin x)$.

34. $\dfrac{d^2y}{dx^2} - 2\dfrac{dy}{dx} + y = x\sin x$.

Ans. $y = (c_1 + c_2 x)e^x + \dfrac{x}{2}\cos x - \dfrac{1}{2}(\sin x - \cos x)$.

Verify the condition of integrability for the following total differential equations.

35. $(6x^2 y - e^x z)dx + (2x^3 + \sin z)dy +$
$(y\cos z - e^x)dz = 0$.

Ans. Integrable.

36. $ydx + dy + dz = 0$.

Ans. Not integrable.

Solve the total differential equations described in problems 37–50.

37. $(y + 3z)dx + (x + 2z)dy + (3x + 2y)dz = 0.$ Ans. $xy + 2yz + 3xz = c.$

38. $(y + z)dx + (z + x)dy + (x + y)dz = 0.$ Ans. $xy + yz + zx = c.$

39. $(2x^3y + 1)dx + x^4dy + x^2 \tan z\,dz = 0.$ Ans. $x^2y - \dfrac{1}{x} + \log \sec z = c.$

40. $2(y + z)dx - (x + z)dy + (2y - x + z)dz = 0.$ Ans. $y + z = a(x + z)^2.$

41. $(2y - z)dx + 2(x - z)dy - (x + 2y)dz = 0.$ Ans. $2xy - xz - 2yz = c.$

42. $(x + z)^2 dy + y^2(dx + dz) = 0.$ Ans. $y(x + z) = a(x + y + z).$

43. $(e^xy + e^z)dx + (e^yz + e^x)dy + (e^y - e^xy - e^yz)dz = 0.$ Ans. $e^xy + e^yz + e^zx = ce^z.$

44. $(y^2 + yz)dx + (z^2 + zx)dy + (y^2 - xy)dz = 0.$ Ans. $y(z + x) = c(y + z).$

45. $(1 + yz)dx + x(z - x)dy - (1 + xy)dz = 0.$ Ans. $x - z = c\,(1 + xy).$

46. $(y^2 + z^2 - x^2)dx - 2xydy - 2xzdz = 0.$ Ans. $x^2 + y^2 + z\,(x + y) = cxy.$

47. $(x^2y - y^3 - y^2z)dx + (xy^2 - x^2z - x^3)dy + (xy^2 + x^2y)$
   $dz = 0.$ (**Hint.** Divide the given equation by x^2y^2)   Ans. $x^2 + y^2 + z(x + y) = cxy.$

48. $xz^3dx - zdy + 2ydz = 0.$ Ans. $\dfrac{x^2}{2} - \dfrac{y}{z^2} = c.$

49. $(2x^2 + 2xy + 2xz^2 + 1)dx + dy + 2zdz = 0.$
   (**Hint.** $2x(x + y + z^2)dx + dx + dy + d(z^2) = 0$ or
   $(x + y + z^2)d(x^2) + d(x + y + z^2) = 0)$ Ans. $x^2 + \log(x + y + z^2) = c.$

50. $yz^2dx - xz^2dy - (2xyz + x^2)dz = 0.$ Ans. $(x + yz)z = cx.$

Solve the following simultaneous total differential equations described in problems 51–64.

51. $\dfrac{dx}{x^2} = \dfrac{dy}{y^2} = \dfrac{dz}{nxy}.$

   Ans. $\dfrac{1}{x} = \dfrac{1}{y} + c_1, \; z = c_2 - \dfrac{nxy}{y - x}\log\dfrac{x}{y}.$

52. $\dfrac{dx}{mz - ny} = \dfrac{dy}{nx - lz} = \dfrac{dz}{ly - mx}.$

   Ans. $lx + my + nz = c_1, \; x^2 + y^2 + z^2 = c_2.$

53. $\dfrac{dx}{y^2} = \dfrac{dy}{x^2} = \dfrac{dz}{x^2y^2z^2}.$

   Ans. $x^3 - y^3 = c_1, \; y^3 + \dfrac{1}{z^3} = c_2.$

54. $\dfrac{dx}{yz} = \dfrac{dy}{zx} = \dfrac{dz}{xy}.$

   Ans. $x^2 - y^2 = c_1, \; x^2 - z^2 = c_2.$

55. $\dfrac{dx}{y + z} = \dfrac{dy}{z + x} = \dfrac{dz}{x + y}.$

   Ans. $y - x = c_1(z - y), \; (x - y)^2(x + y + z) = c_2.$

56. $\dfrac{dx}{y^2 + z^2 - x^2} = \dfrac{dy}{-2xy} = \dfrac{dz}{-2xz}.$

   Ans. $y = c_1z, \; x^2 + y^2 + z^2 = c_2z.$

57. $\dfrac{dx}{xz(z^2 + xy)} = \dfrac{dy}{-yz(z^2 + xy)} = \dfrac{dz}{x^4}.$

   Ans. $xy = c_1, \; (z^2 + xy)^2 - x^4 = c_2.$

58. $\dfrac{dx}{z(x + y)} = \dfrac{dy}{z(x - y)} = \dfrac{dz}{x^2 + y^2}.$

   Ans. $x^2 - y^2 - z^2 = c_1, \; 2xy - z^2 = c_2.$

59. $\dfrac{dx}{x^2 + y^2 + yz} = \dfrac{dy}{x^2 + y^2 - xz} = \dfrac{dz}{z(x + y)}.$

   Ans. $x - y - z = c_1, \; x^2 + y^2 = c_2z^2.$

60. $\dfrac{dx}{x^2 + y^2} = \dfrac{dy}{2xy} = \dfrac{dz}{(x+y)z}$.

Ans. $x + y = c_1 z$, $2y = c_2(x^2 - y^2)$.

61. $\dfrac{dx}{1} = \dfrac{dy}{3} = \dfrac{dz}{5z + \tan(y - 3x)}$.

Ans.
$y - 3x = c_1$, $5z + \tan(y - 3x) = c_2 e^{5x}$.

62. $\dfrac{dx}{\cos(x+y)} = \dfrac{dy}{\sin(x+y)} = \dfrac{dz}{z}$.

Ans. $[\sin(x+y) + \cos(x+y)]e^{y-x} = c_1$, $z^{\sqrt{2}} + \cot\dfrac{1}{2}(x+y+\dfrac{\pi}{4}) = c_2$.

63. $\dfrac{dx}{y-x} = \dfrac{dy}{y+x} = \dfrac{zdz}{x^2 - 2xy - y^2}$.

Ans.
$x^2 + y^2 + z^2 = c_1$, $x^2 - y^2 + 2xy = c_2$.

64. $\dfrac{dx}{y - zx} = \dfrac{dy}{yz + x} = \dfrac{dz}{x^2 + y^2}$.

Ans.
$x^2 - y^2 - 2xy = c_1$, $x^2 - y^2 - z^2 = c_2$.

2

Solution, Classification, and Formation of Partial Differential Equations

We now introduce the idea of partial differential equations formally as well as technically. A partial differential equation (often abbreviated in the sequel as PDE) is defined as an equation involving one or more partial derivatives of an unknown function of several variables. Thus, for an unknown function u of several independent variables $x, y, z, t, \ldots$, the general form of PDE will be

$$F(x, y, z, t, \ldots, u, u_x, u_y, \ldots, u_{xx}, u_{xy}, \ldots, u_{xxx}, \ldots) = 0. \tag{2.1}$$

In Eq. (2.1), we assume that at least one of the partial derivatives of the unknown function u must be involved. We also assume that the mixed partial derivatives of u are independent of the order of differentiation, for example, $u_{xyx} = u_{xxy} = u_{yxx}$.

The following examples of well-known PDEs naturally evolve in various physical considerations.

(i) Linear Transport Equation: $u_t + cu_x = 0$

(ii) Inviscid Burgers' Equation (Shock Waves): $u_t + uu_x = 0$

(iii) Eikonal Equation: $u_x^2 + u_y^2 = 1$

(iv) Laplace Equation or Potential Equation: $u_{xx} + u_{yy} + u_{zz} = 0$

(v) Heat Equation: $u_t = \alpha^2(u_{xx} + u_{yy} + u_{zz})$

(vi) Wave Equation: $u_{tt} = c^2(u_{xx} + u_{yy} + u_{zz})$

(vii) Poisson Equation: $u_{xx} + u_{yy} + u_{zz} = f(x, y, z)$

(viii) Helmholtz Equation: $u_{xx} + u_{yy} + u_{zz} + \lambda u = 0$

(ix) Telegraph Equation: $u_{tt} + au_t + bu = c^2 u_{xx}$

(x) Burgers' Equation: $u_t + uu_x = \mu u_{xx}$

(xi) Minimal Surface Equation: $(1 + u_y^2)u_{xx} - 2u_x u_y u_{xy} + (1 + u_x^2)u_{yy} = 0$

(xii) Born–Infeld Equation: $(1 - u_t^2)u_{xx} + 2u_x u_t u_{xt} - (1 + u_x^2)u_{tt} = 0$

(xiii) Korteweg-de Vries (KdV) Equation: $u_t + uu_x + u_{xxx} = 0$

(xiv) Biharmonic Equation: $u_{xxxx} + 2u_{xxyy} + u_{yyyy} = 0.$

All these PDEs are classical and each of them is profoundly significant in theoretical physics. Here, generally, u remains a function of time t and one/two/three rectangular coordinate(s) of a point.

The theory of PDEs is quite different as compared to that of ordinary differential equations (ODEs) and is relatively difficult and cumbersome in every respect. PDEs form a subject of vigorous mathematical research for over three centuries and still continue to be so. It remains to be a very active and young area for critical investigations and research for mathematicians, scientists, and engineers. In Mathematics, PDEs naturally occur in the study of geometry of surfaces and a wide variety of problems in mechanics whenever involved function is of more than one variables. In Physics, PDEs come from such areas of applications as electric fields, electromagnetism, fluid mechanics, diffusion, wave motion, non-linear optics, quantum field theory, geophysics (seismic wave propagation), and plasma physics (ionised liquids and gases), whose studies have stimulated the development of many important mathematical ideas. In recent years, we have observed an abrupt increase in the use of PDEs in other areas such as Biology, Bio-mathematics, Chemistry, Computer Science (particularly in relation to image processing and graphics), and Economics (finance). Thus, to solve the real-world problems occurring in various disciplines of pure and applied sciences, PDEs will continue to flourish in years to come.

The aim of this chapter is to discuss some relevant notions, basic concepts, and elementary remarks concerning PDEs. At the end of this chapter, the formations of PDEs have been discussed.

2.1 Order and Degree of a PDE

The order of a PDE is the order of highest order partial derivative appearing in the equation. The degree of a PDE is the greatest power of highest order partial derivative occurring in the equation after making it free from radicals and fractions.

Illustrations:

(i) $x\dfrac{\partial u}{\partial x} + y\dfrac{\partial u}{\partial y} = z^2$ is a PDE of first order and first degree.

(ii) $\left(\dfrac{\partial u}{\partial x}\right)^2 + \left(\dfrac{\partial u}{\partial y}\right)^2 = z$ is a PDE of first order and second degree.

(iii) $\left(\dfrac{\partial u}{\partial x}\right)^3 + \left(\dfrac{\partial u}{\partial y}\right)^4 = z$ is a PDE of first order and fourth degree.

(iv) $x\dfrac{\partial^2 u}{\partial x^2} = \dfrac{\partial u}{\partial x}$ is a PDE of second order and first degree.

(v) $\dfrac{\partial^2 u}{\partial x^2} = \sqrt{1 + \dfrac{\partial u}{\partial y}}$ is a PDE of second order and second degree.

(vi) $\left(\dfrac{\partial u}{\partial x}\right)^2 + \dfrac{\partial^3 u}{\partial x \partial y^2} = 2x\dfrac{\partial u}{\partial x}$ is a PDE of third order and first degree.

(vii) $\left(\dfrac{\partial^3 u}{\partial x^3}\right)^2 + \left(\dfrac{\partial^2 u}{\partial y^2}\right)^3 + 8z = 0$ is a PDE of third order and second degree.

2.2 Solution of a PDE

A solution (or integral) of a PDE is an explicit or implicit relation among the variables, which does not contain derivatives and also satisfies the equation identically in the domain of the independent variables.

Therefore, if a function $u(x, y, z, t, ...)$ is sufficiently differentiable, then we can verify whether it is a solution of Eq. (2.1) simply by differentiating u the appropriate number of times w.r.t. the appropriate variables and then substituting these expressions into (2.1). If an identity is obtained, then u solves Eq. (2.1). Consequently, a possible solution u of an nth-order PDE is required to have the property that all partial derivatives of u up to order n must exist and are also continuous.

The notions of "general solution" as well as "complete solution" coincide in the context of ODEs, whereas in the case of PDEs, these two concepts are different. To appreciate these two ideas better, we adopt the following three differential equations of order one:

$$\frac{dy}{dx} = 0 \tag{2.2}$$

$$\frac{\partial z}{\partial x} = 0 \tag{2.3}$$

$$\frac{\partial z}{\partial y} = 0. \tag{2.4}$$

In the last two equations, z has been considered as a function of two independent variables x and y. The general solution (or the complete solution) of ODE represented by (2.2) is $y = c$ (c is an arbitrary constant), which illustrates that the general solution of an ODE is a family of functions depending on arbitrary constants. Although $z = c$ satisfies both PDEs represented by (2.3) and (2.4), yet $z = \phi(y)$ and $z = \psi(x)$ (ϕ, ψ being arbitrary functions) are general solutions of (2.3) and (2.4), respectively, as any constant may be included in arbitrary functions. Thus, the general solution of a PDE is the family of functions, which depends on the choice of arbitrary functions rather than arbitrary constants. Sometimes (especially in case of nonlinear equations), it is not always possible to obtain the solution containing arbitrary functions but instead, it is easy to obtain a solution involving arbitrary constants like ODE. Such a solution is termed as complete solution.

Concretely speaking, there are mainly the following types of the solution of a PDE.

(i) General Solution: A solution of a PDE involving arbitrary functions connecting known functions is called the *general solution* or *general integral*.

(ii) Complete Solution: A solution of a PDE, which contains arbitrary constants is called the *complete solution* or *complete integral*.

(iii) Particular Solution: A solution of a PDE obtained from its complete solution (general solution) by assigning particular values of arbitrary constants (arbitrary functions) is called the *particular solution* or *particular integral*.

(iv) Singular Solution: Sometimes a PDE has yet another solution termed as *singular solution,* which is free from arbitrary constants as well as arbitrary functions but it is not obtainable from general solution or complete solution like as a particular solution. We discuss, in detail, singular solutions of first-order PDEs in Chapter 4.

For instance, consider the following PDE:

$$\frac{\partial z}{\partial x} - \frac{\partial z}{\partial y} = 2\sqrt{z}.$$

The general solution of this equation is $\phi(x + y, x - \sqrt{z}) = 0$ and the complete solution is $\sqrt{z} = \frac{ax + y}{a - 1} + b$. Also, $z = (2x + y)^2$ is a particular solution as it is obtained by putting $a = 2$ and $b = 0$ in complete solution (or by taking $\phi(u, v) = u + v$ in general solution). Singular solution of the given PDE is $z = 0$, which still satisfies the equation but cannot be obtained from the general solution or complete solution.

Here, it is worth noting that we sometimes also have to deal with a solution, which is not everywhere differentiable. Such a solution is called a **weak solution** (or **generalised solution**). In this situation, the solution enjoying differentiability property is often referred as a **classical solution** (or alternately, **regular/genuine solution**). The discussion of weak solutions is out of the scope of this book. Henceforth, by solution, we always mean classical solution (with no qualification) throughout this book.

2.3 PDEs Involving Two Independent Variables

Throughout the text, generally, we will be dealing with PDEs involving two independent variables. However, we can study PDEs with as many independent variables as we wish. We shall now specify some notations for a PDE with two independent variables.

Consider z as an unknown function of two independent variables x and y. As usual, two basic partial derivative operators $\frac{\partial}{\partial x}$ and $\frac{\partial}{\partial y}$ shall often be denoted by D and D', respectively, that is, $D := \frac{\partial}{\partial x}$ and $D' := \frac{\partial}{\partial y}$. Thus, we have

$$D^n z = \frac{\partial^n z}{\partial x^n}$$

$$D'^n z = \frac{\partial^n z}{\partial y^n}$$

and

$$D^m D'^n z = \frac{\partial^{m+n} z}{\partial x^m \partial y^n} = \frac{\partial^{m+n} z}{\partial y^n \partial x^m}.$$

For brevity, in the context of PDEs of order up to two, we shall adopt the following notations:

$$p = \frac{\partial z}{\partial x}, \ q = \frac{\partial z}{\partial y}, \ r = \frac{\partial^2 z}{\partial x^2}, \ s = \frac{\partial^2 z}{\partial x \partial y}, \ t = \frac{\partial^2 z}{\partial y^2}.$$

Therefore, the most general PDEs of first and second order are, respectively, represented by

$$F(x, y, z, p, q) = 0$$

and

$$F(x, y, z, p, q, r, s, t) = 0.$$

Integral Surface: For a PDE involving two independent variables x and y, the possible solution $z = f(x, y)$ (or in implicit form, $G(x, y, z)=0$) remains a surface geometrically in xyz-space, which is also referred as a *solution surface* or an *integral surface* of the given equation.

2.4 Classification of PDEs

PDEs are divided into two classes, namely, linear and non-linear.

1. Linear PDE: A PDE is called *fully linear* or simply, *linear* if the dependent variable and all its appeared partial derivatives are of the first degree and the coefficients of the dependent variable and its derivatives are functions only of the independent variables.

The most general linear PDE of order n can be presented symbolically as

$$\begin{aligned}[(A_0 D^n + A_1 D^{n-1} D' + \cdots + A_n D'^n) + (B_0 D^{n-1} + \cdots + B_{n-1} D'^{n-1}) \\ + \cdots + (PD + QD') + Z] z = f(x, y)\end{aligned} \tag{2.5}$$

where the coefficients $A_0, A_1, \ldots, A_n, B_0, \ldots, P, Q, Z$ are functions of x and y and $A_0, A_1, \ldots, A_n$ do not vanish simultaneously.

In particular, the most general linear PDEs of first and second order have the following forms, respectively,

$$P(x, y)p + Q(x, y)q + Z(x, y)z = R(x, y)$$

where the coefficients P and Q do not vanish simultaneously[1] and

$$R(x, y)r + S(x, y)s + T(x, y)t + P(x, y)p + Q(x, y)q + Z(x, y)z = V(x, y)$$

where the coefficients $R, S,$ and T do not vanish simultaneously[2].

---

[1] This condition is equivalent to the relation $P^2 + Q^2 \neq 0$.

[2] This condition is equivalent to the relation $R^2 + S^2 + T^2 \neq 0$.

Some illustrative examples of linear PDEs are as follows:

(i) $y^2p - xyq = 3z + \sin(x - 5y)$

(ii) $xyr - (x^2 - y^2)s - xyt + px - qy = 2(x^2 - y^2)$

(iii) $(xy^2D^3 - DD' + x^3D'^2)z = e^{x^2+y^2}$.

Linear PDEs are further classified into the following two subclasses.

(a) Homogeneous Linear PDE: A linear PDE of form (2.5), in which f is identically zero, is called a *homogeneous linear PDE*.

(b) Non-homogeneous Linear PDE: A linear PDE that is not homogeneous is called a *non-homogeneous* (or an *inhomogeneous*) *linear PDE*.

2. Non-linear PDE: A PDE that is not linear is called *non-linear*. In other words, a PDE is said to be non-linear if at least one of the appearing partial derivatives is of degree higher than one or the dependent variable occurs in the power of greater than one or the product of some partial derivatives occurs or any partial derivative multiplied with dependent variable.

Some illustrative examples of non-linear PDEs are as follows:

(i) $pq = z$

(ii) $p^2 + q^2 = 1$

(iii) $rt - s^2 + 2s = 1$

(iv) $\dfrac{\partial^5 z}{\partial x^5} \cdot \dfrac{\partial^3 z}{\partial y^3} = \log xy$.

Non-linear PDEs are also classified into the following two subclasses.

(a) Uniformly Non-linear PDE: A non-linear PDE that is linear in highest order derivatives is called uniformly non-linear PDE.

(b) Fully Non-linear PDE: A non-linear PDE that is not linear in highest order derivatives is called fully non-linear PDE.

2.5 Further Classification of PDEs

On the basis of linearity of higher order, some authors defined two more classes of PDEs, namely, semilinear and quasilinear equations. These are indeed super classes of linear PDEs, which also cover some non-linear PDEs.

Semilinear PDE: A PDE is called *semilinear* or *almost linear* if all appearing partial derivatives of the highest order have the degree one and the coefficients of these highest order partial derivatives are independent of the dependent variable as well as its derivatives.

Thus, the most general first-order semilinear PDE must be of the form

$$P(x, y)p + Q(x, y)q = R(x, y, z)$$

where the coefficients P and Q do not vanish simultaneously. Also, the most general second-order semilinear PDE must be of the form

$$R(x, y)r + S(x, y)s + T(x, y)t = V(x, y, z, p, q)$$

where the coefficients R, S, and T do not vanish simultaneously.

Some illustrative examples of semilinear PDEs are as follows:

(i) $xp + qy = z^2$

(ii) $x^2 r + y^2 t - zp^3 = x + y - 4$

(iii) $y\dfrac{\partial^4 z}{\partial x^4} + zx\dfrac{\partial^3 z}{\partial x^2 \partial y} = \dfrac{\partial z}{\partial x} \cdot \dfrac{\partial^2 z}{\partial y^2}.$

Clearly, every linear PDE is semilinear but not conversely. A semilinear PDE that is not linear is called *fully semilinear*.

Quasilinear PDE: A PDE is called *quasilinear* if all appearing partial derivatives of the highest order have the degree one. In other words, a PDE that is linear in all the highest order derivatives involved is called quasilinear. Thus, the coefficients of highest order partial derivatives involved in a quasilinear PDE may be the functions of dependent variable, independent variables and partial derivatives of remaining orders.

Henceforth, the most general first-order quasilinear PDE can be expressed in the form

$$P(x, y, z)p + Q(x, y, z)q = R(x, y, z),$$

where the coefficients P and Q do not vanish simultaneously. Similarly, the most general second-order quasilinear PDE can be expressed in the form

$$R(x, y, z, p, q)r + S(x, y, z, p, q)s + T(x, y, z, p, q)t = V(x, y, z, p, q),$$

where the coefficients R, S, and T do not vanish simultaneously.

Some illustrative examples of quasilinear PDEs are as follows:

(i) $x(y^2 + z)p - y(x^2 + z)q = (x^2 - y^2)z$

(ii) $q(yq + z)r - p(2yq + z)s + yp^2 t + p^2 q = 0$

(iii) $\dfrac{\partial z}{\partial y} = z\dfrac{\partial^2 z}{\partial x^2} \cdot \dfrac{\partial^3 z}{\partial x^3} + \sin x.$

Clearly, every semilinear PDE is quasilinear but not conversely. A quasilinear PDE that is not semilinear is called *non-semilinear*. Moreover, a quasilinear PDE that is not linear must be uniformly non-linear.

Hence, the above-mentioned classes motivate us to classify PDEs in another way rather than linear and non-linear equations. In this way, all PDEs are divided completely into two classes: quasilinear and fully non-linear equations. Quasilinear PDEs are also divided completely into two classes: linear and uniformly non-linear equations. Uniformly non-linear PDEs are further divided completely into two classes: fully semilinear and non-semilinear equations. The class of semilinear PDEs is the combination of the classes of linear as well as fully semilinear PDEs. Also, the class of non-linear PDEs is the combination of the classes of uniformly nonlinear and fully non-linear PDEs.

More precisely, the following relations occur among different classes of PDEs.

$$\text{PDE} = \text{Linear PDE} \cup \text{Nonlinear PDE} = \text{Quasilinear PDE} \cup \text{Fully nonlinear PDE}$$

$$\text{Nonlinear PDE} = \text{Uniformly nonlinear PDE} \cup \text{Fully nonlinear PDE}$$

$$\text{Quasilinear PDE} = \text{Linear PDE} \cup \text{Uniformly nonlinear PDE}$$

$$\text{Semilinear PDE} = \text{Linear PDE} \cup \text{Fully semilinear PDE}$$

$$\text{Uniformly nonlinear PDE} = \text{Fully semilinear PDE} \cup \text{Non} - \text{semilinear PDE}$$

$$\text{Linear PDE} \subsetneq \text{Semilinear PDE} \subsetneq \text{Qusilinear PDE}$$

$$\text{Uniformly nonlinear PDE} \subsetneq \text{Qusilinear PDE}.$$

2.6 Linear Differential Operators

The more precise definitions of linear and semilinear PDEs begin with the concept of linear differential operators. To accomplish this, we need to recall the following relevant notions:

Operators: An operator is a mapping, whose domain and range each consists of a certain class of functions. Thus, an operator A on a class C of functions maps each function $u \in C$ to another function $v = A(u) \in C$.

For example, D and D' are the operators as they convert a differentiable function $z(x, y)$ into new functions $p(x, y)$ and $q(x, y)$, respectively.

Algebra of Operators: Given two operators A and B on a class C and a constant k (real or complex), we have

(i) Sum of A and B, often denoted by $A + B$, defined by

$$(A + B)(u) := A(u) + B(u) \ \forall \ u \in C$$

is an operator on class C.

(ii) Composition of A and B, often denoted by AB, defined by

$$(AB)(u) := A(Bu) \ \forall \ u \in C$$

is an operator on class C.

(iii) Product of A and B, often denoted by $A * B$, defined by

$$(A * B)(u) := A(u)B(u) \ \forall \ u \in C$$

is an operator on class C.

(iv) Scalar multiplication of A with k, often denoted by kA, defined by

$$(kA)(u) := k(Au) \ \forall \ u \in C$$

is an operator on class C.

For instance, $(DD')(z) = \dfrac{\partial^2 z}{\partial x \partial y}$, while $(D * D')(z) = \dfrac{\partial z}{\partial x} \cdot \dfrac{\partial z}{\partial y}$.

As usual, for an operator A on a class C, A^k denotes the composition of k identical copies of A; that is, $A^k := AA \overset{(k)}{\cdots} A$. Conventionally, in the context of basic partial derivative operators, we use $D^0(u) = D'^0(z) = z$; such an operator is called the identity operator and sometimes it is denoted by I.

Differential Operators: An operator assembled by the operations of sum, composition, product, and scalar multiplication on the basic partial derivative operators, with coefficients depending on the independent variables is called differential operator. Some examples of differential operators are as follows:

(i) $A = D^2 + D'^2$
(ii) $B = D * D + D' * D' - I$
(iii) $C = 3xD - 4yD' + (x^2 + y^2)I$
(iv) $E = DD' - I * I.$

Principal and Complementary Parts of Differential Operators: Let A be a differential operator decomposed as $A = \overline{A} + \underline{A}$ such that $\overline{A}$ consists of highest order terms of A, whereas $\underline{A}$ consists of remaining terms of A. Then the differential operators $\overline{A}$ and $\underline{A}$ are, respectively, called principal and complementary parts of A.

Linear Operators: An operator L on a class C is called linear if

$$L(k_1 u + k_2 v) = k_1 L(u) + k_2 L(v)$$

for all $u, v \in C$ and for any constants k_1, k_2.

In the preceding example, A and C are linear differential operators, whereas B and E are non-linear differential operators.

Criteria for Linear and Semilinear PDE: A PDE is linear iff it can be represented by $L(z) = f(x, y)$, where L is a linear differential operator. Similarly, a PDE is semilinear iff it can be represented by $L(z) = f(x, y)$, where L is a differential operator whose principal part is linear.

The justification of the above-mentioned fact is straightforward in virtue of the definitions of linear differential operator and linear/semilinear PDEs.

Theorem 2.1. [Principle of Superposition] *If $z_1, z_2, \ldots, z_k$ are solutions of a homogeneous linear PDE represented by $L(z) = 0$ and if $c_1, c_2, \ldots, c_k$ are arbitrary constants, then the linear combination $c_1 z_1 + c_2 z_2 + \cdots + c_k z_k$ is also a solution of the given PDE.*

Moreover, if $z_1, z_2, \ldots$ is an infinite sequence of solutions of a homogeneous linear PDE represented by $L(z) = 0$, then the infinite series $\displaystyle\sum_{n=1}^{\infty} c_n z_n$ forms a solution of the given PDE for any choice of constants $c_1, c_2, \ldots$ such that the series $\displaystyle\sum_{n=1}^{\infty} c_n z_n$ converges.

Proof. As each z_i $(1 \le i \le k)$ is a solution of the equation $L(z) = 0$, we have

$$L(z_i) = 0 \quad (1 \le i \le k). \tag{2.6}$$

Using the linearity of L, we have

$$L\left(\sum_{n=1}^{k} c_i z_i\right) = \sum_{n=1}^{k} c_i L(z_i)$$

which using Eq. (2.6) reduces to

$$L\left(\sum_{n=1}^{k} c_i z_i\right) = 0.$$

It follows that $z = \displaystyle\sum_{n=1}^{k} c_i z_i$ forms a solution of the given equation. Thus, our result is proved in the case of a finite number of solutions. Similarly, one can prove the result in the case of an infinite number of solutions. $\qquad\square$

2.7 Auxiliary Conditions

The PDEs describing various physical phenomena often have infinitely many solutions. In practice, a particular solution of interest is required for a physical problem satisfying prescribed specific auxiliary conditions that further characterise the system being modelled. The class of auxiliary conditions falls into two categories, namely, *boundary conditions* and *initial conditions*.

Let a PDE of order n be defined in the domain Ω, a subset of space of independent variables. Suppose that $\partial\Omega$ denotes the boundary of Ω and $\overline{\Omega} := \Omega \cup \partial\Omega$, the closure of Ω. The conditions associated with the given PDE, which describe the values of the dependent variable or/and its derivatives of order not greater than $n - 1$, satisfying on $\partial\Omega$ are called **boundary conditions** (abbreviated as BC). Such a PDE together with suitable boundary conditions is called a

boundary value problem (abbreviated as BVP). Following are the three special forms of boundary conditions:

- **Dirichlet conditions** are the values of dependent variable prescribed at each point on $\partial\Omega$.
- **Neumann (or flux) conditions** are the values of normal derivative of dependent variable prescribed at each point on $\partial\Omega$.
- **Robin (or Churchill/radiation/mixed) conditions** are those conditions in which a linear combination of the dependent variable and its normal derivative takes the prescribed values on $\partial\Omega$.

Initial conditions (abbreviated as IC) are the conditions on the dependent variable or/and its derivatives of order not greater than $n - 1$, all given at a single point of Ω. The combination of given PDE along with suitable initial conditions is referred to as an **initial value problem** (abbreviated as IVP). Initial value problem is also known as **Cauchy problem** in honour of French mathematician *Augustin Louis Cauchy* (1789–1857). A PDE associated with both initial and boundary conditions leads to an **initial-boundary value problem** (abbreviated as IBVP).

While considering the problem in an unbounded domain, the solution can be determined uniquely by prescribing initial conditions only, as the solution of such a problem is usually unaffected by the boundary conditions at infinity. However, for problems affected by boundary at infinity, boundedness conditions on behaviour of solutions at infinity must be prescribed. In most of the applications, the dependent variable is a function of space coordinates x, y, z and time t. An initial condition holds at the instant $t = t_0$, when consideration of the physical system begins. Such an initial condition is conveniently regarded as a boundary condition.

For illustration, the PDE

$$u_t - u_{xx} = F(x, t), \quad 0 < x < l, t \geq 0 \tag{2.7}$$

describes heat conduction in a rod of length l. More precisely, the domain in which Eq. (2.7) holds is defined as

$$\Omega = \left\{ (x, t) : 0 < x < l, 0 \leq t < \infty \right\}.$$

Obviously, for each $t \geq 0$, $(0, t) \in \partial\Omega$ and $(l, t) \in \partial\Omega$. Henceforth, the auxiliary conditions

$$u(0, t) = g_1(t), \ u(l, t) = g_2(t), \quad t \geq 0$$

and

$$u_x(0, t) = h_1(t), \ u_x(l, t) = h_2(t), \quad t \geq 0$$

are the boundary conditions on Eq. (2.7), which are Dirichlet conditions and Neumann conditions, respectively.

On the other hand, the auxiliary conditions

$$u(x, 0) = \phi(x), \quad (x, 0) \in \overline{\Omega}$$

and

$$u_t(x, 0) = \psi(x), \quad (x, 0) \in \overline{\Omega}$$

are initial conditions on Eq. (2.7).

2.8 Formation of Partial Differential Equations by Elimination of Arbitrary Constants

Before discussing the different methods for solving the PDEs, it is desirable to know about the initiation of PDEs similar to the cases of ODEs. Partial differential equations can be derived in two ways:

(i) by the elimination of arbitrary constants from a relation among x, y, z, and

(ii) by the elimination of arbitrary functions of these variables.

In this section, we derive the PDEs by the elimination of arbitrary constants, whereas the derivation of PDEs by the elimination of arbitrary functions will be undertaken in the next section.

Theorem 2.2. *Consider the two-parameter family of surfaces*

$$f(x, y, z, a, b) = 0. \tag{2.8}$$

By eliminating arbitrary constants a and b, we get a PDE.

Proof. Differentiating (2.8) partially w.r.t. x and y, we get, respectively,

$$\frac{\partial f}{\partial x} + \frac{\partial f}{\partial z} \cdot \frac{\partial z}{\partial x} = 0 \text{ or } f_x + pf_z = 0 \tag{2.9}$$

and

$$\frac{\partial f}{\partial y} + \frac{\partial f}{\partial z} \cdot \frac{\partial z}{\partial y} = 0 \text{ or } f_y + qf_z = 0. \tag{2.10}$$

Eliminating a and b (if possible) among Eqs (2.8), (2.9), and (2.10), we get an equation of the form

$$F(x, y, z, p, q) = 0 \tag{2.11}$$

which is a PDE of first order.

If it is not possible to eliminate a and b using Eqs (2.8), (2.9), and (2.10), then we determine the second-order (or higher order) partial derivatives so that we can eliminate a and b. In this case, we obtain a PDE of order greater than one. $\qquad\square$

Here, it is also noticed that any family of surfaces does not necessarily derive a unique PDE; it may derive different PDEs. Instead of (2.8), we can consider a one-parameter family of surfaces, a three-parameter family of surfaces, and so on, and hence we may obtain PDEs by eliminating these parameters.

Example 2.1. *Obtain a PDE by eliminating the constants γ and c from*

$$x^2 + y^2 + (z - \gamma)^2 = c^2.$$

Solution: *Differentiating the given equation partially w.r.t. x and y, we get, respectively,*

$$x + p(z - \gamma) = 0 \text{ and } y + q(z - \gamma) = 0.$$

Eliminating γ between these two equations, we obtain the PDE

$$yp - xq = 0.$$

Example 2.2. *Find a PDE by eliminating a and b from*

$$z = ax + by + ab.$$

Solution: *Differentiating the given equation partially w.r.t. x and y, we get, respectively,*

$$p = a \text{ and } q = b.$$

Putting these values of a and b in the given equation, we obtain

$$z = px + qy + pq$$

which is the required PDE.

Example 2.3. *Form a PDE by eliminating a, b, and c from*

$$2z = ax^2 + bxy + cy^2.$$

Solution: *Differentiating the given equation partially w.r.t. x and y, we get, respectively,*

$$2p = 2ax + by \text{ and } 2q = bx + 2cy.$$

It is not possible to eliminate constants using above equations. Thus, differentiating again to get

$$r = a, \ s = b, \text{ and } t = c.$$

Putting the values of a, b, c in the given equation, we get

$$2z = x^2 r + xys + y^2 t$$

which is the required PDE of order two.

Example 2.4. *Form a PDE by eliminating the constant λ from*

$$z = \lambda(x + y).$$

Solution: Differentiating the given equation partially w.r.t. x and y, we get, respectively,

$$p = \lambda \text{ and } q = \lambda.$$

Eliminating λ from the above equations and the given equation, we get the following three distinct PDEs:

$$z = p(x + y), \quad z = q(x + y), \quad \text{and} \quad p = q.$$

Example 2.5. *Eliminate the constants k and ω from*

$$z = ke^{\omega y} \cos \omega x$$

to obtain a PDE.

Solution: Differentiating the given equation partially w.r.t. x and y, we get, respectively,

$$p = -k\omega e^{\omega y} \sin \omega x \text{ and } q = k\omega e^{\omega y} \cos \omega x. \tag{2.12}$$

Now, it is not easy to eliminate the constants between equations and the given equation. Hence, differentiating again, we obtain

$$r = \frac{\partial p}{\partial x} = -k\omega^2 e^{\omega y} \cos \omega x \tag{2.13}$$

$$t = \frac{\partial q}{\partial y} = k\omega^2 e^{\omega y} \cos \omega x \tag{2.14}$$

and

$$s = \frac{\partial p}{\partial y} = k\omega^2 e^{\omega y} \sin \omega x = \frac{\partial q}{\partial x}. \tag{2.15}$$

From (2.13) and (2.14), we get the desired PDE of second order as

$$r + t = 0.$$

Further, from (2.12), (2.13), and (2.15), we obtain

$$s = \omega p \text{ and } r = -\omega q.$$

Eliminating ω between the above equations, we get another PDE given by

$$qs + pr = 0.$$

2.9 Formation of PDEs by Elimination of Arbitrary Functions

In this section, we derive PDEs from the families of surfaces containing arbitrary functions instead of arbitrary constants. For brevity, we divide this section into four subsections.

2.9.1 Formation of First-Order Quasilinear PDEs

We eliminate one arbitrary function to get a first-order quasilinear PDE, which is embodied in the following theorem.

Theorem 2.3. *Let u and v be two known functions of x, y, z. Consider the surface of the form*

$$\phi(u, v) = 0 \tag{2.16}$$

where ϕ is an arbitrary function. By eliminating ϕ from (2.16), we get a PDE of the form: $Pp + Qq = R$, where P, Q, R are functions of x, y, z, respectively.

Proof. Differentiating (2.16) partially w.r.t. *x*, we get

$$\frac{\partial \phi}{\partial u}\left(\frac{\partial u}{\partial x} + \frac{\partial u}{\partial z}\frac{\partial z}{\partial x}\right) + \frac{\partial \phi}{\partial v}\left(\frac{\partial v}{\partial x} + \frac{\partial v}{\partial z}\frac{\partial z}{\partial x}\right) = 0$$

or

$$\frac{\partial \phi}{\partial u}(u_x + pu_z) + \frac{\partial \phi}{\partial v}(v_x + pv_z) = 0. \tag{2.17}$$

Similarly, differentiating (2.16) partially w.r.t. *y*, we get

$$\frac{\partial \phi}{\partial u}(u_y + qu_z) + \frac{\partial \phi}{\partial v}(v_y + qv_z) = 0. \tag{2.18}$$

Eliminating $\dfrac{\partial \phi}{\partial u}$ and $\dfrac{\partial \phi}{\partial v}$ between (2.17) and (2.18), we obtain

$$\begin{vmatrix} u_x + pu_z & v_x + pv_z \\ u_y + qu_z & v_y + qv_z \end{vmatrix} = 0$$

or

$$(v_y u_z - u_y v_z)p + (u_x v_z - v_x u_z)q = u_y v_x - u_x v_y$$

which can be written as

$$Pp + Qq = R$$

where

$$P = v_y u_z - u_y v_z = \frac{\partial(u, v)}{\partial(y, z)}$$

$$Q = u_x v_z - v_x u_z = \frac{\partial(u, v)}{\partial(z, x)}$$

and

$$R = u_y v_x - u_x v_y = \frac{\partial(u, v)}{\partial(x, y)}.$$

Hence, from relation (2.16), we obtain a quasilinear PDE of order one. $\qquad\square$

Notice that the relation (2.16) is also expressed as in the explicit forms: $v = \psi(u)$, for an arbitrary function ψ of u or $u = \theta(v)$, for an arbitrary function θ of v.

Example 2.6. *Form a PDE by eliminating arbitrary function ϕ from*

$$\phi\left(\frac{y}{x}, \frac{z}{x}\right) = 0.$$

Solution: The given equation is of the form

$$\phi(u, v) = 0 \tag{2.19}$$

where $u = \dfrac{y}{x}$ and $v = \dfrac{z}{x}$. Differentiating (2.19) partially w.r.t. x and y, we get, respectively,

$$\frac{\partial\phi}{\partial u}(u_x + pu_z) + \frac{\partial\phi}{\partial v}(v_x + pv_z) = 0 \tag{2.20}$$

and

$$\frac{\partial\phi}{\partial u}(u_y + qu_z) + \frac{\partial\phi}{\partial v}(v_y + qv_z) = 0. \tag{2.21}$$

Here, we have

$$u_x = -\frac{y}{x^2}, u_y = \frac{1}{x}, u_z = 0$$

and

$$v_x = -\frac{z}{x^2}, v_y = 0, v_z = \frac{1}{x}.$$

Putting these values in (2.20) and (2.21), we obtain

$$\frac{\partial\phi}{\partial u}\left(-\frac{y}{x^2}\right) + \frac{\partial\phi}{\partial v}\left(-\frac{z}{x^2} + p.\frac{1}{x}\right) = 0$$

$$\tag{2.22}$$

$$\Rightarrow \qquad \frac{\partial\phi/\partial u}{\partial\phi/\partial v} = -\frac{z - xp}{y}$$

and

$$\frac{\partial \phi}{\partial u}\left(\frac{1}{x}\right) + \frac{\partial \phi}{\partial v}\left(q.\frac{1}{x}\right) = 0$$

(2.23)

$$\Rightarrow \quad \frac{\partial \phi / \partial u}{\partial \phi / \partial v} = -q.$$

Eliminating ϕ from (2.22) and (2.23), we get

$$\frac{z - xp}{y} = q$$

or

$$z = xp + yq$$

which is the required PDE of order one.

Example 2.7. *Obtain a PDE by eliminating arbitrary function ϕ from*

$$lx + my + nz = \phi(x^2 + y^2 + z^2).$$

Solution: *The given equation is of the form $v = \phi(u)$. Differentiating the given equation partially w.r.t. x and y, we get, respectively,*

$$l + np = (2x + 2zp)\phi'(x^2 + y^2 + z^2)$$

and

$$m + nq = (2y + 2zq)\phi'(x^2 + y^2 + z^2).$$

On dividing, we get

$$\frac{l + np}{m + nq} = \frac{x + zp}{y + zq}$$

or

$$(ny - mz)p + (lz - nx)q = mx - ly$$

which is the desired PDE.

2.9.2 Formation of Second-Order Linear PDEs

We eliminate two arbitrary functions from the surface to get a second-order linear PDE, which is embodied in the following theorem.

Theorem 2.4. *Let u, v, w be known functions of x and y. Consider the surface of the form*

$$z = \phi(u) + \psi(v) + w \qquad (2.24)$$

where ϕ and ψ are arbitrary functions of u and v, respectively. By eliminating ϕ and ψ from (2.24), we get a PDE of the form $Rr + Ss + Tt + Pp + Qq = V$, where R, S, T, P, Q, V are functions of x and y.

Proof. Differentiating (2.24) twice partially w.r.t. x and y, we get,

$$p = \phi'(u)u_x + \psi'(v)v_x + w_x$$

$$q = \phi'(u)u_y + \psi'(v)v_y + w_y$$

$$r = \phi''(u)u_x^2 + \psi''(v)v_x^2 + \phi'(u)u_{xx} + \psi'(v)v_{xx} + w_{xx}$$

$$s = \phi''(u)u_x u_y + \psi''(v)v_x v_y + \phi'(u)u_{xy} + \psi'(v)v_{xy} + w_{xy}$$

and

$$t = \phi''(u)u_y^2 + \psi''(v)v_y^2 + \phi'(u)u_{yy} + \psi'(v)v_{yy} + w_{yy}.$$

Above five equations can be written, respectively, as

$$(p - w_x) - \phi'(u)u_x - \psi'(v)v_x = 0 \qquad (2.25)$$

$$(q - w_y) - \phi'(u)u_y - \psi'(v)v_y = 0 \qquad (2.26)$$

$$(r - w_{xx}) - \phi'(u)u_{xx} - \psi'(v)v_{xx} - \phi''(u)u_x^2 - \psi''(v)v_x^2 = 0 \qquad (2.27)$$

$$(s - w_{xy}) - \phi'(u)u_{xy} - \psi'(v)v_{xy} - \phi''(u)u_x u_y - \psi''(v)v_x v_y = 0 \qquad (2.28)$$

and

$$(t - w_{yy}) - \phi'(u)u_{yy} - \psi'(v)v_{yy} - \phi''(u)u_y^2 - \psi''(v)v_y^2 = 0. \qquad (2.29)$$

Eliminating $\phi'(u), \psi'(v), \phi''(u)$, and $\psi''(v)$ among Eqs (2.25)–(2.29), we get

$$\begin{vmatrix} p - w_x & u_x & v_x & 0 & 0 \\ q - w_y & u_y & v_y & 0 & 0 \\ r - w_{xx} & u_{xx} & v_{xx} & u_x^2 & v_x^2 \\ s - w_{xy} & u_{xy} & v_{xy} & u_x u_y & v_x v_y \\ t - w_{yy} & u_{yy} & v_{yy} & u_y^2 & v_y^2 \end{vmatrix} = 0$$

which on expanding gives a PDE of the form

$$Rr + Ss + Tt + Pp + Qq = V$$

where R, S, T, P, Q, V are known functions of x and y. Therefore, the relation (2.24) generates certain linear PDE of order two, in which the dependent variable z does not occur. □

Example 2.8. *Eliminate the arbitrary functions ϕ and ψ from the relation*

$$z = \phi(y + mx) + \psi(y - mx)$$

and find a PDE.

Solution: *Differentiating the given equation partially w.r.t. x and y, we get*

$$p = m\phi'(y + mx) - m\psi'(y - mx) \text{ and } q = \phi'(y + mx) + \psi'(y - mx).$$

Differentiating again, we get

$$r = m^2[\phi''(y + mx) + \psi''(y - mx)] \text{ and } t = \phi''(y + mx) + \psi''(y - mx).$$

Above two equations give rise to

$$r = m^2 t$$

which is a second-order PDE as desired.

Example 2.9. *Find a PDE by eliminating arbitrary functions ϕ and ψ from*

$$z = y\phi(x) + x\psi(y). \tag{2.30}$$

Solution: *Differentiating Eq. (2.30) partially w.r.t. x and y, we get*

$$p = y\phi'(x) + \psi(y) \tag{2.31}$$

and

$$q = \phi(x) + x\psi'(y). \tag{2.32}$$

Differentiating (2.31) partially w.r.t. y (or, (2.32) w.r.t. x), we get

$$s = \phi'(x) + \psi'(y). \tag{2.33}$$

From (2.31) and (2.32), we get

$$\phi'(x) = \frac{p - \psi(y)}{y} \text{ and } \psi'(y) = \frac{q - \phi(x)}{x}.$$

Putting the values of $\phi'(x)$ and $\psi'(y)$ in (2.33), we obtain

$$s = \frac{p - \psi(y)}{y} + \frac{q - \phi(x)}{x}$$

or

$$xys = xp + yq - [y\phi(x) + x\psi(y)]$$

which on using (2.30), becomes the desired PDE given as

$$xys = xp + yq - z.$$

2.9.3 Formation of Second-Order Quasilinear PDEs

Now, we consider the result regarding the elimination of an arbitrary function from a first-order PDE to obtain a second-order quasilinear PDE.

Theorem 2.5. *Let u be a known function of x, y, z; while v be an another known function of x, y, z, p, q. Consider the surface of the form*

$$v = \phi(u) \tag{2.34}$$

where ϕ is an arbitrary function. By eliminating ϕ from (2.34), we get a PDE of the form $Rr + Ss + Tt = V$, where R, S, T, V are functions of x, y, z, p, q.

Proof. Differentiating (2.34) partially w.r.t. x, we get

$$\frac{\partial v}{\partial x} + \frac{\partial v}{\partial z}\frac{\partial z}{\partial x} + \frac{\partial v}{\partial p}\frac{\partial p}{\partial x} + \frac{\partial v}{\partial q}\frac{\partial q}{\partial x} = \phi'(u)\left[\frac{\partial u}{\partial x} + \frac{\partial u}{\partial z}\frac{\partial z}{\partial x}\right]$$

or

$$v_x + pv_z + rv_p + sv_q = \phi'(u)(u_x + pu_z). \tag{2.35}$$

Similarly, differentiating (2.34) partially w.r.t. y, we get,

$$v_y + qv_z + sv_p + tv_q = \phi'(u)(u_y + qu_z). \tag{2.36}$$

Eliminating $\phi'(u)$ from Eqs (2.35) and (2.36), we get

$$(v_x + pv_z + rv_p + sv_q)(u_y + qu_z) = (v_y + qv_z + sv_p + tv_q)(u_x + pu_z)$$

which after rearranging takes the form

$$Rr + Ss + Tt = V$$

where R, S, T, V are known functions of x, y, z, p, q. Therefore, the relation (2.34) generates a quasilinear PDE of order two. $\qquad\square$

Example 2.10. *Form a PDE by eliminating the arbitrary function ϕ from*

$$2xp - yq = \phi(x^2y).$$

Solution: *Differentiating the given equation partially w.r.t. x and y, we get*

$$2p + 2xr - ys = 2xy\phi'(x^2y)$$

and

$$2xs - q - yt = x^2\phi'(x^2y).$$

On dividing, we get

$$\frac{2p + 2xr - ys}{2xs - q - yt} = \frac{2y}{x}$$

or

$$x(2p + 2xr - ys) - 2y(2xs - q - yt) = 0$$

or

$$2x^2r - 5xys + 2y^2t + 2(px + qy) = 0$$

which is a second-order quasilinear PDE as desired.

Example 2.11. *Obtain a PDE by eliminating the arbitrary function ϕ from*

$$p^2 + q = \phi(2x + y).$$

Solution: *Differentiating the given equation partially w.r.t. x and y, we get*

$$2pr + s = 2\phi'(2x + y)$$

and

$$2ps + t = \phi'(2x + y).$$

Combining the above two equations, we obtain

$$2pr + s = 2(2ps + t)$$

or

$$2pr + s = 4ps + 2t$$

which is a quasilinear PDE of order two.

2.9.4 Formation of Second-Order Non-linear PDEs

In the following result, we slightly generalise the idea contained in Theorem 2.5 to obtain a special form of non-linear PDE of order two.

Theorem 2.6. *Let u and v be two known functions of x, y, z, p, q. Consider the surface of the form*

$$v = \phi(u) \tag{2.37}$$

where ϕ is an arbitrary function. By eliminating ϕ from (2.37), we get a PDE of the form $Rr + Ss + Tt + U(rt - s^2) = V$, where R, S, T, U, V are functions of x, y, z, p, q, respectively.

Proof. Differentiating (2.37) partially w.r.t. x, we get

$$\frac{\partial v}{\partial x} + \frac{\partial v}{\partial z}\frac{\partial z}{\partial x} + \frac{\partial v}{\partial p}\frac{\partial p}{\partial x} + \frac{\partial v}{\partial q}\frac{\partial q}{\partial x} = \phi'(u)\left[\frac{\partial u}{\partial x} + \frac{\partial u}{\partial z}\frac{\partial z}{\partial x} + \frac{\partial u}{\partial p}\frac{\partial p}{\partial x} + \frac{\partial u}{\partial q}\frac{\partial q}{\partial x}\right]$$

or

$$v_x + pv_z + rv_p + sv_q = \phi'(u)(u_x + pu_z + ru_p + su_q + ru_p + su_q). \tag{2.38}$$

Similarly, differentiating (2.37) partially w.r.t. y, we get

$$v_y + qv_z + sv_p + tv_q = \phi'(u)(u_y + qu_z + su_p + tu_q). \tag{2.39}$$

Eliminating $\phi'(u)$ from Eqs (2.38) and (2.39), we find that the terms rs and st cancel out, leaving a PDE of the form

$$Rr + Ss + Tt + U(rt - s^2) = V$$

where R, S, T, U, V are known functions of x, y, z, p, q. Therefore, the relation (2.37) generates a certain type of non-linear PDEs of order two. □

Example 2.12. *Form a PDE by eliminating the arbitrary function ϕ from*

$$y - p = \phi(x - q).$$

Solution: Differentiating the given equation partially w.r.t. x and y, we get

$$-r = (1 - s)\phi'(x - q)$$

and

$$1 - s = -t\phi'(x - q).$$

On dividing, we get

$$\frac{-r}{1 - s} = \frac{1 - s}{-t}$$

or

$$rt = (1 - s)^2$$

or

$$2s + (rt - s^2) = 1$$

which is a second-order non-linear PDE as desired.

Example 2.13. *Form a PDE by eliminating the arbitrary function ϕ from*

$$xp + yq = \phi(p^2 + q^2).$$

Solution: Differentiating the given equation partially w.r.t. x and y, we get

$$p + xr + ys = (2pr + 2qs)\phi'(p^2 + q^2)$$

and

$$xs + q + yt = (2ps + 2qt)\phi'(p^2 + q^2).$$

On dividing, we get

$$\frac{p + xr + ys}{xs + q + yt} = \frac{pr + qs}{ps + qt}$$

or

$$(p + xr + ys)(ps + qt) = (xs + q + yt)(pr + qs)$$

or

$$pq(r - t) - (p^2 - q^2)s + (yp - xq)(rt - s^2) = 0$$

which is a second-order non-linear PDE as desired.

▶ **Exercises** ──────────────────────────────

1. Show that the travelling wave $u(x, t) = \phi(x - 3t)$ satisfies the linear transport equation $u_t + 3u_y$, for any arbitrary differentiable function ϕ.

2. Show that each of the following equations has a general solution of the form $z = \phi(ax + by)$ for a proper choice of constants a, b. Find these constants for each.
   (a) $p + 3q = 0$.
   (b) $3p - 7q = 0$.
   (c) $2p + \pi q = 0$.

3. Show that each of the following equations has a complete solution of the form $z = e^{mx+ny}$ for a proper choice of constants m, n. Find these constants for each.
   (a) $p + 3q + z = 0$.
   (b) $r + t = 5e^{x-2y}$.
   (c) $\dfrac{\partial^4 z}{\partial x^4} + \dfrac{\partial^4 z}{\partial y^4} + 2\dfrac{\partial^4 z}{\partial x^2 \partial y^2} = 0$.

4. Verify that $z = e^{mx+m^2 y}$ is a solution of the second-order equation $r = q$. Find a solution of the same equation that vanishes when $y = \infty$.   Ans. $z = e^{-\omega^2 y}(A \cos \omega x + B \sin \omega x)$.

5. Verify that $z = (5x - 6x^5 + x^9)y^6$ satisfies the PDE:

$$x^3 y^2 \frac{\partial^3 z}{\partial x \partial y^2} - 9x^2 y^2 \frac{\partial^2 z}{\partial y^2} = y\frac{\partial^3 z}{\partial x^2 \partial y} + 4\frac{\partial^2 z}{\partial x^2}.$$

6. Verify that $u = \dfrac{1}{t}e^{-x^2/4kt}$ is a solution of the heat equation

$$\frac{\partial^2 u}{\partial x^2} = \frac{1}{k}\frac{\partial u}{\partial t}.$$

7. Verify that any function of the form $\phi(x + ct)$ satisfies the wave equation

$$\frac{\partial^2 u}{\partial x^2} = \frac{1}{c^2}\frac{\partial^2 u}{\partial t^2}.$$

8. Verify that the functions

$$u(x, y) = x^2 - y^2$$
$$u(x, y) = e^x \sin y$$
$$u(x, y) = 2xy$$

are the solutions of the Laplace equation

$$u_{xx} + u_{yy} = 0.$$

Eliminate the arbitrary constants from each of the equations described in problems 9–16 and obtain the corresponding PDEs.

9. $z = (x + a)(y + b)$.   Ans. $z = pq$.
10. $z = ax + by + f(a, b)$.   Ans. $z = xp + yq + f(p, q)$.
11. $2z = \dfrac{x^2}{a^2} + \dfrac{y^2}{b^2}$.   Ans. $2z = xp + yq$.
12. $\log(az - 1) = x + ay + b$.   Ans. $(1 + q)p = zq$.
13. $(x - h)^2 + (y - k)^2 + z^2 = 1$.   Ans. $z^2(p^2 + q^2 + 1) = 1$.
14. $\dfrac{x^2}{a^2} + \dfrac{y^2}{b^2} + \dfrac{z^2}{c^2} = 1$.   Ans. $xzr + xp^2 - zp = 0$.
15. $x^2 + y^2 = (z - c)^2 \tan^2 \alpha$.   Ans. $yp - xq = 0$.
16. $z = ke^{-\omega y} \sin \omega x$.   Ans. $r + t = 0$.

Eliminate the arbitrary functions from each of the equations described in problems 17–24 and obtain the corresponding PDEs.

17. $\phi(x^2 + y^2 + z^2, z^2 - 2xy) = 0$.   Ans. $z(p - q) = y - x$.

18. $\phi(x^2 + y^2, z^2 - xy) = 0$.   Ans. $yp - xq = y^2 - x^2$.

19. $z = \phi(x)\psi(y)$.   Ans. $zs = pq$.

20. $z = \phi(x^2 - y^2)$.   Ans. $yp + qx = 0$.

21. $z = x^n \phi\left(\dfrac{y}{x}\right)$.   Ans. $xp + yq = nz$.

22. $z = x\phi\left(\dfrac{y}{x}\right) + y\psi(x)$.   Ans. $y^2 t + xys = xp + yq - z$.

23. $z = \phi(x + iy) + \psi(x - iy)$, where $i^2 = -1$.   Ans. $r + t = 0$.

24. $z = \dfrac{1}{x}\theta(y - x) + \theta'(y - x)$, where $\theta'(u) = \dfrac{d\theta}{du}$.   Ans. $x^2(r - t) = 2z$.

Eliminate the arbitrary functions from each of the PDEs of first order described in problems 25–30 and obtain the second-order PDEs.

25. $xp - 2yq = \phi(xy^2)$.   Ans.
$2x^2 r - 5xys + 2y^2 t + 2(px + qy) = 0$.

26. $yp - q + 3y^2 = \phi(2x + y^2)$.   Ans. $y^2 - 2ys + t = p + 6y$.

27. $x - \dfrac{1}{q} = \phi(z)$.   Ans. $pt - qs = pq^3$.

28. $p + x - y = \phi(q - 2x + y)$.   Ans. $r + 3s + t + (rt - s^2) = 1$.

29. $p^2 - x = \phi(q^2 - 2y)$.   Ans. $2pr + qt - 2pq(rt - s^2) = 1$.

30. $p + zq = \phi(z)$.   Ans. $qr + (zq - p)s - zpt = 0$.

3

Easily Solvable Partial Differential Equations

There are several ways of solving partial differential equations (PDEs), which are utilised in specific situations depending on the order of the underlying equation. In succeeding chapters, we discuss several PDEs of order one (Chapter 4), two (Chapter 5), and higher (Chapter 6) in a systematic manner. Nevertheless, we begin our journey with some basic methods to solve special types of PDEs, which are solvable easily. These basic methods often do not depend on the order of PDEs. In this chapter, first, we undertake certain types of PDEs, which can be solved merely by integration. Thereafter, we consider the special classes of linear PDEs, which can be solved on the lines of linear ordinary differential equations (ODEs). Finally, we describe the method of separation of variables.

3.1 Direct Integration Method

Roughly speaking, the operations of 'differentiation' and 'integration' are the inverses of each other. To associate an inverse of partial differentiation, consider a function $u(x, y)$, whose partial derivative w.r.t. x is $f(x, y)$, *i.e.*,

$$\frac{\partial}{\partial x} u(x, y) = f(x, y) \text{ or } \partial u(x, y) = f(x, y) \partial x.$$

If the above relation is satisfied, then we say that $u(x, y)$ is a partial integral of $f(x, y)$ w.r.t. x and symbolically, we write

$$\int f(x, y) \partial x = u(x, y).$$

In order to have a substitute of constant of integration, we replace an arbitrary function ϕ of y owing to the fact that an arbitrary constant naturally is absorbed with $\phi(y)$ so that we have

$$\frac{\partial}{\partial x} \left[u(x, y) + \phi(y) \right] = f(x, y)$$

which amounts to saying that if $u_x(x, y) = f(x, y)$, then

$$\int f(x, y) \partial x = u(x, y) + \phi(y).$$

This motivates us to define the concept of partial integration.

Partial Integration: Let $f(x_1, x_2, \ldots, x_n)$ be a function of n independent variables, then the partial integral of f w.r.t. an independent variable x_i ($1 \leq i \leq n$), often denoted by $\int f(x_1, x_2, \ldots, x_n) \partial x_i$, is defined as the integral of f w.r.t. x_i treating the rest $n - 1$ independent variables as constants and taking an arbitrary function $\phi(x_1, x_2, \ldots, x_{i-1}, x_{i+1}, \ldots, x_n)$ of these $n - 1$ variables instead of arbitrary constant of integration.

Generally, a PDE containing single partial derivative can be easily solved by the technique of direct integration. Such PDEs are illustrated by Examples 3.1–3.3. On the other hand, a PDE involving two derivatives of consecutive orders in its distinct terms and not containing dependent variable explicitly, sometimes can also be solved by this method (see Examples 3.4–3.6).

Example 3.1. *Solve $u_{yx} = z^3 + 4xy - 6y^2$, where $u = u(x, y, z)$.*

Solution: *The given equation can be rewritten as*

$$\frac{\partial}{\partial x}\left(\frac{\partial u}{\partial y}\right) = z^3 + 4xy - 6y^2.$$

Integrating partially w.r.t. x, we get

$$\frac{\partial u}{\partial y} = xz^3 + 2x^2 y - 6xy^2 + \phi(y, z)$$

where ϕ is an arbitrary function of y and z. Now, integrating partially w.r.t. y, we get

$$u = xyz^3 + x^2 y^2 - 2xy^3 + \int \phi(y, z) \partial y + \psi(x, z).$$

Denoting $\int \phi(y, z)\partial y = \theta(y, z)$, we obtain the required solution of the given PDE as

$$u = xyz^3 + x^2 y^2 - 2xy^3 + \psi(x, z) + \theta(y, z)$$

where ψ is an arbitrary function of x and z, while θ is an arbitrary function of y and z, differentiable in the variable y.

Example 3.2. *Solve $\dfrac{\partial^3 z}{\partial x \partial y^2} + \cos(x - 2y) - 12xy = 0$.*

Solution: *Integrating the given equation partially w.r.t. y, we get*

$$\frac{\partial^2 z}{\partial x \partial y} - \frac{1}{2}\sin(x - 2y) - 6xy^2 = \phi(x).$$

Integrating again partially w.r.t. y, we get

$$\frac{\partial z}{\partial x} - \frac{1}{4}\cos(x - 2y) - 2xy^3 = y\phi(x) + \psi(x).$$

Finally, integrating partially w.r.t. x, we get

$$z - \frac{1}{4} \sin(x - 2y) - x^2 y^3 = y \int \phi(x) dx + \int \psi(x) dx + \theta(y).$$

Denoting $\int \phi(x) dx = \mu(x)$ and $\int \psi(x) dx = \eta(x)$, we obtain the required solution of the given PDE as

$$z = \frac{1}{4} \sin(x - 2y) + x^2 y^3 + y\mu(x) + \eta(x) + \theta(y).$$

Example 3.3. *Solve $q = 2xyz$, whenever $z(x, 0) = e^{x^2}$.*

Solution: *The given equation can be written as*

$$\frac{\partial z}{\partial y} = 2xyz$$

or

$$\frac{\partial z}{z} = 2xy \partial y.$$

On integrating partially, we get

$$\log z = xy^2 + \log \phi(x)$$

or

$$z = \phi(x) e^{xy^2}.$$

Applying $z(x, 0) = e^{x^2}$, we get

$$\phi(x) = e^{x^2}.$$

Hence, the required solution of the given PDE is

$$z = e^{xy^2 + x^2}.$$

Example 3.4. *Solve $ys + p = \cos(x + y) - y \sin(x + y)$.*

Solution: *The given equation can be written as*

$$y \frac{\partial q}{\partial x} + \frac{\partial z}{\partial x} = \cos(x + y) - y \sin(x + y).$$

Integrating partially w.r.t. x, we get

$$yq + z = \sin(x + y) + y \cos(x + y) + \phi(y)$$

or

$$y\frac{\partial z}{\partial y} + z = \sin(x+y) + y\cos(x+y) + \phi(y).$$

Now, integrating partially w.r.t. y, we get

$$yz = y\sin(x+y) + \int \phi(y)dy + \psi(x).$$

Denote $\theta(y) = \int \phi(y)dy$. Then, the required solution of the given PDE is

$$yz = y\sin(x+y) + \theta(y) + \psi(x).$$

Example 3.5. *Solve $y^2 t + 2yq = 1$.*

Solution: The given equation can be written as

$$y^2\frac{\partial q}{\partial y} + 2yq = 1.$$

Integrating partially w.r.t. y, we get

$$y^2 q = y + \phi(x)$$

or

$$\frac{\partial z}{\partial y} = \frac{1}{y} + \frac{1}{y^2}\phi(x).$$

Integrating partially again w.r.t. y, we get

$$z = \log y - \frac{1}{y}\phi(x) + \psi(x)$$

or

$$yz = y\log y - \phi(x) + y\psi(x)$$

which is the required solution of the given PDE.

Example 3.6. *Solve $rx = (k-1)p$.*

Solution: The given equation can be written as

$$\frac{\partial p}{\partial x}x = (k-1)p$$

or

$$\frac{\partial p}{p} = (k-1)\frac{\partial x}{x}.$$

On integrating partially, we get

$$\log p = (k - 1)\log x + \log \theta(y)$$

or

$$\frac{\partial z}{\partial x} = x^{k-1}\theta(y).$$

Integrating partially again w.r.t. x, we get

$$z = \frac{x^k}{k}\theta(y) + \psi(y)$$

or

$$z = x^k\phi(y) + \psi(y)$$

which is the required solution of the given PDE.

Example 3.7. *Show that the integral surface of the equation s = 8xy passing through the circle z = 0, $x^2 + y^2 = 1$ is $z = (x^2 + y^2)^2 - 1$.*

Solution: The given equation can be written as

$$\frac{\partial^2 z}{\partial x \partial y} = 8xy.$$

Integrating partially w.r.t. x, we get

$$\frac{\partial z}{\partial y} = 4x^2 y + \phi(y).$$

Now, integrating partially w.r.t. y, we get

$$z = 2x^2 y^2 + \int \phi(y)\partial y + \psi(x)$$

or

$$z = 2x^2 y^2 + \theta(y) + \psi(x). \tag{3.1}$$

Equation of the given circle is

$$z = 0, \; x^2 + y^2 = 1. \tag{3.2}$$

Putting z = 0 in Eq. (3.1), we get

$$2x^2 y^2 + \theta(y) + \psi(x) = 0. \tag{3.3}$$

Again, on the circle (3.2), we have

$$x^2 + y^2 = 1.$$

Squaring on both sides, we get

$$x^4 + y^4 + 2x^2y^2 = 1. \tag{3.4}$$

Comparing Eqs (3.3) and (3.4), we obtain

$$\theta(y) + \psi(x) = x^4 + y^4 - 1. \tag{3.5}$$

Using (3.5), Eq. (3.1) reduces to

$$z = 2x^2y^2 + x^4 + y^4 - 1$$

or

$$z = (x^2 + y^2)^2 - 1$$

which is the required integral surface of the given PDE.

Example 3.8. *Show that the integral surface of the equation $r = 6x + 2$ touching $z = x^3 + y^3$ along its section by the plane $x + y + 1 = 0$ is $z = x^3 + y^3 + (x + y + 1)^2$.*

Solution: The given equation can be written as

$$\frac{\partial^2 z}{\partial x^2} = 6x + 2.$$

Integrating partially w.r.t. x, we get

$$\frac{\partial z}{\partial x} = 3x^2 + 2x + \phi(y).$$

Integrating again partially w.r.t. x, we get

$$z = x^3 + x^2 + x\phi(y) + \psi(y). \tag{3.6}$$

Equation of the given surface is

$$z = x^3 + y^3. \tag{3.7}$$

Equation of the given plane is

$$x + y + 1 = 0. \tag{3.8}$$

For Eq. (3.6), we have

$$p = 3x^2 + 2x + \phi(y) \text{ and } q = x\phi'(y) + \psi'(y).$$

For Eq. (3.7), we have

$$p = 3x^2 \text{ and } q = 3y^2.$$

If the surface (3.6) touches the surface (3.7) along its section by the plane (3.8), then the values of p and q for any point on the plane (3.8) must be equal. Hence, we have

$$3x^2 + 2x + \phi(y) = 3x^2 \qquad (3.9)$$

$$x\phi'(y) + \psi'(y) = 3y^2. \qquad (3.10)$$

From (3.8) and (3.9), we have

$$\phi(y) = -2x = 2(y + 1)$$

which gives rise

$$\phi'(y) = 2.$$

Putting the value of $\phi'(y)$ in (3.10), we get

$$\psi'(y) = 3y^2 - 2x$$

which using (3.8) becomes

$$\psi'(y) = 3y^2 + 2(y + 1).$$

Integrating w.r.t. y, we get

$$\psi(y) = y^3 + y^2 + 2y + c.$$

Putting the values of $\phi(y)$ and $\psi(y)$ in (3.6), we get

$$z = x^3 + x^2 + 2x(y + 1) + y^3 + y^2 + 2y + c. \qquad (3.11)$$

Equating the two values of z from (3.7) and (3.11), when $y = -(x + 1)$, we get

$$x^3 - (x + 1)^3 = x^3 + x^2 + 2x(-x) - (x + 1)^3 + (x + 1)^2 - 2(x + 1) + c.$$

Comparing the constants on both the sides, we get

$$c = 1.$$

Putting the value of c in (3.11), we obtain

$$z = x^3 + x^2 + 2x(y + 1) + y^3 + y^2 + 2y + 1$$

or

$$z = x^3 + y^3 + (x + y + 1)^2$$

which is the required integral surface of the given PDE.

3.2 Linear Equations of Order One Containing Single Derivative

In Chapter 1, we have seen that a first-order linear ODE with variable coefficients can be solved employing its integration factor (IF). If a first-order linear PDE with variable coefficients contains only one partial derivative, then it may be treated as a linear ODE. Indeed, such types of PDEs have the following two most general forms:

$$P(x, y)p + Z(x, y)z = R(x, y) \quad \text{or} \quad \frac{\partial z}{\partial x} + \frac{Z}{P}z = \frac{R}{P} \tag{3.12}$$

and

$$Q(x, y)q + Z(x, y)z = R(x, y) \quad \text{or} \quad \frac{\partial z}{\partial y} + \frac{Z}{Q}z = \frac{R}{Q}. \tag{3.13}$$

Equation (3.12) can be solved on the lines of linear ODE in dependent variable z and independent variable x, so that y plays the role of a parameter. Consequently, instead of arbitrary constant of integration, we use an arbitrary function of the parameter y.

Similarly, Eq. (3.13) can be solved on the lines of linear ODE in dependent variable z and independent variable y, so that x plays the role of a parameter. Consequently, instead of arbitrary constant of integration, we use an arbitrary function of the parameter x.

The following two examples illustrate the techniques of finding the solutions of such types of equations.

Example 3.9. *Solve* $(1 + x^2)p = \tan^{-1}x - z$.

Solution: *The given equation can be written as*

$$\frac{\partial z}{\partial x} + \frac{z}{1 + x^2} = \frac{\tan^{-1}x}{1 + x^2}$$

which is a linear differential equation of first order in dependent variable z and independent variable x (here y is treated as a parameter). The integrating factor (IF) of the given equation is

$$\text{IF} = e^{\int \frac{1}{1+x^2}\partial x} = e^{\tan^{-1}x}.$$

Thus, the solution of the given PDE is

$$z \cdot e^{\tan^{-1}x} = \int \frac{\tan^{-1}x}{1 + x^2} \cdot e^{\tan^{-1}x}\partial x + \phi(y).$$

Putting $\tan^{-1} x = t$ *so that* $\dfrac{\partial x}{1 + x^2} = \partial t$, *above equation reduces to*

$$ze^t = \int e^t t \, \partial t + \phi(y)$$
$$= te^t - e^t + \phi(y)$$

or

$$z = t - 1 + \phi(y)e^{-t}.$$

Therefore, the required general solution of the given equation is

$$z = \tan^{-1}x - 1 + \phi(y)e^{-\tan^{-1}x}.$$

Example 3.10. *Solve* $(x^2 - y^2)q + 2yz = y\sqrt{x^2 - y^2}.$

Solution: *The given equation can be written as*

$$\frac{\partial z}{\partial y} + \frac{2yz}{x^2 - y^2} = \frac{y}{\sqrt{x^2 - y^2}}$$

which is a linear differential equation of first order in dependent variable z and independent variable y (here x is treated as a parameter). Hence, its IF is

$$\text{IF} = e^{\int \frac{2y}{x^2-y^2}\,dy} = \frac{1}{x^2 - y^2}.$$

Therefore, the solution of the given equation is

$$z \cdot \frac{1}{x^2 - y^2} = \int \frac{y}{\sqrt{x^2 - y^2}} \cdot \frac{1}{x^2 - y^2}\,dy + \phi(x)$$

or

$$\frac{z}{x^2 - y^2} = \int \frac{y}{(x^2 - y^2)^{3/2}}\,dy + \phi(x).$$

Putting $x^2 - y^2 = t$ *so that* $-2y\partial y = \partial t$, *the above equation reduces to*

$$\frac{z}{x^2 - y^2} = -\frac{1}{2}\int t^{-3/2}\partial t + \phi(x)$$

$$= t^{-1/2} + \phi(x)$$

$$= \frac{1}{\sqrt{x^2 - y^2}} + \phi(x)$$

implying thereby

$$z = \sqrt{x^2 - y^2} + (x^2 - y^2)\phi(x)$$

which is the general solution of the given PDE as desired.

3.3 Linear PDEs Solvable as ODEs

A linear PDE, in which the partial derivatives w.r.t. merely one variable are involved and hence are free from partial derivatives w.r.t other variables, can be easily solved. The technique to solve such equations is similar to that of solving linear ODEs. There are two most general forms of such types of equations, which are given as under:

$$\left[a_0(y)D^n + a_1(y)D^{n-1} + \cdots + a_n(y)D\right]z = f(x, y) \tag{3.14}$$

where the coefficients $a_0, a_1, \ldots, a_n$ all are functions of y only and $a_0 \neq 0$.

$$\left[b_0(x)D'^n + b_1(x)D'^{n-1} + \cdots + b_n(x)D'\right]z = f(x, y) \tag{3.15}$$

where the coefficients $b_0, b_1, \ldots, b_n$ all are functions of x only and $b_0 \neq 0$.

Equation (3.14) can be solved on the lines of linear ODE with constant coefficients in dependent variable z and independent variable x, so that y must be treated as a parameter. Consequently, instead of an arbitrary constant, we use an arbitrary function of the parameter y. Analogously, Eq. (3.15) can be solved by interchanging the role of x and y. The methods of solving linear ODEs with constant coefficients have been summarised in Chapter 1.

We present several examples of PDEs of the forms (3.14) and (3.15).

Example 3.11. *Solve* $\dfrac{\partial^2 z}{\partial x^2} - \omega^2 z = 0$.

Solution: *The given PDE can be written as*

$$(D^2 - \omega^2)z = 0.$$

Its auxiliary equation (AE) is

$$m^2 - \omega^2 = 0$$

which roots are $m = \pm\omega$. Hence, the general solution of the given PDE is

$$z = \phi(y)e^{\omega x} + \psi(y)e^{-\omega x}$$

where ϕ and ψ are the arbitrary functions of y.

Example 3.12. *Solve* $\dfrac{\partial^2 z}{\partial y^2} + x^2 z = 0$. *Also, obtain a particular solution satisfying the following conditions:*

$$z = e^x \text{ and } \frac{\partial z}{\partial y} = x^2 \text{ whenever } y = 0.$$

Solution: The given PDE can be written as

$$(D'^2 + x^2)z = e^y$$

Its AE is

$$m^2 + x^2 = 0$$

whose roots are $m = \pm ix$. *Hence, the general solution of the given PDE is*

$$z = \phi(x)\cos xy + \psi(x)\sin xy.$$

Differentiating partially above equation w.r.t. y, we get

$$\frac{\partial z}{\partial y} = -x\phi(x)\sin xy + x\psi(x)\cos xy.$$

Applying the condition $z(x, 0) = e^x$, *we get* $\phi(x) = e^x$. *Also, using the condition* $\frac{\partial z}{\partial y}(x, 0) = x^2$, *we obtain* $x\psi(x) = x^2$, *which yields that* $\psi(x) = x$. *Therefore, the required particular solution is*

$$z = e^x \cos xy + x \sin xy.$$

Example 3.13. *Solve* $\dfrac{\partial^2 z}{\partial y^2} - 3\dfrac{\partial z}{\partial y} - 4z = e^y.$

Solution: The given PDE can be written as

$$(D'^2 - 3D' - 4)z = e^y$$

Its AE is

$$m^2 - 3m - 4 = 0$$

whose roots are −1 *and* 4. *Hence, complementary function (CF) is*

$$z_c = \phi(x)e^{-y} + \psi(x)e^{4y}.$$

Further, particular integral (PI) of the given equation is

$$z_p = \frac{1}{D'^2 - 3D' - 4}e^y = \frac{e^y}{1^2 - 3.1 - 4}$$
$$= -\frac{1}{6}e^y.$$

Therefore, the general solution of the given PDE is

$$z = z_c + z_p = \phi(x)e^{-y} + \psi(x)e^{4y} - \frac{1}{6}e^y.$$

Example 3.14. *Solve* $\dfrac{\partial^2 z}{\partial x^2} - y\dfrac{\partial z}{\partial x} = e^{xy}$.

Solution: *The given PDE can be written as*

$$(D^2 - yD)z = e^{xy}.$$

Its AE is

$$m^2 - ym = 0$$

which gives $m = 0, y$. Hence, CF is

$$z_c = \phi(y) + \psi(y)e^{xy}.$$

Further, PI of the given equation is

$$
\begin{aligned}
z_p &= \frac{1}{D^2 - yD}e^{xy} \\
&= \frac{1}{D}\left[\frac{1}{D-y}e^{xy}\right] = \frac{1}{D}\left[e^{xy}\int e^{-xy} \cdot e^{xy}\partial x\right] \\
&= \frac{1}{D}xe^{xy} = \int xe^{xy}\partial x
\end{aligned}
$$

which integrating by parts gives

$$z_p = \frac{1}{y^2}(xy - 1)e^{xy}.$$

Therefore, the general solution of the given PDE is

$$z = z_c + z_p = \phi(y) + \psi(y)e^{xy} + \frac{1}{y^2}(xy - 1)e^{xy}.$$

Example 3.15. *Solve* $(D^3 + 8)z = x^4y^2 + xy^3 + 1$.

Solution: *AE of the given PDE is*

$$m^3 + 8 = 0$$

or

$$(m + 2)(m^2 - 2m + 4) = 0$$

whose roots are

$$m = -2, m = 1 \pm i\sqrt{3}.$$

Hence, CF is

$$z_c = \phi(y)e^{-2x} + e^x \left[\psi(y) \cos \sqrt{3}x + \theta(y) \sin \sqrt{3}x \right].$$

Further, PI of the given equation is

$$z_p = \frac{1}{(D^3 + 8)}(x^4 y^2 + xy^3 + 1)$$

$$= \frac{1}{8} \left[1 + \frac{D^3}{8} \right]^{-1} (x^4 y^2 + xy^3 + 1)$$

$$= \frac{1}{8} \left[1 - \frac{D^3}{8} + \frac{D^6}{64} + \cdots \right] (x^4 y^2 + xy^3 + 1)$$

$$= \frac{1}{8} \left[(x^4 y^2 + xy^3 + 1) - \frac{1}{8}D^3(x^4 y^2 + xy^3 + 1) + 0 \right]$$

$$= \frac{1}{8} \left[x^4 y^2 + xy^3 + 1 - \frac{1}{8}y^2.24x + 0 \right]$$

$$= \frac{1}{8} \left[x^4 y^2 + xy^3 - 3xy^2 + 1 \right].$$

Therefore, the general solution of the given PDE is

$$z = \phi(y)e^{-2x} + e^x \left[\psi(y) \cos \sqrt{3}x + \theta(y) \sin \sqrt{3}x \right] + \frac{1}{8} \left[x^4 y^2 + xy^3 - 3xy^2 + 1 \right].$$

Example 3.16. *Solve* $(D'^3 + D'^2 - D' - 1)z = (x^2 + 1)^2 \cos xy.$

Solution: AE of the given PDE is

$$m^3 + m^2 - m - 1 = 0$$

or

$$(m^2 - 1)(m + 1) = 0$$

which has three roots, viz. $m = 1, -1, -1$. *Hence, CF is*

$$z_c = \phi(x)e^y + (\psi(x) + y\theta(x))e^{-y}.$$

Further, PI of the given equation is

$$z_p = \frac{1}{D'^3 + D'^2 - D' - 1}(x^2 + 1)^2 \cos xy$$

$$= (x^2 + 1)^2 \frac{1}{(-x^2)D' + (-x^2) - D' - 1} \cos xy$$

$$= (x^2 + 1)^2 \frac{1}{(-x^2 - 1)(D' + 1)} \cos xy = -\frac{(x^2 + 1)}{D' + 1} \cos xy$$

$$= -\frac{(x^2 + 1)(D' - 1)}{D'^2 - 1} \cos xy$$

$$= -\frac{(x^2 + 1)(D' - 1)}{-x^2 - 1} \cos xy = (D' - 1) \cos xy$$

$$= -x \sin xy - \cos xy.$$

Therefore, the general solution of the given PDE is

$$z = \phi(x)e^y + (\psi(x) + y\theta(x))e^{-y} - x \sin xy - \cos xy.$$

Example 3.17. *Solve* $\dfrac{\partial^2 z}{\partial x^2} - 2y\dfrac{\partial z}{\partial x} + y^2 z = xe^{xy}\sin(x + y).$

Solution: *Here CF is*

$$z_c = \left(\phi(y) + x\psi(y)\right)e^{xy}.$$

Further, PI of the given equation is

$$z_p = \frac{1}{D^2 - 2yD + y^2}xe^{xy}\sin(x + y)$$

$$= e^{xy}\frac{1}{(D + y)^2 - 2y(D + y) + y^2}x\sin(x + y)$$

$$= e^{xy}\frac{1}{D^2}x\sin(x + y) = e^{xy}\int\left(\int x\sin(x + y)\partial x\right)\partial x$$

which integrating by parts gives rise

$$z_p = -e^{xy}\left[x\sin(x + y) + 2\cos(x + y)\right].$$

Therefore, the general solution of the given PDE is

$$z = \left[\phi(y) + x\psi(y) - x\sin(x + y) - 2\cos(x + y)\right]e^{xy}.$$

3.4 Method of Separation of Variables

In this section, we discuss a common and basic method called the **method of separation of variables**. This method is very old but a natural systematic method for solving certain types of linear PDEs. This classical method has undergone several refinements and improvements in the long course of the last two centuries. This technique applies to a wide range of problems from different domains such as mathematical physics, applied mathematics, and engineering sciences.

In fact, this method transforms the underlying linear PDE into two or more linear ODEs, which can be applied when the solution remains separable in independent variables, which is equivalent to saying that the dependent variable can be written as the product of functions of the individual independent variables. In the case of homogeneous linear PDEs, more general solution can be expressed as an infinite series in the appropriate separable solutions using the principle of superposition.

Now, we describe the method of separation of variables and examine in two independent variables x and y. We assume that the dependent variable z is expressed in the separable form, *i.e.*, the product of two functions, one of them is a function of x alone whereas the other is a function of y alone, that is,

$$z = X(x) \cdot Y(y) \neq 0. \tag{3.16}$$

Now, if Eq. (3.16) is reduced to the form

$$\frac{1}{X} P(D)X = \frac{1}{Y} Q(D')Y \tag{3.17}$$

where $P(D)$ and $Q(D')$ are the functions of $D = \dfrac{d}{dx}$ and $D' = \dfrac{d}{dy}$, respectively, then we say that the given PDE is separable in x and y and thus the method of separation of variables is applicable.

Since LHS of (3.17) is a function of x alone, while RHS remains a function of y alone and x and y are independent variables, it follows that (3.17) can be true only if each side is equal to the same constant value k, which is called a separation constant. Therefore, we obtain

$$P(D)X = kX$$

and

$$Q(D')Y = kY.$$

By solving these two ODEs, we can find the values X and Y, which using (3.16) gives rise to the solution of the given PDE. The value of the constants of integration can be obtained by applying the conditions given in the problem. A similar technique can be applied to PDEs in three or more independent variables.

Example 3.18. *Solve $p = q$.*

Solution: *Let the solution of the given PDE be of the form*

$$z = X(x) \cdot Y(y). \tag{3.18}$$

Differentiating (3.18) partially w.r.t. x and y, we get, respectively,

$$\frac{\partial z}{\partial x} = \frac{\partial}{\partial x}(XY) = Y\frac{\partial X}{\partial x} \Rightarrow p = X'Y$$

and

$$\frac{\partial z}{\partial y} = \frac{\partial}{\partial y}(XY) = X\frac{\partial Y}{\partial y} \Rightarrow q = XY'.$$

Putting these values in the given equation, we have

$$X'Y = XY'$$

which on separating the variables reduces to

$$\frac{X'}{X} = \frac{Y'}{Y} = k$$

where k is a separation constant. Thus, we get the following two ODEs:

$$X' = kX \qquad\qquad (3.19)$$

and

$$Y' = kY. \qquad\qquad (3.20)$$

From (3.19), we have

$$\frac{dX}{dx} = kX \text{ or } \frac{dX}{X} = kdx.$$

On integrating, we obtain

$$\log X = kx + \log c_1$$

or

$$X(x) = c_1 e^{kx}$$

which is the solution of (3.19). Similarly, the solution of (3.20) is

$$Y(y) = c_2 e^{ky}.$$

Putting the values of X and Y in (3.18), we get

$$z = c_1 c_2 e^{kx} e^{ky}$$

or

$$z = c e^{k(x+y)} \quad (where \; c = c_1 c_2)$$

which is the required solution of the given PDE.

Example 3.19. *Solve $p = 2q + z$, given that*

(i) $z(x, 0) = 8e^{-3x}$.

(ii) $z(x, 0) = 3e^{-5x} + 2e^{-3x}$.

Solution: Let the solution of the given PDE be of the form

$$z = X(x) \cdot Y(y). \tag{3.21}$$

Differentiating partially (3.21) w.r.t. x and y, we get, respectively,

$$p = X'Y$$

and

$$q = XY'.$$

Putting these values in the given equation, we have

$$X'Y = 2XY' + XY.$$

On separating the variables, the above equation reduces to

$$\frac{X'}{X} = \frac{2Y' + Y}{Y} = k$$

where k is a separation constant. Thus, we get the following two ODEs:

$$X' = kX \tag{3.22}$$

and

$$Y' = \frac{1}{2}(k - 1)Y. \tag{3.23}$$

The solutions of (3.22) and (3.23) are, respectively,

$$X(x) = c_1 e^{kx}$$

and

$$Y(y) = c_2 e^{\frac{1}{2}(k-1)y}.$$

Putting the values of X and Y in (3.21), we get

$$z = c_1 c_2 e^{kx} e^{\frac{1}{2}(k-1)y}$$

or

$$z = c_1 c_2 e^{kx+(k-1)y/2}. \tag{3.24}$$

(i) Given

$$z(x, 0) = 8e^{-3x}.$$

From (3.24), we have

$$z(x, 0) = c_1 c_2 e^{kx}.$$

On comparing, we get $c_1 c_2 = 8$ and $k = -3$. Putting these values in (3.24), the required solution is

$$z = 8e^{-(3x+2y)}.$$

(ii) Given

$$z(x, 0) = 3e^{-5x} + 2e^{-3x}.$$

In view of (3.24), the solutions of the given PDE may be rewritten as

$$z = C_k e^{kx+(k-1)y/2}$$

where C_k are constants depending on k. Then, by principle of superposition, the most general solution of the given equation is

$$z(x, y) = \sum_k C_k e^{kx+(k-1)y/2}. \tag{3.25}$$

Now, we have

$$z(x, 0) = \sum_k C_k e^{kx}$$

or

$$3e^{-5x} + 2e^{-3x} = \sum_k C_k e^{kx}.$$

Comparing coefficients, we get

$$C_k = \begin{cases} 3 & \text{if } k = -5 \\ 2 & \text{if } k = -3 \\ 0 & \text{otherwise.} \end{cases}$$

Putting these values in (3.25), the required solution is

$$z = 3e^{-(5x+3y)} + 2e^{-(3x+2y)}.$$

Example 3.20. *Solve $r - 2p + q = 0$.*

Solution: *Let the solution of the given PDE be of the form*

$$z = X(x) \cdot Y(y). \tag{3.26}$$

Then, we have

$$p = X'Y, \ q = XY' \text{ and } r = X''Y.$$

Putting these values in the given equation, we have

$$X''Y - 2X'Y + XY' = 0$$

or

$$(X'' - 2X')Y + XY' = 0.$$

On separating the variables, the above equation reduces to

$$\frac{X'' - 2X'}{X} = -\frac{Y'}{Y} = k$$

where k is a separation constant. Thus, we get the following two ODEs:

$$X'' - 2X' - kX = 0 \tag{3.27}$$

and

$$Y' + kY = 0. \tag{3.28}$$

AE of (3.27) is

$$m^2 - 2m - k = 0.$$

Its roots are

$$m = \frac{2 \pm \sqrt{4 + 4k}}{2} = 1 \pm \sqrt{1 + k}.$$

Therefore, the solution of (3.27) is

$$X(x) = c_1 e^{(1+\sqrt{1+k})x} + c_2 e^{(1-\sqrt{1+k})x}.$$

Also, the solution of (3.28) is

$$Y(y) = c_3 e^{-ky}.$$

Putting the values of X and Y in (3.26), we get

$$z = (c_1 e^{(1+\sqrt{1+k})x} + c_2 e^{(1-\sqrt{1+k})x}) c_3 e^{-ky}.$$

Thus, the required solution is

$$z = \left(a e^{(1+\sqrt{1+k})x} + b e^{(1-\sqrt{1+k})x} \right) e^{-ky}$$

where $a = c_1 c_3$ and $b = c_2 c_3$.

Example 3.21. *Using the method of separation of variables, find the product solution, if possible, for the PDE*

$$r + s + t = 0.$$

Solution: *Let the solution of the given PDE be of the form*

$$z = X(x) \cdot Y(y).$$

Then, we have

$$r = X''Y, \ s = X'Y' \text{ and } t = XY''.$$

Putting these values in the given equation, we have

$$X''Y + X'Y' + XY'' = 0$$

which is not separable and hence we cannot apply the method of separation of variables.

Example 3.22. *Use the method of separation of variables to solve the PDE*

$$q = r$$

satisfying $z = 0$ whenever $y \to \infty$. Also, obtain a particular solution of the given equation under the following boundary conditions (BCs):

$$z(0, y) = z(\pi, y) = 0, z(x, 0) = 5 \sin 2x + 7 \sin 6x.$$

Solution: *Let the solution of the given PDE be of the form*

$$z = X(x) \cdot Y(y).$$ (3.29)

Then, we have

$$q = XY' \text{ and } r = X''Y.$$

Putting these values in the given equation, we have

$$XY' = X''Y$$

which on separating the variables reduces to

$$\frac{X''}{X} = \frac{Y'}{Y} = k$$

where k is a separation constant. Thus, we get the following two ODEs:

$$X'' - kX = 0 \text{ and } Y' - kY = 0.$$ (3.30)

Now, there are the following three cases arising:

Case-I: *If $k = 0$, then Eq. (3.30) becomes*

$$X'' = 0 \text{ and } Y' = 0$$

whose solutions are

$$X = c_1 x + c_2 \text{ and } Y = c_3.$$

Hence, we obtain

$$z = (c_1 x + c_2)c_3.$$

Since this solution does not satisfy the condition $z = 0$ whenever $y \to \infty$, we reject it.

Case-II: *If $k > 0$, then we can take $k = \lambda^2, \lambda \neq 0$, so that Eq. (3.30) reduces to*

$$X'' - \lambda^2 X = 0 \text{ and } Y' - \lambda^2 Y = 0$$

whose solutions are

$$X = c_4 e^{\lambda x} + c_5 e^{-\lambda x} \text{ and } Y = c_6 e^{\lambda^2 y}.$$

Hence, we obtain

$$z = (c_4 e^{\lambda x} + c_5 e^{-\lambda x})c_6 e^{\lambda^2 y}.$$

Taking $y \to \infty$ in the above solution, we have $z \to \infty$. The above solution is also rejected as it also does not satisfy the given condition.

Case-III: *If $k < 0$, then we can take $k = -\lambda^2, \lambda \neq 0$, so that Eq. (3.30) reduces to*

$$X'' + \lambda^2 X = 0 \text{ and } Y' + \lambda^2 Y = 0$$

whose solutions are

$$X = c_7 \cos \lambda x + c_8 \sin \lambda x \text{ and } Y = c_9 e^{-\lambda^2 y}.$$

Hence, we get

$$z = (c_7 \cos \lambda x + c_8 \sin \lambda x)c_9 e^{-\lambda^2 y}.$$

Clearly, this solution satisfies the condition $z = 0$ whenever $y \to \infty$. Therefore, among the three possible solutions, the last one is the suitable solution of the given problem. Putting $c_7 c_9 = a$ and $c_8 c_9 = b$, this solution can be written as

$$z = (a \cos \lambda x + b \sin \lambda x)e^{-\lambda^2 y}. \tag{3.31}$$

Now, applying BC $z(0, y) = 0$ in (3.31), we get

$$ae^{-\lambda^2 y} = 0 \Rightarrow a = 0.$$

Thus, the solution (3.31) becomes

$$z = b \sin \lambda x \cdot e^{-\lambda^2 y}. \tag{3.32}$$

Again, applying BC $z(\pi, y) = 0$ in (3.32), we get

$$b \sin \lambda \pi \cdot e^{-\lambda^2 y} = 0 \Rightarrow \sin \lambda \pi = 0$$

which gives rise

$$\lambda \pi = n\pi \text{ or } \lambda = n$$

where n is any natural number. Hence, the solution (3.32) becomes

$$z = b \sin nx \cdot e^{-n^2 y},$$

which is a sequence of solutions. As for each solution, the constant b also depends on n, we take b_n instead of b. Thus, the solutions of the given PDE satisfying BC $z(0, y) = z(\pi, y) = 0$ are

$$z_n = b_n \sin nx \cdot e^{-n^2 y}.$$

Using the principle of superposition, the most general solution is

$$z = \sum_{n=1}^{\infty} z_n = \sum_{n=1}^{\infty} b_n \sin nx \cdot e^{-n^2 y}. \tag{3.33}$$

Finally, applying BC $z(x,0) = 5 \sin 2x + 7 \sin 6x$ in (3.33), we get

$$\sum_{n=1}^{\infty} b_n \sin nx = 5 \sin 2x + 7 \sin 6x.$$

Comparing coefficients, we get

$$b_n = \begin{cases} 5 & \text{if } n = 2 \\ 7 & \text{if } n = 6 \\ 0 & \text{otherwise.} \end{cases}$$

Putting the values of b_n in (3.33), the required solution is

$$z = 5e^{-4y} \sin 2x + 7e^{-36y} \sin 6x.$$

Sometimes a PDE has the solution, which is separable by dint of the sum of functions of the individual independent variables rather than their product. In this case, we use an analogue to the preceding technique, which has been demonstrated by the following example:

Example 3.23. *Use the method of separation of variables, to find the solution of the form $z(x,y) = X(x) + Y(y)$ of the PDE*

$$p^2 + q + x^2 = 0.$$

Solution: *Given that the solution of the given PDE is of the form*

$$z = X(x) + Y(y). \tag{3.34}$$

Differentiating the above equation partially w.r.t. x and y, we get respectively,

$$p = X' \text{ and } q = Y'.$$

Putting these values in the given equation, we have

$$X'^2 + Y' + x^2 = 0$$

which on separating the variables reduces to

$$X'^2 + x^2 = -Y' = a^2$$

where a^2 is a separation constant. Thus, we get the following two ODEs:

$$X' = \sqrt{a^2 - x^2} \text{ and } Y' = -a^2.$$

The solution of these ODEs are

$$X = \frac{a^2}{2}\sin^{-1}\left(\frac{x}{a}\right) + \frac{x}{2}\sqrt{a^2 - x^2} + c_1 \text{ and } Y = -a^2 y + c_2.$$

Putting the values of X and Y in (3.34), we get

$$z = \frac{a^2}{2}\sin^{-1}\left(\frac{x}{a}\right) + \frac{x}{2}\sqrt{a^2 - x^2} - a^2 y + c_1 + c_2.$$

Therefore, the required solution of the given PDE is

$$2z = a^2\sin^{-1}\left(\frac{x}{a}\right) + x\sqrt{a^2 - x^2} - 2a^2 y + b$$

where $b = 2c_1 + 2c_2$ is an arbitrary constant.

▶ Exercises

Find the general solution of the linear PDEs described in problems 1–14 using the direct integration method.

1. $p = x^2 + y^2$.
Ans. $z = \frac{x^3}{3} + xy^2 + \phi(y)$.

2. $q = \cos y$.
Ans. $z = \sin y + \phi(x)$.

3. $p = 2xy + 1, z(0, y) = \cosh y$.
Ans. $z = x^2 y + x + \cosh y$.

4. $u_x = 0$, where $u = u(x, y, z)$.
Ans. $u = \phi(y, z)$.

5. $s = 0$.
Ans. $z = \phi(x) + \psi(y)$.

6. $ar = xy$.
Ans. $z = \frac{1}{6a}x^3 y + \phi(y) + x\psi(y)$.

7. $r = x^2 e^y$.
Ans. $z = \frac{1}{12}x^4 e^y + \phi(y) + x\psi(y)$.

8. $xys = 1$.
Ans. $z = \log x \log y + \phi(y) + \psi(x)$.

9. $\dfrac{\partial^3 z}{\partial x \partial y \partial x} = -2$.
Ans. $z = -x^2 y + \phi(x) + x\psi(y)$.

10. $\dfrac{\partial^3 z}{\partial x \partial y^2} = \cos(3x + 2y)$.
Ans. $z = \phi(y) + y\psi(x) + \theta(x) - \frac{1}{12}\sin(3x + 2y)$.

11. $\dfrac{\partial^5 z}{\partial x^5} = x + y$. Ans.
$$z = \frac{1}{6!}x^6 + \frac{1}{5!}yx^5 + \frac{1}{4!}\phi_1(y)x^4 + \frac{1}{3!}\phi_2(y)x^3 + \frac{1}{2!}\phi_3(y)x^2 + \frac{1}{1!}\phi_4(y)x + \phi_5(y).$$

12. $xr = p$.
Ans. $z = \frac{1}{2}x^2\phi(y) + \psi(y)$.

13. $ty = q$.
Ans. $z = y^2\phi(x) + \psi(x)$.

14. $xs + q = 4x + 2y + 2$.

Ans.
$$xz = 2x^2y + xy^2 + 2xy + \phi(x) + \psi(y).$$

15. Find a surface passing through the parabolas $z = 0$, $y^2 = 4ax$ and $z = 1$, $y^2 = -4ax$ and satisfying $xr + 2p = 0$.

Ans. $8axz = 4ax - y^2$.

16. Find a surface satisfying $t = 6x^3y$ containing the two lines $y = 0 = z$ and $y = 1 = z$.

Ans. $z = x^3y^3 + y(1 - x^3)$.

Solve the linear PDEs described in problems 17–30.

17. $y(1 - y^2)q + 2zy^2 - z - x^2y^3 = 0$.

Ans. $z = x^2y + \phi(x)y\sqrt{y^2 - 1}$.

18. $(x + 1)p - yz = e^x(x + 1)^{y+1}$.

Ans. $z = (e^x + \phi(y))(x + 1)^y$.

19. $(D^4 - y^4)z = 0$.

Ans. $z = \phi_1(y)e^{xy} + \phi_2(y)e^{-xy} + \phi_3(y)\cos xy + \phi_4(y)\sin xy$.

20. $(D^2 + \omega^2y^2)^2z = 0$.

Ans. $z = [\phi_1(y) + x\phi_2(y)]\cos \omega xy + [\phi_3(y) + x\phi_4(y)]\sin \omega xy$.

21. $(D'^3 + 3D'^2 + 3D' + 1)z = 0$.

Ans.
$$z = [\phi_1(x) + y\phi_2(x) + y^2\phi_3(x)]e^{-x}.$$

22. $(D'^2 + x^2)z = \cot xy$.

Ans. $z = \phi(x)\cos xy + \psi(x)\sin xy + \dfrac{1}{x^2}\sin xy \log \tan \dfrac{xy}{2}$.

23. $(D^3 - 1)z = e^x + e^y$.

Ans. $z = \phi_1(y)e^x$
$$+ e^{-x/2}\left[\phi_2(y)\cos \frac{\sqrt{3}x}{2} + \phi_3(y)\sin \frac{\sqrt{3}x}{2}\right]$$
$$+ \frac{1}{7}e^{2x} + \frac{2}{3}xe^{x+y} - e^y$$

24. $(D^2 - 5D + 6)z = \sin 3x$.

Ans. $z = \phi(y)e^{2x}\psi(y)e^{3x} + \dfrac{1}{78}(5\cos 3x - \sin 3x)$.

25. $\dfrac{\partial^3 z}{\partial y^3} + x^2\dfrac{\partial z}{\partial y} = \sin xy$.

Ans. $z = $
$\phi_1(x) + \phi_2(x)\sin xy + \phi_3(x)\cos xy - \dfrac{y}{2x^2}\sin xy - \dfrac{1}{2x^3}\cos xy$.

26. $t + 4q + 4z = e^{2y} - e^{-2y}$.

Ans. $z = (\phi(x) + y\psi(x) + 18y^2)e^{3y/2}$.

27. $r - y^2z = \cos ax$.

Ans. $z = $
$\phi(y)e^{xy} + \psi(y)e^{-xy} - \dfrac{1}{a^2 + y^2}\cos ax$.

28. $t - x^4z = y^4$.

Ans. $z = $
$\phi_1(x)e^{xy} + \phi_2(x)e^{-xy} + \phi_3(x)\cos xy + \phi_4(x)\sin xy - \dfrac{1}{x^4}(y^4 + \dfrac{2y}{x^4})$.

29. $(D^2 + D)z = (x + y)^2$.

Ans. $z = \phi(y) + \psi(y)e^{-x} + \dfrac{1}{3}x^3 + x^2y + xy^2 - 2xy - x^2 + 2x$.

30. $(D'^4 - 1)z = y \sin y$.

Ans.
$$z = \phi_1(x)e^y + \phi_2(x)e^{-y} + \phi_3(x)\cos y$$
$$+\phi_4(x)\sin y + \frac{1}{8}(y^2 \cos y - 3y \sin y)$$

Using the method of separation of variables, find the solution of the PDEs described in problems 31–45.

31. $3p + 2q = 0$.

Ans. $z = Ae^{k(\frac{x}{3} - \frac{y}{2})}$.

32. $xp + yq = 0$.

Ans. $z = ce^{\frac{1}{2}(x^2 - y^2)k}$.

33. $xp = yq$.

Ans. $z = c(xy)^k$.

34. $p = yq$.

Ans. $z = cy^k e^{kx}$.

35. $p + q = 2(x + y)z$.

Ans. $z = ce^{x^2 + y^2 - k(x-y)}$.

36. $p = 4q$, $z(0, y) = 8e^{-3y}$.

Ans. $z = 6e^{-3(4x+y)}$.

37. $p - 3q = 0$, $x > 0$, $z(0, y) = \frac{1}{2}e^{-2y}$.

Ans. $z = \frac{1}{2}e^{-6x-2y}$.

38. $p + 2q = 3z$, $z(0, y) = 3e^{-y} - e^{-5y}$.

Ans. $z = e^{x-y} - e^{2x-5y}$.

39. $ar = xy$.

Ans. $z = \frac{1}{6a}x^3 y + \phi(y) + x\psi(y)$.

40. $y^2 s + 3x^2 z = 0$.

Ans. $z = e^{\frac{x^3}{k} - \frac{k}{y}}$.

41. $a^2 r = t$. Ans. $z = (c_1 x + c_2)((c_3 x + c_4))$,
$\qquad z = (c_5 e^{\lambda x} + c_6 e^{-\lambda x})(c_7 e^{\lambda a y} + c_8 e^{-\lambda a y})$,
$\qquad z = (c_9 \cos \lambda x + c_{10} \sin \lambda x)$.
$\qquad (c_{11} \cos \lambda a y + c_{12} \sin \lambda a y)$

42. $r + t = z$.

43. $r = q + 2z$, $z(0, y) = 0$, $p(0, y) = 1 + e^{-3y}$.

Ans. $z = \frac{1}{\sqrt{2}} \sinh \sqrt{2}x + e^{-3y} \sin x$.

44. $s = e^{-y} \cos x$, $z(x, 0) = 0$, $q(0, y) = 0$.

Ans. $z = (1 - e^{-y}) \sin x$.

45. $q = r$, $z(x, 0) = \sin \pi x$.

Ans. $z = \sin \pi x . e^{-ky}$.

46. Use $w = \log z$ and $w = X(x) + Y(y)$
to solve $x^2 p^2 + y^2 q^2 = z^2$.

Ans. $z = bx^a y^{\sqrt{1-a^2}}$.

47. Find the solution of the form $z(x, y) = X(x) + Y(y)$
of the equation $p^2 q^2 = 1$.

Ans. $z = ax + \sqrt{1 - a^2}y + b$.

4

First-Order Partial Differential Equations

Many problems in mathematical, physical, and engineering sciences deal with the formulation and the solution of first-order partial differential equations (PDEs), for example, Brownian motion, glacier motion, mechanical transport of solvents in fluids, propagation of wavefronts in optics, stochastic process, waves in shallow water, thermal efficiency of heat exchangers, heat propagation between two superconducting cables, radioactive disintegration, flood waves, acoustics, gas dynamics, traffic flow, noise in communication systems, population growth, telephone traffic, and so on. Nevertheless, first-order PDEs appear less frequently than second-order equations. From a mathematical point of view, first-order PDEs have the advantage of providing a conceptual framework that can be utilised for second- and higher order equations.

In this chapter, we discuss various methods for finding the different types of solutions, namely, general, complete, singular, and particular solutions of linear/semilinear/quasilinear/non-linear PDEs of order one. Recall that the most general PDE of first order can be written in symbolic form as

$$F(x, y, z, p, q) = 0.$$

4.1 Linear PDEs: Reduction to Canonical Form

In Chapter 2, we have seen that the general linear PDE of order one in two independent variables x and y is of the form

$$P(x, y)p + Q(x, y)q + Z(x, y)z = R(x, y). \tag{4.1}$$

Here, the coefficients P and Q are continuously differentiable functions and do not vanish simultaneously. Without loss of generality, we shall assume that $P \neq 0$.

In Chapter 3, we have discussed the method for finding the general solution of standard forms of Eq. (4.1) in which one of P and Q remains identically zero, so that Eq. (4.1) contains a single derivative. Now, we discuss a technique to solve the general form of Eq. (4.1). In this technique, we consider certain transformation of independent variables of the form

$$\xi = \xi(x, y) \text{ and } \eta = \eta(x, y) \tag{4.2}$$

under which Eq. (4.1) is transformed into standard form and hence it can be solved using the technique discussed in Section 3.2. The obtained standard form is called the **canonical form** of

Eq. (4.1). Such new coordinates (ξ, η), under which the original equation reduces to its canonical form, are called the **canonical coordinates**. This method of solving linear PDEs is sometimes referred to as **coordinate method** or **canonical method**.

Under consideration, we assume that the Jacobian of the transformation (4.2) does not vanish at (x, y), *i.e.*,

$$J := \frac{\partial(\xi, \eta)}{\partial(x, y)} = \xi_x \eta_y - \xi_y \eta_x \neq 0$$

so that by inverse function theorem, x and y can be determined uniquely from the system (4.2).

Definition 4.1. *The ordinary differential equation (ODE)*

$$\frac{dy}{dx} = \frac{Q}{P}$$

*is called the **characteristic equation** of Eq. (4.1). The general solution of the characteristic equation is called the **characteristic** of Eq. (4.1).*

Now, we prove that by making a certain change of independent variables, a linear PDE reduces to a new linear equation involving a single partial derivative.

Theorem 4.1. *Let $u(x, y) = c$ be the characteristics of (4.1). Also, assume that $v(x, y)$ is arbitrarily chosen such that the Jacobian $\dfrac{\partial(u, v)}{\partial(x, y)} \neq 0$. Then, under the invertible transformation*

$$\xi = u(x, y) \text{ and } \eta = v(x, y)$$

Equation (4.1) reduces to its canonical form given by

$$z_\eta + a(\xi, \eta)z = b(\xi, \eta).$$

Proof: Using the chain rule, we compute the partial derivatives of $z = z(\xi, \eta)$ as follows:

$$p = \frac{\partial z}{\partial x} = \frac{\partial z}{\partial \xi}\frac{\partial \xi}{\partial x} + \frac{\partial z}{\partial \eta}\frac{\partial \eta}{\partial x} = \xi_x z_\xi + \eta_x z_\eta$$

$$q = \frac{\partial z}{\partial y} = \frac{\partial z}{\partial \xi}\frac{\partial \xi}{\partial y} + \frac{\partial z}{\partial \eta}\frac{\partial \eta}{\partial y} = \xi_y z_\xi + \eta_y z_\eta.$$

Substituting the values of these partial derivatives in Eq. (4.1) and simplifying, we get

$$Az_\xi + Bz_\eta + Cz = D \tag{4.3}$$

where

$$A(\xi, \eta) = P\xi_x + Q\xi_y$$
$$B(\xi, \eta) = P\eta_x + Q\eta_y$$
$$C(\xi, \eta) = Z \text{ and } D(\xi, \eta) = R.$$

Along the characteristics $\xi \equiv u(x, y) = c$, we have

$$d\xi = 0$$

which yields that

$$\xi_x dx + \xi_y dy = 0$$

or

$$\frac{dy}{dx} = -\frac{\xi_x}{\xi_y}. \tag{4.4}$$

On the other hand, by the definition of characteristic equation of Eq. (4.1), we have

$$\frac{dy}{dx} = \frac{Q}{P}. \tag{4.5}$$

From Eqs (4.4) and (4.5), we obtain

$$-\frac{\xi_x}{\xi_y} = \frac{Q}{P} \tag{4.6}$$

so that

$$A = P\xi_x + Q\xi_y = 0.$$

Hence, Eq. (4.1) reduces to

$$Bz_\eta + Cz = D. \tag{4.7}$$

We claim that $B \neq 0$. To prove this, we suppose on contrary that $B = 0$, i.e.,

$$P\eta_x + Q\eta_y = 0.$$

Rearranging the above equation and using (4.6), we have

$$\frac{\eta_x}{\eta_y} = -\frac{Q}{P} = \frac{\xi_x}{\xi_y}$$

which contradicts the hypothesis $\dfrac{\partial(\xi, \eta)}{\partial(x, y)} \neq 0$. Hence, $B \neq 0$. Dividing Eq. (4.7) by B, we get

$$z_\eta + az = b$$

where

$$a = \frac{C}{B} \text{ and } b = \frac{D}{B}.$$

$\square$

In practice, it is convenient to choose $\eta \equiv v(x, y) = x$ or $\eta \equiv v(x, y) = y$ so that $J \neq 0$. Now, we furnish several examples, which illustrate the coordinate method.

Example 4.1. *Reduce the equation $yp + q = x$ to canonical form and hence obtain the general solution.*

Solution: *Here*

$$P = y, \ Q = 1.$$

The characteristic equation is

$$\frac{dy}{dx} = \frac{Q}{P} = \frac{1}{y}$$

or

$$ydy - dx = 0$$

which on integrating gives rise to

$$y^2 - 2x = c$$

which form the characteristics of the given PDE. The canonical coordinates are

$$\xi = y^2 - 2x \text{ and } \eta = y.$$

Now, we have

$$\xi_x = -2, \ \xi_y = 2y, \ \eta_x = 0, \ \eta_y = 1.$$

Using the above, we obtain

$$p = -2z_\xi \text{ and } q = 2yz_\xi + z_\eta.$$

Putting these values in the given PDE, we get

$$-2yz_\xi + 2yz_\xi + z_\eta = x$$

or

$$z_\eta = x. \tag{4.9}$$

From the system (4.8), we obtain $x = \dfrac{1}{2}(\eta^2 - \xi)$ *and hence Eq. (4.9) reduces to*

$$z_\eta = \frac{1}{2}(\eta^2 - \xi)$$

which is the canonical form of the given PDE. Integrating it partially w.r.t. η, we obtain

$$z = \frac{1}{6}\eta^3 - \frac{1}{2}\xi\eta + \phi(\xi).$$

Putting the values of ξ and η in the above equation, we get

$$z = xy - \frac{1}{3}y^3 + \phi(y^2 - 2x)$$

which is the required general solution.

Example 4.2. *Using the coordinate method, solve the linear equation*

$$xp - yq + y^2z = y^2, \ x, y \neq 0.$$

Solution: Here

$$P = x, \ Q = -y.$$

The characteristic equation is

$$\frac{dy}{dx} = \frac{Q}{P} = -\frac{y}{x}$$

or

$$xdy + ydx = 0$$

which on integrating gives rise to

$$xy = c$$

which remain the characteristics of the given PDE. The canonical coordinates are

$$\xi = xy \ and \ \eta = x. \tag{4.10}$$

Now, we have

$$\xi_x = y, \ \xi_y = x, \ \eta_x = 1, \ \eta_y = 0.$$

Using the above, we obtain

$$p = yz_\xi + z_\eta \text{ and } q = xz_\xi.$$

Putting these values in the given PDE, we get

$$x(yz_\xi + z_\eta) - y(xz_\xi) + y^2 z = y^2$$

or

$$xz_\eta + y^2 z = y^2. \tag{4.11}$$

Solving the system (4.10) for x and y, we get

$$x = \eta \text{ and } y = \frac{\xi}{\eta}.$$

Using these values, Eq. (4.11) reduces to

$$\eta z_\eta + \frac{\xi^2}{\eta^2} z = \frac{\xi^2}{\eta^2}$$

or

$$z_\eta + \frac{\xi^2}{\eta^3} z = \frac{\xi^2}{\eta^3}$$

which is the canonical form of the given PDE. Its integrating factor (IF) is

$$\text{IF} = e^{\int \frac{\xi^2}{\eta^3} \partial \eta} = e^{-\xi^2/2\eta^2}.$$

Thus, the solution is

$$z \cdot e^{-\xi^2/2\eta^2} = \int \frac{\xi^2}{\eta^3} \cdot e^{-\xi^2/2\eta^2} \partial \eta + \phi(\xi)$$
$$= e^{-\xi^2/2\eta^2} + \phi(\xi)$$

implying thereby

$$z = 1 + \phi(\xi) e^{\xi^2/2\eta^2}.$$

Putting the values of ξ and η in the above equation, we get

$$z = 1 + \phi(xy) e^{y^2/2}$$

which is the required general solution.

4.2 Semilinear PDEs: Geometry of Solutions

As discussed in Chapter 2, the general semilinear PDE of order one in two independent variables x and y is of the form

$$P(x, y)p + Q(x, y)q = R(x, y, z). \tag{4.12}$$

Here, the coefficients P and Q are continuously differentiable functions and do not vanish simultaneously. Without loss of generality, we shall assume that $P \neq 0$.

The idea of characteristics can be extended naturally to semilinear equations from linear equations as in both (linear and semilinear) cases, P and Q are functions of x and y only and independent of z.

Definition 4.2. *The ODE*

$$\frac{dy}{dx} = \frac{Q}{P}$$

is called the **characteristic equation** *of Eq. (4.12). The general solution of the characteristic equation forms a one-parameter family of plane curves, which are called the* **characteristics** *of Eq. (4.12).*

Geometric Interpretation of Characteristics: Equation (4.12) associates with a vector field $\mathbf{T} := (P(x, y), Q(x, y))$, which is called **characteristic direction** of Eq. (4.12). Geometrically speaking, the characteristics of Eq. (4.12) are integral curves[1] of its characteristic direction $\mathbf{T}$. As the components of $\mathbf{T}$ are continuously differentiable functions, at each point (x, y), there exists a unique one-parameter family of integral curves of the direction field $\mathbf{T}$, which are characteristics of Eq. (4.12) and hence it can be determined by its slop (that is, characteristic equation). These characteristics being tangent to the integral surface of Eq. (4.12) must lie entirely within the integral surface of the PDE.

Definition 4.3. *The ODE*

$$\frac{dz}{dx} = \frac{R}{P}$$

is called the **compatible condition** *for Eq. (4.12).*

Notice that Eq. (4.12) can be written as

$$\mathbf{T}.\nabla z = R.$$

It follows that the expression on LHS of Eq. (4.12) represents the directional derivative of z in the direction of $\mathbf{T}$ at each point (x, y). Thus, we conclude that the function $R(x, y, z)$ remains the rate of

---

[1] See Section 1.5.

change of the function $z = z(x, y)$ in the characteristic direction $\mathbf{T}$. Therefore, the characteristics are not enough to construct the required general solution (integral surface) of the PDE. Additionally, it is also required to determine the variation of the function $z = z(x, y)$ along these characteristics. The following result concludes that this variation remains the compatible condition on the characteristics.

Theorem 4.2. *Along the characteristics, every solution of Eq. (4.12) satisfies its compatibility condition.*

Proof. Let $z = z(x, y)$ be a solution of Eq. (4.12). The total derivative of z is

$$dz = p\,dx + q\,dy$$

which gives rise to

$$\frac{dz}{dx} = p + q \left(\frac{dy}{dx} \right).$$

Thus, the variation of z w.r.t. x along the characteristics of Eq. (4.12) is

$$\frac{dz}{dx} = p + q \left(\frac{Q}{P} \right) = \frac{Pp + Qq}{P} = \frac{R}{P}$$

or

$$\frac{dz}{dx} = \frac{R}{P}.$$

which is compatibility condition on these characteristics. $\qquad\square$

Construction of Integral Surface: To find the general solution of a semilinear PDE represented by (4.12), we first solve its characteristic equation to find the characteristics of the form

$$u(x, y) = c. \qquad (4.13)$$

Using implicit function theorem, we can write (4.13) as $y = y(x, c)$. The compatibility condition on the characteristics (4.13) is

$$\frac{dz}{dx} = \frac{R\left(x, y(x, c), z\right)}{P\left(x, y(x)\right)}.$$

whose general solution is of the form

$$v(x, z, c) = \phi(c). \qquad (4.14)$$

Remember that while solving compatibility condition, we write the constant of integration as a function $\phi(c)$. Finally, we eliminate the parameter c between (4.13) and (4.14) to obtain the required integral surface of PDE.

With a view to illustrate the foregoing description, we furnish the following examples.

Example 4.3. *Solve $ap + bq = 0$, where a and b are constants and $a \neq 0$.*

Solution: *The characteristic equation is*

$$\frac{dy}{dx} = \frac{Q}{P} = \frac{b}{a}$$

or

$$ady - bdx = 0$$

which on integrating gives rise to the characteristics

$$ay - bx = c \tag{4.15}$$

which is a family of straight lines. The compatibility condition on these characteristics is

$$\frac{dz}{dx} = \frac{R}{P} = 0$$

or

$$dz = 0.$$

Integrating it, we get

$$z = \phi(c) \tag{4.16}$$

where ϕ is an arbitrary function. This means that z remains a constant on the family of straight lines represented by (4.15). Eliminating the constant c between (4.15) and (4.16), we obtain

$$z = \phi(ay - bx)$$

which forms the general solution of the given PDE as desired.

Example 4.4. *Solve $yp - xq = z$.*

Solution: *The characteristic equation is*

$$\frac{dy}{dx} = -\frac{x}{y}$$

or

$$xdx + ydy = 0$$

which on integrating gives rise to

$$x^2 + y^2 = c^2. \tag{4.17}$$

Thus, the characteristics of given PDE are the circles with centre at origin and radius c. The compatibility condition on these characteristics is

$$\frac{dz}{dx} = \frac{z}{\sqrt{c^2 - x^2}}.$$

On separating the variables, the above equation becomes

$$\frac{dz}{z} = \frac{dx}{\sqrt{c^2 - x^2}}$$

which on integrating gives rise to

$$\log z = \sin^{-1}(x/c) + \phi(c^2) \tag{4.18}$$

where ϕ is an arbitrary function. Eliminating the constant c between (4.17) and (4.18), we obtain

$$\log z = \sin^{-1}\left(x/\sqrt{x^2 + y^2}\right) + \phi(x^2 + y^2)$$

which is the general solution of the given PDE as desired.

Example 4.5. *Find the particular solution of $xp + yq = xe^{-z}$ subject to the condition $z = 0$ on $y = x^2$.*

Solution: *The characteristic equation is*

$$\frac{dy}{dx} = \frac{y}{x}$$

which on separating the variables becomes

$$\frac{dy}{y} = \frac{dx}{x}.$$

On integrating, we obtain

$$y = cx \tag{4.19}$$

which is the characteristics of the given PDE. The compatibility condition on these characteristics is

$$\frac{dz}{dx} = \frac{xe^{-z}}{x} = e^{-z}$$

or

$$e^z dz = dx.$$

On integrating, we obtain

$$e^z = x + \phi(c) \tag{4.20}$$

where ϕ is an arbitrary function. Eliminating the constant c between (4.19) and (4.20), we obtain

$$e^z = x + \phi(y/x)$$

or

$$z(x,y) = \log\left[x + \phi(y/x)\right].$$

Applying the condition $z(x, x^2) = 0$, we get $\phi(x) = 1 - x$. Therefore, the required particular solution of the given problem is

$$z = \log\left[1 + x - y/x\right].$$

4.3 Quasilinear PDEs: Lagrange's Method

In Chapter 2, we have already undertaken quasilinear PDE of order one, which can be expressed symbolically as

$$P(x,y,z)p + Q(x,y,z)q = R(x,y,z). \tag{4.21}$$

Here, the coefficients P and Q are continuously differentiable functions and do not vanish simultaneously. A systematic method to obtain the general solution of such type of equations is known as **Lagrange's method,** named after the great Italian mathematician *Joseph-Louis Lagrange* (1736–1813). To describe Lagrange's method comprehensively, we need the following notion.

Definition 4.4. *The simultaneous total differential equations*

$$\frac{dx}{P} = \frac{dy}{Q} = \frac{dz}{R}$$

are called Lagrange's auxiliary equations or Lagrange's subsidiary equations or Lagrange's characteristic equations of Eq. (4.21).

Lagrange's method is embodied in the following result.

Theorem 4.3. *The general solution of Eq. (4.21) is*

$$\phi(u, v) = 0$$

where ϕ is an arbitrary differentiable function of u and v and

$$u(x, y, z) = c_1 \quad \text{and} \quad v(x, y, z) = c_2$$

are two independent solutions of Lagrange's auxiliary equations of Eq. (4.21).

Proof. In view of Theorem 2.3, we have seen that the functional relation

$$\phi(u, v) = 0$$

leads to the quasilinear PDE

$$Pp + Qq = R$$

where

$$P = \frac{\partial(u, v)}{\partial(y, z)}, \quad Q = \frac{\partial(u, v)}{\partial(z, x)}, \quad R = \frac{\partial(u, v)}{\partial(x, y)}. \tag{4.22}$$

We now proceed to determine u and v. Taking differentials of the equations $u(x, y, z) = c_1$ and $v(x, y, z) = c_2$, we get, respectively,

$$du = 0 \quad \text{and} \quad dv = 0.$$

Using the chain rule, the above equations become

$$u_x dx + u_y dy + u_z dz = 0$$

and

$$v_x dx + v_y dy + v_z dz = 0.$$

Solving these two equations for dx, dy and dz, we get

$$\frac{dx}{u_y v_z - u_z v_y} = \frac{dy}{u_z v_x - u_x v_z} = \frac{dz}{u_x v_y - u_y v_x}$$

which in view of (4.22) reduces to

$$\frac{dx}{P} = \frac{dy}{Q} = \frac{dz}{R}.$$

Thus, we conclude that $u = c_1$ and $v = c_2$ form the solutions of Lagrange's auxiliary equations of Eq. (4.21). This completes the proof. □

Notice that herein at least one of u and v must contain z. The general solution $\phi(u, v) = 0$ can also alternately be represented by the following equivalent form:

$$v = \psi(u)$$

where ψ is an arbitrary differentiable function.

The following examples are utilised to explain the Lagrange's method.

Example 4.6.   *Solve* $yzp + zxq = xy$.

Solution: *Here* $P = yz$, $Q = zx$, $R = xy$. *Lagrange's auxiliary equations are*

$$\frac{dx}{yz} = \frac{dy}{zx} = \frac{dz}{xy}$$

which multiplying by xyz become

$$xdx = ydy = zdz. \tag{4.23}$$

Taking the first two fractions of (4.23), we obtain

$$xdx = ydy$$

which on integrating gives rise to

$$x^2 - y^2 = c_1. \tag{4.24}$$

Taking the last two fractions of (4.23), we obtain

$$ydy = zdz$$

which on integrating gives rise to

$$y^2 - z^2 = c_2. \tag{4.25}$$

Hence, from (4.24) and (4.25), the required general solution of the given PDE is

$$\phi(x^2 - y^2, y^2 - z^2) = 0.$$

Example 4.7.   *Solve* $xyp + y^2 q = xyz - 2x^2$.

Solution: *Here* $P = xy$, $Q = y^2$, $R = xyz - 2x^2$. *Lagrange's auxiliary equations are*

$$\frac{dx}{xy} = \frac{dy}{y^2} = \frac{dz}{xyz - 2x^2}. \tag{4.26}$$

Taking the first two fractions of (4.26) and cancelling y, we obtain

$$\frac{dx}{x} = \frac{dy}{y}.$$

Integrating, we obtain

$$\log x = \log y + \log c_1$$

or

$$\frac{x}{y} = c_1. \tag{4.27}$$

Taking the second and third fractions of (4.26), we obtain

$$\frac{dy}{y^2} = \frac{dz}{xyz - 2x^2}$$

which using (4.27) reduces to

$$\frac{dy}{y^2} = \frac{dz}{(c_1 y)yz - 2(c_1 y)^2}$$
$$= \frac{dz}{c_1 y^2 (z - 2c_1)}.$$

Cancelling y^2 on both sides, the above equation becomes

$$c_1 dy = \frac{dz}{z - 2c_1}.$$

On integrating, we get

$$c_1 y - \log(z - 2c_1) = c_2$$

which using (4.27) reduces to

$$x - \log\left(z - \frac{2x}{y}\right) = c_2. \tag{4.28}$$

Hence, from (4.27) and (4.28), the required general solution of the given PDE is

$$x = \log\left(z - \frac{2x}{y}\right) + \phi\left(\frac{x}{y}\right).$$

Example 4.8. *Solve $(y + z)p + (z + x)q = x + y$.*

Solution: *Lagrange's auxiliary equations are*

$$\frac{dx}{y + z} = \frac{dy}{z + x} = \frac{dz}{x + y}.$$

Using multipliers $1, 1, 1$, we obtain

$$each \ fraction = \frac{dx + dy + dz}{(y + z) + (z + x) + (x + y)} = \frac{d(x + y + z)}{2(x + y + z)}. \tag{4.29}$$

Further, using multipliers $1, -1, 0$, *we obtain*

$$each\,fraction = \frac{dx - dy}{(y + z) - (z + x)} = \frac{d(x - y)}{-(x - y)}. \tag{4.30}$$

Finally, using multipliers $1, 0, -1$, *we obtain*

$$each\,fraction = \frac{dx - dz}{(y + z) - (x + y)} = \frac{d(x - z)}{-(x - z)}. \tag{4.31}$$

From (4.29), (4.30), and (4.31), we have

$$\frac{d(x + y + z)}{2(x + y + z)} = \frac{d(x - y)}{-(x - y)} = \frac{d(x - z)}{-(x - z)}. \tag{4.32}$$

Taking the first two fractions of (4.32), we obtain

$$\frac{d(x + y + z)}{x + y + z} + 2\frac{d(x - y)}{x - y} = 0.$$

Integrating, we obtain

$$\log(x + y + z) + 2\log(x - y) = \log c_1$$

or

$$(x + y + z)(x - y)^2 = c_1. \tag{4.33}$$

Taking the last two fractions of (4.32), we obtain

$$\frac{d(x - y)}{x - y} = \frac{d(x - z)}{x - z}.$$

Integrating, we obtain

$$\log(x - y) = \log(x - z) + \log c_2$$

or

$$\frac{x - y}{x - z} = c_2. \tag{4.34}$$

Hence, from (4.33) and (4.34), the required general solution of the given PDE is

$$\phi\left[(x + y + z)(x - y)^2, \frac{x - y}{x - z}\right] = 0.$$

Example 4.9. *Solve $x^2 p + y^2 q = (x+y)z$.*

Solution: *Lagrange's auxiliary equations are*

$$\frac{dx}{x^2} = \frac{dy}{y^2} = \frac{dz}{(x+y)z}. \tag{4.35}$$

Taking the first two fractions of (4.35), we obtain

$$\frac{dx}{x^2} - \frac{dy}{y^2} = 0.$$

Integrating, we obtain

$$-\frac{1}{x} + \frac{1}{y} = c_1$$

or

$$\frac{x-y}{xy} = c_1. \tag{4.36}$$

Taking multipliers $1, -1, 0$ in (4.35), we obtain

$$each\,fraction = \frac{dx - dy}{x^2 - y^2} = \frac{d(x-y)}{x^2 - y^2}. \tag{4.37}$$

Now, from (4.35) and (4.37), we have

$$\frac{d(x-y)}{x^2 - y^2} = \frac{dz}{(x+y)z}$$

or

$$\frac{d(x-y)}{x-y} = \frac{dz}{z}$$

which on integrating gives rise to

$$\frac{x-y}{z} = c_2. \tag{4.38}$$

Hence, from (4.36) and (4.38), the required general solution of the given PDE is

$$\phi \left[\frac{x-y}{xy}, \frac{x-y}{z} \right] = 0.$$

Example 4.10. *Solve $(z^2 - 2yz - y^2)p + (xy + zx)q = xy - zx$.*

Solution: Lagrange's auxiliary equations corresponding to the given PDE are

$$\frac{dx}{z^2 - 2yz - y^2} = \frac{dy}{xy + zx} = \frac{dz}{xy - zx}$$

which on multiplying by x becomes

$$\frac{xdx}{z^2 - 2yz - y^2} = \frac{dy}{y + z} = \frac{dz}{y - z}. \tag{4.39}$$

Taking multipliers $1, y, z$, we obtain

$$each\, fraction = \frac{xdx + ydy + zdz}{(z^2 - 2yz - y^2) + (y^2 + yz) + (yz - z^2)} = \frac{xdx + ydy + zdz}{0}$$

which gives rise to

$$xdx + ydy + zdz = 0.$$

Integrating, we obtain

$$x^2 + y^2 + z^2 = c_1. \tag{4.40}$$

Taking the last two fractions of (4.39), we obtain

$$\frac{dy}{y + z} = \frac{dz}{y - z}$$

or

$$ydy - (zdy + ydz) - zdz = 0$$

or

$$2ydy - 2d(yz) - 2zdz = 0.$$

Integrating, we obtain

$$y^2 - 2yz - z^2 = c_2. \tag{4.41}$$

Hence, from (4.40) and (4.41), the required general solution of the given equation is

$$y^2 = 2yz + z^2 + \phi(x^2 + y^2 + z^2).$$

4.3.1 Geometric Interpretation of Lagrange's Method

In Theorem 4.3, an analytic proof of Lagrange's method has been presented. We are now interested to verify the Lagrange's method using the geometrical approach. Suppose that $z = f(x, y)$ is an integral surface of quasilinear equation (4.21), which can be expressed explicitly as

$$F(x, y, z) \equiv f(x, y) - z = 0. \tag{4.42}$$

We know that the vector $\mathbf{N} := \nabla F = (p, q, -1)$ represents the direction of normal to the surface (4.42) at any point (x, y, z). Denoting $\mathbf{T} := (P, Q, R)$, Eq. (4.21) can be written as

$$\mathbf{T}.\mathbf{N} = 0$$

which shows that $\mathbf{T}$ and $\mathbf{N}$ are orthogonal. As $\mathbf{N}$ is the normal to the integral surface (4.42), the vector $\mathbf{T}$ must be a tangent vector to the surface (4.42) and hence must lie in the tangent plane to the surface at every point. Thus, our task of finding a solution is equivalent to finding a surface such that at every point on the surface, the vector $\mathbf{T}$ is tangent to the surface. To construct such a surface, the integral curves of the vector field $\mathbf{T}$ play an essential role (see Section 1.5). To understand better, we define the following notions.

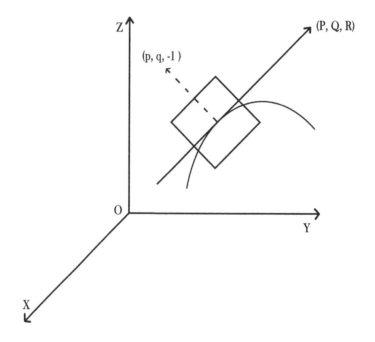

FIGURE 4.1   A geometric interpretation of Lagrange's method

Characteristic Direction: The vector field $\big(P(x, y, z), Q(x, y, z), R(x, y, z)\big)$ is called the *characteristic direction* or *Monge direction* of Eq. (4.21).

Characteristic Curves: The integral curves of the characteristic direction of Eq. (4.21) are called *characteristic curves* or *Monge curves* or *space characteristics* of Eq. (4.21). Thus, the characteristic curve of Eq. (4.21) is a space curve, whose tangent at every point (x, y, z) has direction ratios P, Q, R.

Characteristic: The projection of a characteristic curve on xy-plane is called *projected characteristic curve* or *characteristic base curve* or *characteristic ground curve* or simply, *characteristic*.

Therefore, we conclude that the characteristic curves of Eq. (4.21) lie wholly on the integral surface of (4.21). Once we have found the characteristic curves, the required integral surface is the union of these characteristic curves. By the definition of integral curves of the vector field $\mathbf{T}$, the general solution of simultaneous total differential equation

$$\frac{dx}{P} = \frac{dy}{Q} = \frac{dz}{R}$$

forms the characteristic curves of (4.21). Due to this fact, these equations are referred to as Lagrange's characteristic equations.

Thus in all, if

$$u(x, y, z) = c_1 \text{ and } v(x, y, z) = c_2$$

represent the two-parameter family of characteristic curves of (4.21) and ϕ is an arbitrary function, then in lieu of Theorem 1.1, the required integral surface of (4.21) can be represented by the functional relation

$$\phi(u, v) = 0$$

which is indeed a union of these characteristic curves.

Peculiarities of Semilinear PDE: Lagrange's auxiliary equations for a semilinear equation $P(x, y)p + Q(x, y)q = R(x, y, z)$ are

$$\frac{dx}{P(x, y)} = \frac{dy}{Q(x, y)} = \frac{dz}{R(x, y, z)}.$$

First two fractions taken together give the characteristic equation

$$\frac{dy}{dx} = \frac{Q(x, y)}{P(x, y)}.$$

Unlike general case of quasilinear equations, this equation does not contain z, therefore, the characteristics are completely determined from the characteristic equation. These characteristics are indeed a one-parameter family of plane curves of the form $u(x, y) = c$.

On associating last fraction with first fraction, we obtain the compatibility condition

$$\frac{dz}{dx} = \frac{R(x, y, z)}{P(x, y)}.$$

A general solution of the above ODE is a one-parameter family of surface of the form $v(x, y, z) = c'$. The characteristics $u(x, y) = c$ intersects the surface $v(x, y, z) = c'$ at a two-parameter family of space curves, which are indeed characteristic curves or Monge curves of the equation. Due to this fact, the characteristics $u(x, y) = c$ being projection of these characteristic curves on xy-plane are sometimes referred to as *projected characteristic curves*.

4.3.2 An Application to Determine Surfaces Orthogonal to a Given Family of Surfaces

An interesting application of the theory of quasilinear PDE in solid geometry can be recorded in the form of the following result.

Theorem 4.4. *Consider a one-parameter family of surfaces of the form*

$$f(x, y, z) = c. \tag{4.43}$$

Then, the general solution of the equation

$$Pp + Qq = R \tag{4.44}$$

where $P = \dfrac{\partial f}{\partial x}, Q = \dfrac{\partial f}{\partial y}, R = \dfrac{\partial f}{\partial z},$ *forms a system of surfaces, which cuts each member of family (4.43) orthogonally.*

Proof. Suppose that the required surface, which cuts each surface of (4.43) at right angles, is

$$z = \phi(x, y). \tag{4.45}$$

Rewriting Eq. (4.45) in implicit form as

$$\Phi(x, y, z) \equiv \phi(x, y) - z = 0.$$

The vector $\nabla\Phi = (p, q, -1)$ is normal vector to surface (4.45), that is, the direction ratios of the normal to the surface (4.45) are $p, q, -1$. On the other hand, P, Q, R are direction ratios of the normal to the surface (4.43). Also, we know that two surfaces are orthogonal if and only if their normals are mutually perpendicular. Therefore, we have

$$Pp + Qq + R(-1) = 0$$

or,

$$Pp + Qq = R.$$

It follows that Eq. (4.45) remains an integral surface of (4.44). Conversely, it can be easily proved that every integral surface of (4.44) is orthogonal to each surface of the family (4.43). Hence, the general solution of (4.44) forms a system of surfaces orthogonal to family (4.43). $\qquad\square$

The following example demonstrates Theorem 4.4.

Example 4.11. *Find the equation of the surface, which intersects the surfaces of system $z(x + y) = c(3z + 1)$ orthogonally and which passes through the circle $x^2 + y^2 = 1$, $z = 1$.*

Solution: *The given family of surfaces is*

$$f(x, y, z) \equiv \frac{z(x + y)}{3z + 1} = c.$$

Then

$$P = \frac{\partial f}{\partial x} = \frac{z}{3z + 1}$$

$$Q = \frac{\partial f}{\partial y} = \frac{z}{3z + 1}$$

$$R = \frac{\partial f}{\partial z} = \frac{x + y}{(3z + 1)^2}.$$

*Therefore, the **PDE** of required orthogonal surfaces is*

$$Pp + Qq = R$$

or

$$\frac{z}{3z + 1}p + \frac{z}{3z + 1}q = \frac{x + y}{(3z + 1)^2}$$

or

$$z(3z + 1)p + z(3z + 1)q = x + y.$$

Its Lagrange's auxiliary equations are

$$\frac{dx}{z(3z + 1)} = \frac{dy}{z(3z + 1)} = \frac{dz}{x + y}.$$

Taking the first two fractions, we get $dx - dy = 0$ so that

$$x - y = c_1.$$

Using the multipliers $x, y, -z(3z + 1)$, we obtain

$$each\ fraction = \frac{xdx + ydy - z(3z + 1)dz}{0}$$

which gives rise to

$$xdx + ydy - 3z^2 dz - zdz = 0.$$

On integrating, we get

$$x^2 + y^2 - z^2 - 2z^3 = c_2.$$

Hence, the system of surfaces orthogonal to the given family of surfaces is

$$x^2 + y^2 - z^2 - 2z^3 = \phi(x - y).$$

Putting $x^2 + y^2 = 1$, $z = 1$ in the above equation, we get

$$\phi(x - y) = -2.$$

Thus, the required particular surface is

$$x^2 + y^2 - z^2 - 2z^3 + 2 = 0.$$

4.3.3 Quasilinear Equations in n Independent Variables

The idea of Lagrange's method can be easily extended to the case of first-order quasilinear equations containing more than two independent variables. Hence, the general solution of PDE

$$Q_1 \frac{\partial u}{\partial x_1} + Q_2 \frac{\partial u}{\partial x_2} + \cdots + Q_n \frac{\partial u}{\partial x_n} = R$$

where $Q_i = Q_i(x_1, x_2, ..., x_n, u)$, $i = 1, 2, ..., n$ and $R = R(x_1, x_2, ..., x_n, u)$ is

$$\phi(v_1, v_2, ..., v_n) = 0$$

so that $v_i(x_1, x_2, ..., x_n, u) = c_i$ $(i = 1, 2, ..., n)$ are independent solutions of the corresponding auxiliary equations

$$\frac{dx_1}{Q_1} = \frac{dx_2}{Q_2} = \cdots = \frac{dx_n}{Q_n} = \frac{du}{R}.$$

The following example is adopted to illustrate the foregoing description.

Example 4.12. *Solve $yzu \dfrac{\partial u}{\partial x} + zux \dfrac{\partial u}{\partial y} + uxy \dfrac{\partial u}{\partial z} = xyz.$*

Solution: *The auxiliary equations are*

$$\frac{dx}{yzu} = \frac{dy}{zux} = \frac{dz}{uxy} = \frac{du}{xyz}.$$

Taking the first and last fractions, we get

$$xdx - udu = 0.$$

Integrating, we obtain

$$x^2 - u^2 = c_1.$$

Similarly, taking the second and last fractions, we obtain

$$y^2 - u^2 = c_2.$$

Also, from the third and last fractions, we obtain

$$z^2 - u^2 = c_3.$$

Hence, the general solution of the given PDE is

$$\phi(x^2 - u^2, y^2 - u^2, z^2 - u^2) = 0.$$

4.4 Certain Standard Forms of First-Order PDEs

Now, we discuss the techniques of finding the solution of first-order PDEs of the form $F(x, y, z, p, q) = 0$, which are not necessarily linear. It is easier to find the complete solution rather than the general solution, especially whenever such an equation is fully non-linear. The complete solution of such an equation involves merely two arbitrary constants. There is a general method to solve such equations, which will be given in the next section. Nevertheless, before considering the general method, we discuss some simplest standard forms. The methods of solving such forms are shorter than the general method, which is advantageous to see whether an equation is included under one of them.

4.4.1 Standard Form I: Equations Containing Partial Derivatives Only

This form is comprised of all those equations, which do not contain x, y, z explicitly. Specifically speaking, such equations are of the form

$$f(p, q) = 0.$$

Let $z = ax + by + c$ be a solution of this equation. Then, we have

$$p = a \ \text{ and } \ q = b.$$

Putting these values of p and q in the given PDE, we obtain

$$f(a, b) = 0$$

which on solving for b gives rise to $b = \phi(a)$. Thus, $z = ax + \phi(a)y + c$ is the complete solution of the given PDE.

Now, we take up some examples of this form.

Example 4.13. *Find the complete solution of PDE*

$$pq = 1.$$

Solution: As the given PDE is of the form $f(p, q) \equiv pq - 1 = 0$, *its solution is*

$$z = ax + by + c$$

so that

$$ab = 1 \ or \ b = \frac{1}{a}.$$

Hence, the required complete solution is

$$z = ax + \frac{1}{a}y + c.$$

Example 4.14.   *Find the complete solution of PDE*

$$p + q + pq = 0.$$

Solution: *Since the given PDE is of the form* $f(p, q) \equiv p + q + pq = 0$, *therefore, its solution is*

$$z = ax + by + c$$

so that

$$a + b + ab = 0$$

or

$$b = -\frac{a}{1 + a}.$$

Hence, the required complete solution is

$$z = ax - \frac{ay}{1 + a} + c.$$

4.4.2 Standard Form II: Equations Containing None of the Independent Variables Explicitly

This form is the class of those equations, which do not contain x and y, that is, these are of the form

$$f(z, p, q) = 0.$$

Suppose that z is a function of $u = x + ay$, where a is an arbitrary constant, that is, $z = \phi(u) = \phi(x + ay)$. Then by chain rules, we obtain

$$p = \frac{\partial z}{\partial x} = \frac{dz}{du}\frac{\partial u}{\partial x} = \frac{dz}{du}$$

$$q = \frac{\partial z}{\partial y} = \frac{dz}{du}\frac{\partial u}{\partial y} = a\frac{dz}{du}.$$

Putting these values of p and q in the given PDE, we obtain

$$f\left(z, \frac{dz}{du}, a\frac{dz}{du}\right) = 0$$

which being a first-order ODE can be solved to give z as a function of u. Finally, we replace u by $x + ay$ to get the complete solution of the given PDE.

For the sake of clarity, we furnish the following examples.

Example 4.15. *Find the complete solution of PDE*

$$p(1 + q) = zq.$$

Solution: *The given PDE is of the form*

$$F(z, p, q) \equiv p(1 + q) - zq = 0.$$

Assume that $u = x + ay$. Putting $p = \dfrac{dz}{du}$ and $q = a\dfrac{dz}{du}$ in the given equation, we get

$$\frac{dz}{du}\left[1 + a\frac{dz}{du}\right] = za\frac{dz}{du}$$

or

$$1 + a\frac{dz}{du} = az$$

or

$$\frac{adz}{az - 1} = du.$$

Integrating, we get

$$\log(az - 1) = u + b.$$

Hence, the required complete solution is

$$\log(az - 1) = x + ay + b.$$

Example 4.16. *Find the complete solution of PDE*

$$z = p^2 + q^2.$$

Solution: *The given PDE is of the form*

$$F(z, p, q) \equiv z - p^2 - q^2 = 0.$$

Assume that $u = x + ay$. Putting $p = \dfrac{dz}{du}$ and $q = a\dfrac{dz}{du}$ in the given equation, we get

$$z = \left(\frac{dz}{du}\right)^2 + a^2 \left(\frac{dz}{du}\right)^2$$

or

$$\sqrt{z} = \sqrt{1 + a^2} \left(\frac{dz}{du}\right)$$

or

$$\sqrt{1 + a^2} \frac{dz}{\sqrt{z}} = du.$$

On integrating, we obtain

$$2\sqrt{1 + a^2}\sqrt{z} = u + b$$

or

$$4(1 + a^2)z = (x + ay + b)^2$$

which is the required complete solution.

4.4.3 Standard Form III: Separable Equations

A first-order PDE, in which the dependent variable z does not appear and the terms containing x and p can be separated from those containing y and q, is called a **separable equation**. Thus, these equations are of the form

$$f(x, p) = g(y, q).$$

Put each of these expressions equal to an arbitrary constant a so that

$$f(x, p) = g(y, q) = a.$$

Solving these equations for p and q, we obtain

$$p = \phi(x, a) \quad \text{and} \quad q = \psi(y, a).$$

The total differential of z is

$$dz = \frac{\partial z}{\partial x} dx + \frac{\partial z}{\partial y} dy = p dx + q dy$$
$$= \phi(x, a) dx + \psi(y, a) dy.$$

On integrating, we obtain

$$z = \int \phi(x, a) dx + \int \psi(y, a) dy + b$$

which is the complete solution of PDE as desired.

We furnish the following examples of separable equations to understand the ideas outlined above.

Example 4.17. *Find the complete solution of PDE*

$$p^2 + q = x + y.$$

Solution: *The given PDE can be written as*

$$p^2 - x = y - q = a$$

which gives rise to

$$p = \sqrt{x + a}$$

and

$$q = y - a.$$

Putting these values in the equation $dz = pdx + qdy$, we get

$$dz = \sqrt{x + a} dx + (y - a) dy.$$

Integrating the above, we get

$$z = \frac{2}{3}(x + a)^{3/2} + \frac{1}{2}(y - a)^2 + b$$

which is the complete solution of the given equation as desired.

Example 4.18. *Find the complete solution of PDE*

$$2yp^2 = q.$$

Solution: *The given PDE can be expressed in the separable form as*

$$p^2 = \frac{q}{2y} = a^2$$

which gives rise to

$$p = a$$

and

$$q = 2a^2 y.$$

Putting these values in the equation $dz = pdx + qdy$, we get

$$dz = adx + 2a^2 ydy.$$

Integrating the above, we get

$$z = ax + a^2 y^2 + b$$

which is the required complete solution of the given equation.

4.4.4 Standard Form IV: Clairaut's Equation

A first-order PDE is called **Clairaut's equation** (or **Clairaut's form**) if it can be written as

$$z = xp + yq + f(p, q).$$

We assert that the complete solution of this equation is

$$z = ax + by + f(a, b) \tag{4.46}$$

which is obtained by replacing p by a and q by b in the given PDE. To establish our claim, differentiating Eq. (4.46) partially w.r.t. x and y, we get, respectively,

$$p = a \text{ and } q = b.$$

Using the above to eliminate a and b in Eq. (4.46), we obtain the given Clairaut's equation. It yields that Eq. (4.46) satisfies the given PDE.

We now present some illustrative examples of Clairaut's form.

Example 4.19. *Find the complete solution of PDE*

$$z = xp + yq + p^2 q^2.$$

Solution: The given PDE is Clairaut's equation. Therefore, its complete solution is

$$z = ax + by + a^2b^2.$$

Example 4.20. *Find the complete solution of PDE*

$$(p + q)(z - xp - yq) = 1.$$

Solution: The given PDE can be written as

$$z = xp + yq + \frac{1}{p + q}$$

which is a Clairaut's equation. Hence, its complete solution is

$$z = ax + by + \frac{1}{a + b}.$$

4.4.5 Equations Reducible to Standard Forms

In this subsection, we undertake some non-linear PDEs of order one, which do not fall in any of the standard forms discussed earlier. However, these equations can be reduced in one of the standard forms by a suitable change of variables. For the sake of simplicity, we distinguish such PDEs into five categories as under.

Type I (Equations Containing $x^m p$ or $y^n q$): (i) The PDE of the form $F(x^m p, y, z, q) = 0$ reduces to the form $G(X, y, z, P, q) = 0$ so that $P = \dfrac{\partial z}{\partial X}$, using the transformation

$$X = \begin{cases} \log x, & \text{if } m = 1 \\ x^{1-m}, & \text{if } m \neq 1. \end{cases}$$

To verify this claim, we use the chain rule to obtain

$$p = \frac{\partial z}{\partial x} = \frac{\partial z}{\partial X}\frac{dX}{dx} = P\frac{dX}{dx}.$$

Hence, whenever $m = 1$, we have $p = Px^{-1}$ so that $xp = P$. Otherwise, if $m \neq 1$, then we have $p = P(1 - m)x^{-m}$ so that $x^m p = P(1 - m)$. Thus, in both the cases, the conclusion is immediate.

(ii) The PDE of the form $F(x, y^n q, z, p) = 0$ reduces to the form $G(x, Y, z, p, Q) = 0$ so that $Q = \dfrac{\partial z}{\partial Y}$ using the transformation

$$Y = \begin{cases} \log y, & \text{if } n = 1 \\ y^{1-n}, & \text{if } n \neq 1. \end{cases}$$

The verification of (ii) is similar to that of (i).

(iii) The PDE of the form $F(x^m x, y^n q, z) = 0$ reduces to the form $G(X, Y, z, P, Q) = 0$ so that $P = \dfrac{\partial z}{\partial X}$ and $Q = \dfrac{\partial z}{\partial Y}$ using the transformations

$$X = \begin{cases} \log x, & \text{if } m = 1 \\ x^{1-m}, & \text{if } m \neq 1 \end{cases}$$

and

$$Y = \begin{cases} \log y, & \text{if } n = 1 \\ y^{1-n}, & \text{if } n \neq 1. \end{cases}$$

The verification of (iii) can be carried out by combining (i) and (ii).

Now, we present some illustrative examples of PDEs of Type I, whose reducible forms are in fact standard forms.

Example 4.21. *Find the complete solution of PDE*

$$y^2 q^2 = z(z - xp).$$

Solution: *The given PDE can be written as*

$$(yq)^2 = z(z - xp). \tag{4.47}$$

Here, $m = 1$ and $n = 1$. Put $X = \log x$ and $Y = \log y$ so that $xp = \dfrac{\partial z}{\partial X} = P$ and $yq = \dfrac{\partial z}{\partial Y} = Q$. Therefore, Eq. (4.47) becomes

$$Q^2 = z^2 - zP \tag{4.48}$$

which is of the form $f(z, P, Q) = 0$. Hence, putting $P = \dfrac{dz}{dt}$ and $Q = a\dfrac{dz}{dt}$, where $u = X + aY$, in Eq. (4.50), we get

$$a^2 \left(\frac{dz}{du}\right)^2 + z\frac{dz}{du} - z^2 = 0$$

which yields that

$$\frac{dz}{du} = \left[\frac{-1 \pm \sqrt{1 + 4a^2}}{2a^2}\right] z = kz$$

or

$$\frac{1}{k}\frac{dz}{z} = du.$$

Integrating the above, we get

$$\frac{1}{k} \log z = u + \log b = X + aY + \log b = \log x + a \log y + \log b$$

or

$$\log z^{1/k} = \log(bxy^a)$$

or

$$z^{1/k} = bxy^a$$

which is the required complete solution, where $k = \dfrac{-1 \pm \sqrt{1 + 4a^2}}{2a^2}$.

Example 4.22. *Find the complete solution of PDE*

$$yp + xq = pq.$$

Solution: The given PDE can be written as

$$(y^{-1}q)^{-1} + (x^{-1}p)^{-1} = 1. \tag{4.49}$$

Here, $m = -1$ *and* $n = -1$. *Put* $X = x^{1-m} = x^2$ *and* $Y = y^{1-n} = y^2$ *so that* $x^{-1}p = 2\dfrac{dz}{dX} = 2P$ *and* $y^{-1}q = 2\dfrac{dz}{dY} = 2Q$. *Therefore, Eq. (4.49) becomes*

$$Q^{-1} + P^{-1} = 2 \tag{4.50}$$

which is of the form $f(P, Q) = 0$. *Hence, its solution is* $z = aX + bY + c$, *where*

$$b^{-1} + a^{-1} = 2$$

or

$$a + b = 2ab$$

or

$$b = \frac{a}{2a - 1}.$$

Thus, we have

$$z = aX + \frac{a}{2a - 1}Y + c$$

or

$$z = ax^2 + \frac{ay^2}{2a-1} + c.$$

which is the required complete solution of the given equation.

Example 4.23. *Find the complete solution of PDE*

$$2(y + zq) = q(xp + yq).$$

Solution: *The given PDE can be written as*

$$z = \frac{1}{2}xp + \frac{1}{2}yq - \frac{y}{q}$$

or

$$z = \frac{1}{2}x^2(x^{-1}p) + \frac{1}{2}y^2(y^{-1}q) - (y^{-1}q)^{-1}. \tag{4.51}$$

Here $m = -1$ and $n = -1$. Put $X = x^{1-m} = x^2$ and $Y = y^{1-m} = y^2$ so that $x^{-1}p = 2\dfrac{\partial z}{\partial X} = 2P$ and $y^{-1}q = 2\dfrac{\partial z}{\partial Y} = 2Q$. Therefore, Eq. (4.51) becomes

$$z = PX + QY - \frac{1}{2Q}$$

which is Clairaut's form and hence its solution is

$$z = aX + bY - \frac{1}{2b}.$$

Therefore, the complete solution of the given equation is

$$z = ax^2 + by^2 - \frac{1}{2b}.$$

Type II (Equations Containing $z^k p$ and $z^k q$): The PDE of the form $F(x, y, z^k p, z^k q) = 0$ reduces to the form $G(x, y, Z, P, Q) = 0$ so that $P = \dfrac{\partial Z}{\partial x}$ and $Q = \dfrac{\partial Z}{\partial y}$, using the transformation

$$Z = \begin{cases} \log z, & \text{if } k = -1 \\ z^{k+1}, & \text{if } k \neq -1. \end{cases}$$

To verify our claim, we use the chain rule to obtain

$$p = \frac{\partial z}{\partial x} = \frac{dz}{dZ}\frac{\partial Z}{\partial x} = \frac{P}{dZ/dz}$$

and

$$q = \frac{\partial z}{\partial y} = \frac{dz}{dZ}\frac{\partial Z}{\partial y} = \frac{Q}{dZ/dz}.$$

For $k = -1$, above relations reduce to $p = Pz$ and $q = Qz$ so that $z^{-1}p = P$ and $z^{-1}q = Q$. On the other hand, whenever $k \neq -1$, then we have $p = \dfrac{P}{(k+1)z^k}$ and $q = \dfrac{Q}{(k+1)z^k}$ so that $z^k p = \dfrac{P}{k+1}$ and $z^k q = \dfrac{Q}{k+1}$. Hence, in both the cases, the conclusion is immediate.

In the following lines, we adopt some examples of PDEs of Type II, whose reducible forms are in fact standard forms.

Example 4.24. *Find the complete solution of PDE*

$$p^2 + q^2 = z^2(x + y).$$

Solution: *The given PDE can be written as*

$$(z^{-1}p)^2 + (z^{-1}q)^2 = x + y. \tag{4.52}$$

Here $k = -1$. Put $Z = \log z$ so that $P = \dfrac{\partial Z}{\partial x} = z^{-1}p$ and $Q = \dfrac{\partial Z}{\partial y} = z^{-1}q$. Therefore, Eq. (4.52) becomes

$$P^2 + Q^2 = x + y.$$

This equation can be expressed in the form $f(x, P) = g(y, Q)$ as

$$P^2 - x = y - Q^2 = a$$

which gives rise to

$$P = \sqrt{x + a}$$

and

$$Q = \sqrt{y - a}.$$

Putting these values of P and Q in the relation $dZ = Pdx + Qdy$, we get

$$dZ = \sqrt{x + a}\,dx + \sqrt{y - a}\,dy.$$

Integrating, we get

$$Z = \frac{2}{3}(x + a)^{3/2} + \frac{2}{3}(y - a)^{3/2} + \frac{b}{3}.$$

Therefore, the complete solution of the given equation is

$$3 \log z = 2(x+a)^{3/2} + 2(y-a)^{3/2} + b.$$

Example 4.25. *Find the complete solution of PDE*

$$z^2(p^2 + q^2) = x^2 + y^2.$$

Solution: *The given PDE can be written as*

$$(zp)^2 + (zq)^2 = x^2 + y^2. \tag{4.53}$$

Here $k = 1$. Put $Z = z^{k+1} = z^2$ so that $P = \dfrac{\partial Z}{\partial x} = 2zp$ and $Q = \dfrac{\partial Z}{\partial y} = 2zq$. Therefore, Eq. (4.53) becomes

$$\frac{P^2}{4} + \frac{Q^2}{4} = x^2 + y^2$$

which is of the form $f(x, P) = g(y, Q)$. Let

$$\frac{P^2}{4} - x^2 = y^2 - \frac{Q^2}{4} = a^2$$

which gives rise to

$$P = 2\sqrt{x^2 + a^2}$$

and

$$Q = 2\sqrt{y^2 - a^2}.$$

Putting these values of P and Q in the relation $dZ = Pdx + Qdy$, we get

$$dZ = 2\sqrt{a^2 + x^2}dx + 2\sqrt{y^2 - a^2}dy.$$

Integrating, we get

$$Z = x\sqrt{x^2 + a^2} + y\sqrt{y^2 - a^2} + a^2 \sinh^{-1}\left(\frac{x}{a}\right) - a^2 \cosh^{-1}\left(\frac{y}{a}\right) + b.$$

Therefore, the complete solution of the given equation is

$$z^2 = x\sqrt{x^2 + a^2} + y\sqrt{y^2 - a^2} + a^2 \sinh^{-1}\left(\frac{x}{a}\right) - a^2 \cosh^{-1}\left(\frac{y}{a}\right) + b.$$

Type III (Equations Containing $x^m z^k p$ and $y^n z^k q$): The PDE of the form $F(x^m z^k p, y^n z^k q) = 0$ reduces to the form $G(X, Y, Z, P, Q) = 0$ so that $P = \dfrac{\partial Z}{\partial X}$ and $Q = \dfrac{\partial Z}{\partial Y}$, using the transformations

$$X = \begin{cases} \log x, & \text{if } m = 1 \\ x^{1-m}, & \text{if } m \neq 1 \end{cases}$$

$$Y = \begin{cases} \log y, & \text{if } n = 1 \\ y^{1-n}, & \text{if } n \neq 1 \end{cases}$$

and

$$Z = \begin{cases} \log z, & \text{if } k = -1 \\ z^{k+1}, & \text{if } k \neq -1. \end{cases}$$

The following examples are utilised to demonstrate the PDEs of Type III, whose reducible forms are indeed standard forms.

Example 4.26. *Find the complete solution of PDE*

$$x^2 p^2 + y^2 q^2 = z^2.$$

Solution: *The given PDE can be written as*

$$(xz^{-1}p)^2 + (yz^{-1}q)^2 = 1. \tag{4.54}$$

Here $m = n = 1$ and $k = -1$. Put $X = \log x$, $Y = \log y$, $Z = \log z$ so that $P = \dfrac{\partial Z}{\partial X} = xz^{-1}p$ and $Q = \dfrac{\partial Z}{\partial Y} = yz^{-1}q$. Therefore, Eq. (4.54) becomes

$$P^2 + Q^2 = 1$$

which is of the form $f(P, Q) = 0$. Hence, its solution is $Z = aX + bY + c$, wherein $a^2 + b^2 = 1$, i.e., $b = \sqrt{1 - a^2}$. Therefore, we have

$$Z = aX + \sqrt{1 - a^2}\, Y + c$$

or

$$\log z = a \log x + \sqrt{1 - a^2} \log y + \log k = \log(kx^a y^{\sqrt{1-a^2}}), \quad (\text{where } c = \log k)$$

or

$$z = kx^a y^{\sqrt{1-a^2}}$$

which is the complete solution of the given equation as desired.

Example 4.27. *Find the complete solution of PDE*

$$z^2(x^2p^2 + q^2) = 1.$$

Solution: The given PDE can be written as

$$(xzp)^2 + (zq)^2 = 1. \tag{4.55}$$

Here $m = 1$, $n = 0$ and $k = 1$. Put $X = \log x$, $Z = z^{k+1} = z^2$ so that $P = \dfrac{\partial Z}{\partial X} = 2xzp$ and $Q = \dfrac{\partial Z}{\partial y} = 2zq$. Therefore, Eq. (4.54) becomes

$$P^2 + Q^2 = 4$$

which is of the form $f(P, Q) = 0$. Hence, its solution is $Z = aX + by + c$, wherein $a^2 + b^2 = 4$, i.e., $b = \sqrt{4 - a^2}$. Therefore, we have

$$Z = aX + \sqrt{4 - a^2}\, y + c$$

or

$$z^2 = a \log x + \sqrt{4 - a^2}\, y + c$$

which is the complete solution of the given equation as desired.

Example 4.28. *Find the complete solution of PDE*

$$xp^2 + yq^2 = z.$$

Solution: The given PDE can be written as

$$\left(x^{1/2}z^{-1/2}p\right)^2 + \left(y^{1/2}z^{-1/2}q\right)^2 = 1. \tag{4.56}$$

Here $m = \dfrac{1}{2}$, $n = \dfrac{1}{2}$ and $k = -\dfrac{1}{2}$. Put $X = x^{1-m} = \sqrt{x}$, $Y = y^{1-n} = \sqrt{y}$, $Z = z^{k+1} = \sqrt{z}$ so that $P = \dfrac{\partial Z}{\partial X} = x^{1/2}z^{-1/2}p$ and $Q = \dfrac{\partial Z}{\partial Y} = y^{1/2}z^{-1/2}q$. Therefore, Eq. (4.56) becomes

$$P^2 + Q^2 = 1$$

which is of the form $f(P, Q) = 0$. Hence, its solution is $Z = aX + bY + c$, wherein $a^2 + b^2 = 1$, i.e., $b = \sqrt{1 - a^2}$. Therefore, we have

$$Z = aX + \sqrt{1 - a^2}\, Y + c$$

or

$$\sqrt{z} = a\sqrt{x} + \sqrt{1 - a^2}\sqrt{y} + c$$

which is the complete solution of the given equation as desired.

Type IV (Changing Variables to Polar Coordinates): Sometimes the substitution $x = r\cos\theta$ and $y = r\sin\theta$ transforms a given PDE into one of the standard forms. This idea can be demonstrated through the following example.

Example 4.29. *Find the complete solution of PDE*

$$(x^2 + y^2)(p^2 + q^2) = 1.$$

Solution: Put $x = r\cos\theta$ *and* $y = r\sin\theta$ *so that* $r^2 = x^2 + y^2$ *and* $\theta = \tan^{-1}(y/x)$. *Using chain rules, we obtain*

$$
\begin{aligned}
p = \frac{\partial z}{\partial x} &= \frac{\partial z}{\partial r}\frac{\partial r}{\partial x} + \frac{\partial z}{\partial \theta}\frac{\partial \theta}{\partial x} \\
&= \frac{\partial z}{\partial r}\frac{x}{r} + \frac{\partial z}{\partial \theta}\left[\frac{-y}{x^2 + y^2}\right] \\
&= \cos\theta\frac{\partial z}{\partial r} - \frac{\sin\theta}{r}\frac{\partial z}{\partial \theta}
\end{aligned}
$$

and

$$
\begin{aligned}
q = \frac{\partial z}{\partial y} &= \frac{\partial z}{\partial r}\frac{\partial r}{\partial y} + \frac{\partial z}{\partial \theta}\frac{\partial \theta}{\partial y} \\
&= \frac{\partial z}{\partial r}\frac{y}{r} + \frac{\partial z}{\partial \theta}\left[\frac{x}{x^2 + y^2}\right] \\
&= \sin\theta\frac{\partial z}{\partial r} + \frac{\cos\theta}{r}\frac{\partial z}{\partial \theta}.
\end{aligned}
$$

Therefore, the given equation reduces to

$$r^2\left[\left(\frac{\partial z}{\partial r}\right)^2 + \frac{1}{r^2}\left(\frac{\partial z}{\partial \theta}\right)^2\right] = 1$$

or

$$r^2\left(\frac{\partial z}{\partial r}\right)^2 = 1 - \left(\frac{\partial z}{\partial \theta}\right)^2 = a^2$$

which gives rise to

$$\frac{\partial z}{\partial r} = \frac{a}{r}$$

and

$$\frac{\partial z}{\partial \theta} = \sqrt{1 - a^2}.$$

Now, we have

$$dz = \frac{\partial z}{\partial r} dr + \frac{\partial z}{\partial \theta} d\theta = \frac{a}{r} dr + \sqrt{1 - a^2} d\theta.$$

Integrating, we get

$$z = a \log r + \sqrt{1 - a^2} \theta + b.$$

Hence, the complete solution of the given equation is

$$z = \frac{a}{2} \log(x^2 + y^2) + \sqrt{1 - a^2} \tan^{-1} \left(\frac{y}{x}\right) + b.$$

Type V (Changing Variables by Inspection): Apart from types I to IV, we present several miscellaneous examples for which comprehensive rules of change of variables are not possible. However, we choose the change of variables by inspection.

Example 4.30. *Find the complete solution of PDE*

$$(x + y)(p + q)^2 + (x - y)(p - q)^2 = 1.$$

Solution: Put $x + y = u^2$ and $x - y = v^2$. Using chain rule, we get

$$p = \frac{\partial z}{\partial x} = \frac{\partial z}{\partial u}\frac{\partial u}{\partial x} + \frac{\partial z}{\partial v}\frac{\partial v}{\partial x}$$

$$= \frac{1}{2u}\frac{\partial z}{\partial u} + \frac{1}{2v}\frac{\partial z}{\partial v}$$

and

$$q = \frac{\partial z}{\partial y} = \frac{\partial z}{\partial u}\frac{\partial u}{\partial y} + \frac{\partial z}{\partial v}\frac{\partial v}{\partial y}$$

$$= \frac{1}{2u}\frac{\partial z}{\partial u} - \frac{1}{2v}\frac{\partial z}{\partial v}.$$

Therefore, the given equation reduces to

$$\left(\frac{\partial z}{\partial u}\right)^2 + \left(\frac{\partial z}{\partial v}\right)^2 = 1.$$

Hence, its solution is $z = au + bv + c$, where $a^2 + b^2 = 1$, i.e., $b = \sqrt{1 - a^2}$. Thus, we have

$$z = au + \sqrt{1 - a^2} v + c$$

or

$$z = a\sqrt{x+y} + \sqrt{1-a^2}\sqrt{x-y} + c$$

which is the complete solution of the given equation as desired.

Example 4.31. *Find the complete solution of PDE*

$$(y-x)(yq-xp) = (p-q)^2.$$

Solution: Put $x+y = u$ *and* $xy = v$. *Using chain rule, we get*

$$p = \frac{\partial z}{\partial x} = \frac{\partial z}{\partial u}\frac{\partial u}{\partial x} + \frac{\partial z}{\partial v}\frac{\partial v}{\partial x}$$

$$= \frac{\partial z}{\partial u} + y\frac{\partial z}{\partial v}$$

and

$$q = \frac{\partial z}{\partial y} = \frac{\partial z}{\partial u}\frac{\partial u}{\partial y} + \frac{\partial z}{\partial v}\frac{\partial v}{\partial y}$$

$$= \frac{\partial z}{\partial u} + x\frac{\partial z}{\partial v}.$$

Therefore, the given equation reduces to

$$\frac{\partial z}{\partial u} = \left(\frac{\partial z}{\partial v}\right)^2.$$

Hence, the required complete solution is

$$z = au + \sqrt{a}v + c$$

or

$$z = a(x+y) + \sqrt{a}xy + c.$$

4.5 Charpit's Method: A General Method of Finding the Complete Solution

Consider the most general PDE of order one

$$F(x, y, z, p, q) = 0. \tag{4.57}$$

A general method of solving Eq. (4.57) is termed as **Charpit's method**. This method was initially carried out partly by *Joseph Louis Lagrange* (1736–1813), which is contained in his memoir published in 1776. Later, it was accomplished by the French mathematician *Paul Charpit de Villecourt* (18th century). The method was contained in a memoir, which was presented to the Paris Academy of Sciences on June 30, 1784. But Charpit died young prematurely (in 1785) and the memoir was never published. In 1814, a French mathematician *Sylvestre François Lacroix* (1775–1843) published some information about his results and finally Charpit's memoir was found at the beginning of the 20th century.

To describe Charpit's method comprehensively, we need the following notion.

Definition 4.5. *The simultaneous total differential equations*

$$\frac{dx}{-F_p} = \frac{dy}{-F_q} = \frac{dz}{-pF_p - qF_q} = \frac{dp}{F_x + pF_z} = \frac{dq}{F_y + qF_z}$$

are called Charpit's auxiliary equations of (4.57).

Charpit's method is embodied in the following result.

Theorem 4.5. *A complete solution of Eq. (4.57) is the integral of total differential equation*

$$dz = pdx + qdy$$

where p and q are determined from Eq. (4.57) and an integral of its Charpit's auxiliary equations.

Proof. We have

$$dz = pdx + qdy. \tag{4.58}$$

Let us find another equation

$$\Phi(x, y, z, p, q) = 0 \tag{4.59}$$

such that p and q can be determined from (4.57) and (4.59), which make (4.58) integrable. The integral of Eq. (4.58) will satisfy the given equation (4.57) as the values of p and q are obtained from it.

The proof is complete if we show that an integral of Charpit's auxiliary equations of (4.57) forms such a relation (4.59). Differentiating (4.57) and (4.59) partially w.r.t. x and y, we get

$$\frac{\partial F}{\partial x} + \frac{\partial F}{\partial z}p + \frac{\partial F}{\partial p}\frac{\partial p}{\partial x} + \frac{\partial F}{\partial q}\frac{\partial q}{\partial x} = 0$$

$$\frac{\partial \Phi}{\partial x} + \frac{\partial \Phi}{\partial z}p + \frac{\partial \Phi}{\partial p}\frac{\partial p}{\partial x} + \frac{\partial \Phi}{\partial q}\frac{\partial q}{\partial x} = 0$$

$$\frac{\partial F}{\partial y} + \frac{\partial F}{\partial z}q + \frac{\partial F}{\partial p}\frac{\partial p}{\partial y} + \frac{\partial F}{\partial q}\frac{\partial q}{\partial y} = 0$$

$$\frac{\partial \Phi}{\partial y} + \frac{\partial \Phi}{\partial z}q + \frac{\partial \Phi}{\partial p}\frac{\partial p}{\partial y} + \frac{\partial \Phi}{\partial q}\frac{\partial q}{\partial y} = 0.$$

Eliminating $\dfrac{\partial p}{\partial x}$ between the first pair of these equations, we get

$$\left[\frac{\partial F}{\partial x}\frac{\partial \Phi}{\partial p} - \frac{\partial F}{\partial p}\frac{\partial \Phi}{\partial x}\right] + p\left[\frac{\partial F}{\partial z}\frac{\partial \Phi}{\partial p} - \frac{\partial F}{\partial p}\frac{\partial \Phi}{\partial z}\right] + \frac{\partial q}{\partial x}\left[\frac{\partial F}{\partial q}\frac{\partial \Phi}{\partial p} - \frac{\partial F}{\partial p}\frac{\partial \Phi}{\partial q}\right] = 0.$$

Similarly, eliminating $\dfrac{\partial q}{\partial y}$ between the second pair, we get

$$\left[\frac{\partial F}{\partial y}\frac{\partial \Phi}{\partial q} - \frac{\partial F}{\partial q}\frac{\partial \Phi}{\partial y}\right] + q\left[\frac{\partial F}{\partial z}\frac{\partial \Phi}{\partial q} - \frac{\partial F}{\partial q}\frac{\partial \Phi}{\partial z}\right] + \frac{\partial p}{\partial y}\left[\frac{\partial F}{\partial p}\frac{\partial \Phi}{\partial q} - \frac{\partial F}{\partial q}\frac{\partial \Phi}{\partial p}\right] = 0.$$

Adding these two equations and using $\dfrac{\partial q}{\partial x} = \dfrac{\partial p}{\partial y}$, we get

$$(-F_p)\frac{\partial \Phi}{\partial x} + (-F_q)\frac{\partial \Phi}{\partial y} + (-pF_p - qF_q)\frac{\partial \Phi}{\partial z} + (F_x + pF_z)\frac{\partial \Phi}{\partial p}$$
$$+(F_y + qF_z)\frac{\partial \Phi}{\partial q} = 0 \tag{4.60}$$

which is a quasilinear PDE of order one in dependent variable Φ and independent variables x, y, z, p, q. Hence, the corresponding Lagrange's auxiliary equations are

$$\frac{dx}{-F_p} = \frac{dy}{-F_q} = \frac{dz}{-pF_p - qF_q} = \frac{dp}{F_x + pF_z} = \frac{dq}{F_y + qF_z} = \frac{d\Phi}{0} \tag{4.61}$$

which are indeed Charpit's auxiliary equations of Eq. (4.57). Therefore, any integral of (4.61) will satisfy Eq. (4.60) also. Such an integral involving p or q or both can be taken as the additional PDE of the form (4.59). □

In practice, there is no need to solve all Charpit's auxiliary equations. We take the simplest possible integral of (4.61), which with the help of (4.57) will provide the values of p and q easily. Substituting these values in Eq. (4.58) and integrating it, we obtain a complete solution of the given PDE of the form

$$f(x, y, z, a, b) = 0.$$

Here, a and b are constants of integration. One of these constants is obtained from the integration of Charpit's auxiliary equations, whereas another is obtained from the integration of $dz = pdx + qdy$.

The following examples are adopted to demonstrate the Charpit's method.

Example 4.32. *Find the complete solution of PDE*

$$p^2 + q^2 - 2px - 2qy + 1 = 0.$$

Solution: *The given PDE can be written as*

$$F \equiv p^2 + q^2 - 2px - 2qy + 1 = 0. \tag{4.62}$$

The Charpit's auxiliary equations are

$$\frac{dx}{-(2p-2x)} = \frac{dy}{-(2q-2y)} = \frac{dz}{-p(2p-2x) - q(2q-2y)} = \frac{dp}{-2p} = \frac{dq}{-2q}.$$

Taking the last two fractions, we get

$$\frac{dp}{p} = \frac{dq}{q}$$

which on integrating gives rise to

$$p = aq.$$

Putting $p = aq$ in (4.62), we get

$$a^2 q^2 + q^2 - 2aqx - 2qy + 1 = 0$$

or

$$(a^2 + 1)q^2 - 2(ax + y)q + 1 = 0$$

which yields that

$$q = \frac{(ax+y) \pm \sqrt{(ax+y)^2 - (a^2+1)}}{a^2+1}.$$

Also, we have

$$p = aq = a\frac{(ax+y) \pm \sqrt{(ax+y)^2 - (a^2+1)}}{a^2+1}.$$

Putting these values of p and q in $dz = pdx + qdy$, we get

$$dz = \frac{(ax+y) \pm \sqrt{(ax+y)^2 - (a^2+1)}}{a^2+1}(adx + dy). \tag{4.63}$$

Put $ax + y = t$ so that $adx + dy = dt$. Hence, Eq. (4.63) reduces to

$$(a^2 + 1)dz = [t \pm \sqrt{t^2 - (a^2+1)}]dt.$$

Integrating, we obtain

$$(a^2 + 1)z = \frac{t^2}{2} \pm \frac{1}{2}\left\{t\sqrt{t^2 - (a^2 + 1)} - (a^2 + 1)\log[t + \sqrt{t^2 - (a^2 + 1)}]\right\} + \frac{b}{2}.$$

Therefore, the required complete solution of the given PDE is

$$2(a^2 + 1)z = (ax + y)^2 \pm \left\{(ax + y)\sqrt{(ax + y)^2 - (a^2 + 1)}\right.$$
$$\left. - (a^2 + 1)\log\left[(ax + y) + \sqrt{(ax + y)^2 - (a^2 + 1)}\right]\right\} + b.$$

Example 4.33. *Find the complete solution of PDE*

$$2(z + xp + yq) = yp^2.$$

Solution: *The given PDE can be written as*

$$F \equiv 2(z + xp + yq) - yp^2 = 0. \tag{4.64}$$

The Charpit's auxiliary equations are

$$\frac{dx}{-2x + 2yp} = \frac{dy}{-2y} = \frac{dz}{p(-2x + 2yp) - 2qy} = \frac{dp}{2p + 2p} = \frac{dq}{2q - p^2 + 2q}.$$

Taking the second and fourth fractions, we get

$$2\frac{dy}{y} + \frac{dp}{p} = 0.$$

Integrating, we get

$$y^2 p = a$$

or

$$p = \frac{a}{y^2}.$$

Putting this value of p in Eq. (4.64), we get

$$z + \frac{ax}{y^2} + yq = \frac{a^2}{2y^3}$$

or

$$q = -\frac{z}{y} - \frac{ax}{y^3} + \frac{a^2}{2y^4}.$$

Putting the values of p and q in the relation dz = pdx + qdy, we obtain

$$dz = \frac{a}{y^2}dx - \frac{z}{y}dy - \frac{ax}{y^3}dy + \frac{a^2}{2y^4}dy$$

or

$$ydz + zdy = a\left[\frac{ydx - xdy}{y^2}\right] + \frac{a^2}{2y^3}dy$$

or

$$d(yz) = ad\left(\frac{x}{y}\right) + \frac{a^2}{2}y^{-3}dy = 0.$$

Integrating, we get

$$yz = \frac{ax}{y} - \frac{a^2}{4y^2} + b$$

or

$$4y^3z = 4by^2 + 4axy - a^2$$

which is the required complete solution.

Example 4.34. *Find the complete solution of PDE*

$$xp^2 + yq^2 = z.$$

Solution: *The given PDE can be written as*

$$F \equiv xp^2 + yq^2 - z = 0. \tag{4.65}$$

The Charpit's auxiliary equations are

$$\frac{dx}{-2px} = \frac{dy}{-2qy} = \frac{dz}{-2(p^2x + q^2y)} = \frac{dp}{p^2 - p} = \frac{dq}{q^2 - q}.$$

Taking the first and fourth fractions and using multipliers p^2 and 2px, we have

$$each\ fraction = \frac{p^2dx + 2pxdp}{-2p^2x}. \tag{4.66}$$

Taking the second and fifth fractions and using multipliers q^2 and 2qy, we have

$$each\ fraction = \frac{q^2dy + 2qydq}{-2q^2y}. \tag{4.67}$$

From (4.66) and (4.67), we get

$$\frac{p^2 dx + 2pxdp}{p^2 x} = \frac{q^2 dy + 2qydq}{q^2 y}$$

or

$$\frac{d(p^2 x)}{p^2 x} = \frac{d(q^2 y)}{q^2 y}.$$

Integrating, we get

$$p^2 x = aq^2 y$$

where a is an arbitrary constant. Putting $p^2 x = aq^2 y$ in Eq. (4.65), we obtain

$$aq^2 y + q^2 y = z$$

which gives rise to

$$q = \sqrt{\frac{z}{(1 + a)y}}.$$

Putting this value of q in Eq. (4.65), we get

$$p = \sqrt{\frac{az}{(1 + a)x}}.$$

Making use of these values of p and q, the relation $dz = pdx + qdy$ becomes

$$dz = \sqrt{\frac{az}{(1 + a)x}}dx + \sqrt{\frac{z}{(1 + a)y}}dy$$

or

$$\sqrt{1 + a}\frac{dz}{\sqrt{z}} = \sqrt{a}\frac{dx}{\sqrt{x}} + \frac{dy}{\sqrt{y}}.$$

Integrating the above, we get

$$\sqrt{(1 + a)z} = \sqrt{ax} + \sqrt{y} + b$$

which is the complete solution as desired.

4.6 Singular Solution of a Non-linear PDE

The idea of singular solution is essentially motivated by the following result.

Theorem 4.6. *Let $f(x, y, z, a, b) = 0$ be the complete solution of a first-order PDE of the form*

$$F(x, y, z, p, q) = 0. \tag{4.68}$$

Then the envelope of the family $f(x, y, z, a, b) = 0$, if exists, satisfies Eq. (4.68).

Proof. Recall that the envelope of $f(x, y, z, a, b) = 0$ is obtained by eliminating 'a' and 'b' among

$$f(x, y, z, a, b) = 0 \tag{4.69}$$

$$f_a(x, y, z, a, b) = 0 \tag{4.70}$$

$$f_b(x, y, z, a, b) = 0. \tag{4.71}$$

Hence, the envelope of the given family is of the form

$$g(x, y, z) \equiv f\big(x, y, z, a(x, y, z), b(x, y, z)\big) = 0 \tag{4.72}$$

where $a(x, y, z)$ and $b(x, y, z)$ are determined by Eqs (4.70) and (4.71).

By the definition, the envelope (4.72) is touched at each of its points by a member of the family (4.69). Now, we show that at these points, this envelope has same partial derivatives as those of that particular surface of (4.69). To substantiate this, differentiating the function g partially w.r.t. x, y, z and using chain rule so that

$$g_x = f_x + f_a a_x + f_b b_x$$
$$g_y = f_y + f_a a_y + f_b b_y$$
$$g_z = f_z + f_a a_z + f_b b_z.$$

Using (4.70) and (4.71), the above three equations reduce to

$$g_x = f_x, \ g_y = f_y, \ g_z = f_z.$$

As $f(x, y, z, a, b) = 0$ being a complete solution satisfies $F(x, y, z, p, q) = 0$, we have $p = -\dfrac{f_x}{f_z}$ and $q = -\dfrac{f_y}{f_z}$. Thus in all, the values of x, y, z, p, q at any point of envelope $g(x, y, z) = 0$ coincide with those on some members of the family $f(x, y, z, a, b) = 0$, which is itself a solution of Eq. (4.68). It follows that the envelope (4.72) remains a solution of the PDE. $\square$

Thus, in lieu of Theorem 4.6, we define the notion of singular solution as follows:

Singular Solution: The envelope of the two-parameter family of surfaces represented by the complete solution of a first-order PDE, if it exists, remains a solution of the given PDE. Such a solution is called the *singular solution* or *singular integral*.

On and around the idea of singular solutions, the following observations are natural.

- The singular solution and particular solution both never contain any arbitrary constants. But the main difference between them is that the singular solution generally cannot be obtained from the complete solution by giving the particular values of the constants a and b.
- The eliminant besides containing the envelope sometimes may also produce the extraneous loci, which do not satisfy the PDE. Hence, the resulting eliminant to be a singular solution must satisfy the underlying PDE. An eliminant failing to do so will not be a singular solution.
- Obviously, all first-order PDEs may not have singular solutions. Especially, the quasilinear equations never admit singular solutions. To substantiate this, let $v = \phi(u)$ be the general solution of a quasilinear PDE: $Pp + Qq = R$ so that $u = c_1$ and $v = c_2$ are two independent solutions of corresponding Lagrange's auxiliary equations. If a and b are arbitrary constants, then

$$v = au + b \tag{4.73}$$

forms the complete solution. As the constants enter linearly, therefore, the two-parameter family of surfaces represented by (4.73) has no envelope. Only fully non-linear PDEs may have singular solutions.

Now, we discuss non-existence of singular solutions of two standard forms, which have been introduced in Section 4.4.

Proposition 4.1. *The PDEs of standard forms*

  (i) $f(p, q) = 0$
  (ii) $f(x, p) = g(y, q)$

have no singular solutions.

Proof. (i) The complete solution of the form $f(p, q) = 0$ is

$$z = ax + \phi(a)y + c \tag{4.74}$$

where $f(a, \phi(a)) = 0$. To obtain the singular solution, we differentiate (4.74) partially w.r.t. a and c so that

$$0 = x + \phi'(a)y$$
$$0 = 1.$$

The relation $0 = 1$ is absurd. Hence, the form $f(p, q) = 0$ has no singular solution.

(ii) The complete solution of the form $f(x, p) = g(y, q)$ is

$$z = \int \phi(x, a)dx + \int \psi(y, a)dy + b \tag{4.75}$$

where $p = \phi(x, a)$ and $q = \psi(y, a)$ are obtained from $f(x, p) = a$ and $g(y, q) = a$, respectively. To obtain the singular solution, we differentiate (4.75) partially w.r.t. a and b so that

$$0 = \int \frac{\partial \phi}{\partial a}dx + \int \frac{\partial \psi}{\partial a}dy$$

$$0 = 1.$$

The relation $0 = 1$ is absurd. Thus, there is no singular solution of the form $f(x, p) = g(y, q)$. $\qquad\square$

Theorem 4.7. *The singular solution of $F(x, y, z, p, q) = 0$ is determined by eliminating p and q among*

$$F(x, y, z, p, q) = 0$$
$$F_p(x, y, z, p, q) = 0$$
$$F_q(x, y, z, p, q) = 0.$$

Proof. Let $z = f(x, y, a, b)$ be the complete solution of the given PDE. Then, we have

$$F(x, y, f(x, y, a, b), f_x(x, y, a, b), f_y(x, y, a, b)) = 0.$$

Differentiating partially the above w.r.t. a and b, we get

$$\left.\begin{array}{l} F_z f_a + F_p f_{xa} + F_q f_{ya} = 0 \\ F_z f_b + F_p f_{xb} + F_q f_{yb} = 0. \end{array}\right\}$$

For the singular solution, we have

$$f_a = 0 \text{ and } f_b = 0. \tag{4.77}$$

Hence, Eq. (4.76) reduces to

$$\left.\begin{array}{l} F_p f_{xa} + F_q f_{ya} = 0 \\ F_p f_{xb} + F_q f_{yb} = 0. \end{array}\right\}$$

As $z = f(x, y, a, b)$ forms the complete solution, due to the availability of (4.77), we have

$$f_{xa} f_{yb} - f_{xb} f_{ya} \neq 0.$$

In lieu of the above relation, the system (4.78) has only a trivial solution for F_p and F_q, that is,

$$F_p = 0 \text{ and } F_q = 0.$$

This concludes the proof. $\qquad\square$

To demonstrate the idea of singular integrals, we furnish several examples.

Example 4.35. *Find the singular solution of PDE*

$$p^2 = zq.$$

Solution: *The given PDE can be written as*

$$F \equiv p^2 - zq = 0.$$

Differentiating the above equation partially w.r.t. p and q, we get

$$2p = 0 \ \ or \ \ p = 0,$$
$$-z = 0 \ \ or \ \ z = 0.$$

Eliminating p and q between these three equations, we get

$$z = 0.$$

Example 4.36. *Find the singular solution of Clairaut's equation*

$$z = px + qy + \sqrt{1 + p^2 + q^2}.$$

Solution: *The given PDE can be written as*

$$F \equiv z - px - qy - \sqrt{1 + p^2 + q^2} = 0. \tag{4.79}$$

Differentiating (4.79) partially w.r.t. 'p', we get

$$0 = -x - \frac{p}{\sqrt{1 + p^2 + q^2}}$$

$$x = -\frac{p}{\sqrt{1 + p^2 + q^2}}. \tag{4.80}$$

Similarly, differentiating (4.79) partially w.r.t. 'q', we get

$$y = -\frac{q}{\sqrt{1 + p^2 + q^2}}. \tag{4.81}$$

Squaring and adding (4.80) and (4.81), we get

$$x^2 + y^2 = \frac{p^2}{1 + p^2 + q^2} + \frac{q^2}{1 + p^2 + q^2}$$

$$= \frac{p^2 + q^2}{1 + p^2 + q^2} = 1 - \frac{1}{1 + p^2 + q^2}$$

so that

$$1 + p^2 + q^2 = \frac{1}{1 - x^2 - y^2}.$$

Using the above equation, (4.80) and (4.81) give rise to

$$p = -\frac{x}{\sqrt{1 - x^2 - y^2}} \quad and \quad q = -\frac{y}{\sqrt{1 - x^2 - y^2}}.$$

Putting the values p and q in (4.79), we get

$$x^2 + y^2 + z^2 = 1. \tag{4.82}$$

To verify, differentiating (4.82) partially w.r.t. x and y, we get

$$p = -\frac{x}{z} \quad and \quad q = -\frac{y}{z}.$$

Using these values in left-hand side of (4.79), we get

$$\begin{aligned}
\text{LHS} &= z + \frac{x^2}{z} + \frac{y^2}{z} - \sqrt{1 + \frac{x^2}{z^2} + \frac{y^2}{z^2}} \\
&= \frac{1}{z}(z^2 + x^2 + y^2 - \sqrt{z^2 + x^2 + y^2}) \\
&= \frac{1}{z}(1 - \sqrt{1}) \quad (using\ Eq.(4.82)) \\
&= 0 = \text{RHS}
\end{aligned}$$

Hence, $x^2 + y^2 + z^2 = 1$ is the required singular solution of the given PDE.

Example 4.37. *Find the complete solution and singular solution of PDE*

$$z^2(z^2p^2 + q^2) = 1.$$

Solution: *The given PDE can be written as*

$$z^2(z^2p^2 + q^2) = 1. \tag{4.83}$$

Complete Solution: Equation (4.83) is of the form $f(z, p, q) = 0$. Let $u = x + ay$ so that $p = \dfrac{dz}{du}$ and $q = a\dfrac{dz}{du}$, then Eq. (4.83) reduces to

$$z^2 \left[z^2 \left(\frac{dz}{du} \right)^2 + a^2 \left(\frac{dz}{du} \right)^2 \right] = 1$$

or

$$z^2(z^2 + a^2)\left(\frac{dz}{du}\right)^2 = 1$$

or

$$\pm z(z^2 + a^2)^{1/2}dz = du.$$

On integrating it, we get

$$\pm \frac{1}{3}(z^2 + a^2)^{3/2} = u + b$$

or

$$(z^2 + a^2)^3 = 9(u + b)^2$$

or

$$(z^2 + a^2)^3 = 9(x + ay + b)^2 \tag{4.84}$$

which is the complete solution of the given equation.

Singular Solution: Differentiating Eq. (4.84) partially w.r.t. 'a' and 'b', we obtain

$$(z^2 + a^2)^2 a = 3(x + ay + b)y \tag{4.85}$$

$$0 = x + ay + b. \tag{4.86}$$

From (4.85) and (4.86), we get $x + ay + b = 0$ and $a = 0$. Putting these values in (4.84), we obtain

$$z = 0.$$

Hence, we have $p = 0$ and $q = 0$. But $z = 0, p = 0$, and $q = 0$ do not satisfy the given PDE. Therefore, the given PDE has no singular solution.

Example 4.38. *Find the complete solution and singular solution of PDE*

$$xp + yq = pq.$$

Solution: *The given PDE can be written as*

$$F(x, y, z, p, q) \equiv xp + yq - pq = 0. \tag{4.87}$$

Complete Solution: The Charpit's auxiliary equations are

$$\frac{dx}{-(x-q)} = \frac{dy}{-(y-p)} = \frac{dz}{-p(x-q) - q(y-p)} = \frac{dp}{p} = \frac{dq}{q}.$$

Taking the last two fractions, we get $\dfrac{dp}{p} = \dfrac{dq}{q}$ so that

$$p = aq. \tag{4.88}$$

Putting this value of p in (4.87), we get $aqx + qy = aq^2$, i.e.,

$$q = \frac{y + ax}{a}.$$

From (4.88), we have

$$p = y + ax.$$

Putting these values of p and q in $dz = pdx + qdy$, we get

$$dz = (y + ax)dx + \frac{y + ax}{a}dy$$

or

$$adz = (y + ax)(dy + adx).$$

Integrating it, we obtain

$$az = \frac{1}{2}(y + ax)^2 + b \tag{4.89}$$

which is a complete solution of the given equation.

Singular Solution: Differentiating (4.89) partially w.r.t. 'a' and 'b', we get

$$z = x(y + ax)$$
$$0 = 1.$$

The last relation $0 = 1$ is absurd. Hence, the given PDE has no singular solution.

4.7 General Solution of a Non-linear PDE

As discussed earlier, Lagrange's method provides the general solution of a quasilinear PDE. However, in the case of non-linear PDEs, one can obtain only the complete solutions using Charpit's method/standard methods. In this section, from the complete solution of a non-linear PDE, we endeavour to compute the general solution. To accomplish this, consider the general form of first-order PDE

$$F(x, y, z, p, q) = 0. \tag{4.90}$$

Let $f(x, y, z, a, b) = 0$ be the complete solution of Eq. (4.90). Put $b = \phi(a)$, where ϕ is an arbitrary function, then envelope of the one-parameter subfamily $f\big(x, y, z, a, \phi(a)\big) = 0$ is indeed the surface

$$h(x, y, z) \equiv f\big(x, y, z, a(x, y, z), \phi(a(x, y, z))\big) = 0$$

which is obtained by eliminating 'a' between

$$f\big(x, y, z, a, \phi(a)\big) = 0$$
$$f_a\big(x, y, z, a, \phi(a)\big) = 0.$$

Proceeding along the lines of the proof of Theorem 4.6, we can show that the coordinates of any point (x, y, z) on the envelope $h(x, y, z) = 0$ along with the values of p and q at that point, being identical with x, y, z, p, q on some members of family $f(x, y, z, a, b) = 0$, must satisfy the given PDE. This new solution contains the arbitrary function ϕ, therefore, it remains the general solution.

General Solution: From the two-parameter family of surfaces represented by the complete solution of a first-order PDE, if we obtain a one-parameter subfamily by assuming an arbitrary functional relation between two parameters, then the envelope of this one-parameter subfamily, if it exists, forms the general solution or general integral.

It is usually not possible to perform the elimination of 'a' between two equations giving envelope, on account of the arbitrary function ϕ and its differential coefficient. If we choose $\phi(a)$ suitably, then the general solution reduces to a particular solution, which at the same time can cover many geometrical situations. Similar to the case of singular solutions, a particular solution obtained from the general solution is generally different from a particular solution obtained from the complete solution.

Peculiarities of the Quasilinear PDE: A complete solution of a quasilinear PDE

$$Pp + Qq = R \tag{4.91}$$

is of the form $v = au + b$, wherein $u(x, y, z) = c_1$ and $v(x, y, z) = c_2$ are two independent solutions of Lagrange's auxiliary equations of (4.91). Hence, the general solution of Eq. (4.91) can be obtained from the equations

$$v = au + \phi(a) \tag{4.92}$$

$$0 = u + \phi'(a). \tag{4.93}$$

Solving Eq. (4.93) for a, we get a relation of the form $a = \lambda(u)$. Putting this value in (4.92), we get

$$v = \lambda(u)u + \phi(\lambda(u)).$$

Thus, v remains a function of u, say

$$v = \psi(u). \tag{4.94}$$

If $\phi(a)$ is taken as an arbitrary function of a, then $\phi(u)$ is an arbitrary function of u. Consequently, the relation (4.94) is equivalent to the general solution of Eq. (4.91) available in Section 4.3 (in the context of Lagrange's method).

Intending to illustrate the foregoing description, we furnish the following examples.

Example 4.39. *Find the general solution of PDE*

$$p^2 + q^2 = m^2.$$

Also, deduce a particular solution from general solution.

Solution: *The given PDE can be written as*

$$p^2 + q^2 = m^2 \tag{4.95}$$

which is of the form $f(p, q) = 0$. Its solution is

$$z = ax + by + c$$

so that

$$a^2 + b^2 = m^2 \quad or \quad b = \sqrt{m^2 - a^2}.$$

Hence, the complete solution of the given equation is

$$z = ax + \sqrt{m^2 - a^2}\, y + c. \tag{4.96}$$

Putting $c = \phi(a)$ in Eq. (4.96), we get

$$z = ax + \sqrt{m^2 - a^2}\, y + \phi(a). \tag{4.97}$$

Differentiating Eq. (4.97) partially w.r.t. 'a', we obtain

$$0 = x - \frac{a}{\sqrt{m^2 - a^2}} y + \phi'(a). \tag{4.98}$$

The general solution is obtained by eliminating a from (4.97) and (4.98). If we take $\phi(a) = 0$, then a particular solution obtained by eliminating a is given by

$$z^2 = m^2(x^2 + y^2)$$

which represents a cone.

Example 4.40. *Determine two different particular solutions of PDE*

$$z = xp + yq + p^2 + q^2$$

from its general solution.

Solution: *The given equation is Clairaut's equation. Hence, its complete solution is*

$$z = ax + by + a^2 + b^2. \tag{4.99}$$

Putting $b = \phi(a)$ in Eq. (4.99), we get

$$z = ax + \phi(a)y + a^2 + \big(\phi(a)\big)^2. \tag{4.100}$$

Differentiating Eq. (4.100) partially w.r.t. 'a', we obtain

$$0 = x + \phi'(a)y + 2a + 2\phi(a)\phi'(a). \tag{4.101}$$

The general solution is obtained by eliminating a from (4.100) and (4.101). If we define $\phi(a) = 0$, then by the elimination of a, we get

$$4z + x^2 = 0$$

which is a parabolic cylinder. Let us now take $\phi(a) = \sqrt{1 - a^2}$. Then on eliminating a, we obtain

$$(z - 1)^2 = x^2 + y^2$$

which represents a right circular cone.

Example 4.41. *Find the complete solution, singular solution, and general solution of PDE*

$$(p^2 + q^2)y = qz.$$

Solution: *The given PDE can be written as*

$$F(x, y, z, p, q) \equiv (p^2 + q^2)y - qz = 0. \tag{4.102}$$

Complete Solution: The Charpit's auxiliary equations are

$$\frac{dx}{-2py} = \frac{dy}{-2qy + z} = \frac{dz}{-2p^2y + qz - 2q^2y} = \frac{dp}{-pq} = \frac{dq}{p^2}.$$

Taking the last two fractions, we get $pdp + qdq = 0$ so that

$$p^2 + q^2 = a^2. \tag{4.103}$$

From (4.102) and (4.103), we have

$$q = \frac{a^2 y}{z} \quad \text{and} \quad p = \frac{a}{z}\sqrt{z^2 - a^2 y^2}.$$

Putting these values of p and q in $dz = pdx + qdy$, we get

$$dz = \frac{a}{z}\sqrt{z^2 - a^2 y^2}dx + \frac{a^2 y}{z}dy$$

or

$$\frac{zdz - a^2 ydy}{\sqrt{z^2 - a^2 y^2}} = adx.$$

On integrating it, we obtain

$$\sqrt{z^2 - a^2 y^2} = ax + b$$

or

$$z^2 = (ax + b)^2 + a^2 y^2 \tag{4.104}$$

which is the complete solution of the given equation.

Singular Solution: Differentiating Eq. (4.104) partially w.r.t. 'a' and 'b', we obtain

$$0 = 2(ax + b)x + 2ay^2 \tag{4.105}$$

$$0 = 2(ax + b). \tag{4.106}$$

Eliminating a and b between (4.104), (4.105), and (4.106), we get

$$z = 0$$

which is the required singular solution.

General Solution: Replacing $b = \phi(a)$, where ϕ is some function of a, in Eq. (4.104), we get

$$z^2 = [ax + \phi(a)]^2 + a^2 y^2. \tag{4.107}$$

Differentiating Eq. (4.107) partially w.r.t. 'a', we obtain

$$0 = 2[ax + \phi(a)][x + \phi'(a)] + 2ay^2$$

or

$$0 = [ax + \phi(a)][x + \phi'(a)] + ay^2. \tag{4.108}$$

The general solution is obtained by eliminating a from (4.107) and (4.108).

Example 4.42. *Find the complete solution, singular solution, and general solution of PDE*

$$p^2 + q^2 = z.$$

Solution: *The given PDE can be written as*

$$p^2 + q^2 = z. \tag{4.109}$$

Complete Solution: Equation (4.109) is of the form $f(z, p, q) = 0$. Let $u = x + ay$ so that $p = \dfrac{dz}{du}$ and $q = a\dfrac{dz}{du}$, then Eq. (4.109) reduces to

$$\left(\frac{dz}{du}\right)^2 + a^2\left(\frac{dz}{du}\right)^2 = z$$

or

$$\frac{dz}{du} = \pm\frac{\sqrt{z}}{\sqrt{1+a^2}}$$

or

$$\pm z^{-1/2}(1+a^2)^{1/2}dz = du.$$

On integrating it, we get

$$\pm 2z^{1/2}(1+a^2)^{1/2} = u + b$$

or

$$4z(1+a^2) = (x+ay+b)^2 \tag{4.110}$$

which is the complete solution of the given equation.

Singular Solution: Differentiating Eq. (4.110) partially w.r.t. 'a' and 'b', we obtain

$$8az = 2y(x+ay+b) \tag{4.111}$$

$$0 = 2(x+ay+b). \tag{4.112}$$

From (4.111) and (4.112), we get

$$z = 0$$

which is the required singular solution.

General Solution: Putting $b = \phi(a)$ in Eq. (4.110), we get

$$4z(1 + a^2) - \left[x + ay + \phi(a)\right]^2 = 0. \tag{4.113}$$

Differentiating Eq. (4.113) partially w.r.t. 'a', we obtain

$$8az - 2\left[x + ay + \phi(a)\right]\left[y + \phi'(a)\right] = 0$$

or

$$4az - \left[x + ay + \phi(a)\right]\left[y + \phi'(a)\right] = 0. \tag{4.114}$$

The general solution is obtained by eliminating a from (4.113) and (4.114).

4.8 Cauchy Problem and Particular Solution: Integral Surface through a Given Curve

We have discussed several classical methods for finding the complete solution, general solution, and singular solution of a first-order PDE. But from an application point of view, a particular solution has importance as for a physical problem, only one solution satisfying prescribed conditions is desired. In the case of a first-order PDE, we determine a particular solution by formulating a Cauchy problem (or an initial value problem). Geometrically speaking, a Cauchy problem of a first-order PDE is equivalent to the problem of finding an integral surface of PDE passing through a space curve. For illustration, consider the Cauchy problem $zp + q = 1$, $z(x, x) = 0$. Then, Initial conditions (IC) is $z = 0$ on the line $y = x$, which can be parameterised as $x = s$, $y = s$, $z = 0$. Thus, the problem reduces to finding an integral surface of PDE: $zp + q = 1$ passing through the line $x = s, y = s, z = 0$.

Consider the general form of a first-order PDE

$$F(x, y, z, p, q) = 0 \tag{4.115}$$

Let Γ be a space curve, whose parametric equations are

$$x = x_0(s), \ y = y_0(s), \ z = z_0(s) \tag{4.116}$$

where x_0, y_0, z_0 are continuously differentiable functions on an interval $I \subset \mathbb{R}$. Then, the Cauchy problem is the problem to determine in a certain neighbourhood N of the plane curve $\Gamma' : x = x_0(s), \ y = y_0(s)$, an integral surface $z = z(x, y)$ of Eq. (4.115) containing the curve Γ, that is,

$$z(x_0(s), y_0(s)) = z_0(s) \quad \forall \, s \in I. \tag{4.117}$$

Here, the space curve Γ is called **initial data curve** or **datum curve** of the problem and Eq. (4.117) is called **initial condition** or **Cauchy condition** of the problem. The curve Γ', which is a projection of Γ on xy-plane, is called the **initial curve**, while $z_0(s)$ is called **initial data** or **Cauchy data**. The initial condition (4.117) thus is equivalent to saying that z takes prescribed values of the initial data on initial curve in the neighbourhood N.

4.8.1 Cauchy Problems Solvable by Lagrange's Method: Particular Solution from General Solution

Consider Cauchy problem governing by quasilinear PDE

$$P(x, y, z)p + Q(x, y, z)q = R(x, y, z) \tag{4.118}$$

with initial data curve Γ. In the context of Lagrange's method, the integral surface of Eq. (4.118) can be written as $\phi(u, v) = 0$, wherein

$$\left. \begin{array}{l} u(x, y, z) = c_1 \\ v(x, y, z) = c_2 \end{array} \right\} \tag{4.119}$$

constitute the general solution of Lagrange's auxiliary equations of Eq. (4.118). Suppose that the parametric equations of the given initial data curve Γ are

$$x = x_0(s), \ y = y_0(s), \ z = z_0(s).$$

As the curve Γ lies on the integral surface, we have

$$u\left(x_0(s), y_0(s), z_0(s)\right) = c_1$$
$$v\left(x_0(s), y_0(s), z_0(s)\right) = c_2.$$

Eliminating s between these two equations, we get a relation $\phi(c_1, c_2) = 0$. Thereafter, we replace $c_1 = u(x, y, z)$ and $c_2 = v(x, y, z)$ in this relation to obtain the particular solution as desired. If the curve Γ is described by its Cartesian form

$$\left. \begin{array}{l} f(x, y, z) = 0 \\ g(x, y, z) = 0 \end{array} \right\} \tag{4.120}$$

then we eliminate x, y, z from Eqs (4.119) and (4.120) to obtain the relation $\phi(c_1, c_2) = 0$. Finally, by replacing $c_1 = u(x, y, z)$ and $c_2 = v(x, y, z)$ in this relation, we obtain the required particular solution. Notice that there exists a unique solution of the Cauchy problem governing by a quasilinear PDE, if the initial data curve Γ is properly chosen. With a view to illustrate the foregoing description, we furnish the following examples.

Example 4.43. *Find the integral surface of the PDE*

$$x(y^2 + z)p - y(x^2 + z)q = (x^2 - y^2)z$$

which contains the straight line $x + y = 0$, $z = 1$.

Solution: Lagrange's auxiliary equations corresponding to the given PDE are

$$\frac{dx}{x(y^2 + z)} = \frac{dy}{-y(x^2 + z)} = \frac{dz}{z(x^2 - y^2)}.$$

Using multipliers $\frac{1}{x}, \frac{1}{y}, \frac{1}{z}$, *we obtain*

$$each\,fraction = \frac{\frac{1}{x}dx + \frac{1}{y}dy + \frac{1}{z}dz}{(y^2 + z) - (x^2 + z) + (x^2 - y^2)} = \frac{\frac{1}{x}dx + \frac{1}{y}dy + \frac{1}{z}dz}{0},$$

which gives rise to

$$\frac{1}{x}dx + \frac{1}{y}dy + \frac{1}{z}dz = 0.$$

Integrating, we obtain

$$\log x + \log y + \log z = \log c_1$$

or

$$xyz = c_1. \tag{4.121}$$

Using multipliers $x, y, -1$, *we obtain*

$$each\,fraction = \frac{xdx + ydy - dz}{x^2(y^2 + z) - y^2(x^2 + z) - z(x^2 - y^2)} = \frac{xdx + ydy - dz}{0}$$

which gives rise to

$$xdx + ydy - dz = 0.$$

Integrating, we obtain

$$x^2 + y^2 - 2z = c_2. \tag{4.122}$$

Parametric equations of the given straight line are

$$x = s, \; y = -s, \; z = 1.$$

Putting these values in Eqs (4.121) and (4.122), we get, respectively,

$$-s^2 = c_1$$
$$2s^2 - 2 = c_2.$$

Eliminating s in the above equations, we get

$$c_2 + 2c_1 + 2 = 0.$$

Putting the values of c_1 and c_2 from (4.121) and (4.122) in the above equation, we obtain

$$2xyz + x^2 + y^2 - 2z + 2 = 0$$

which is the integral surface as desired.

Example 4.44. *Find the integral surface of the PDE*

$$2y(z - 3)p + (2x - z)q = y(2x - 3)$$

which passes through the circle $z = 0$, $x^2 + y^2 = 2x$.

Solution: Lagrange's auxiliary equations corresponding to the given PDE are

$$\frac{dx}{2y(z - 3)} = \frac{dy}{(2x - z)} = \frac{dz}{y(2x - 3)}.$$

Taking the first and last fractions, we get

$$(2x - 3)dx - 2(z - 3)dz = 0.$$

Integrating, we obtain

$$x^2 - 3x - z^2 + 6z = c_1 \tag{4.123}$$

Using multipliers $\frac{1}{2}, y, -1$, we obtain

$$each\ fraction = \frac{\frac{1}{2}dx + ydy - dz}{y(z - 3) + y(2x - z) - y(2x - 3)} = \frac{\frac{1}{2}dx + ydy - dz}{0}$$

which gives rise to

$$dx + 2ydy - 2dz = 0.$$

Integrating, we obtain

$$x + y^2 - 2z = c_2. \tag{4.124}$$

Equation of the given circle is

$$z = 0,\ x^2 + y^2 = 2x. \tag{4.125}$$

Now, we shall eliminate x, y, z *between (4.123), (4.124), and (4.125). Putting* $z = 0$ *in Eqs (4.123) and (4.124), we get, respectively,*

$$x^2 - 3x = c_1 \ \ and \ \ x + y^2 = c_2.$$

Adding the above and using (4.125), we get

$$c_1 + c_2 = 0.$$

Putting the values of c_1 *and* c_2, *we get the required integral surface*

$$x^2 + y^2 - z^2 - 2x + 4z = 0.$$

4.8.2 Cauchy Problems Solvable by Charpit's Method: Particular Solution from Complete Solution

We now deal the problem finding a particular solution of a non-linear PDE

$$F(x, y, z, p, q) = 0. \tag{4.126}$$

passing through the curve Γ: $x = x_0(s)$, $y = y_0(s)$, $z = z_0(s)$. By Charpit's method, we can find the complete solution of Eq. (4.126), which is indeed a two-parameter family of surfaces of the form

$$f(x, y, z, a, b) = 0 \tag{4.127}$$

In Section 4.7, we have derived the general solution of Eq. (4.126) from its complete solution as an envelope of one-parameter subfamily of the family (4.127). We have to find a function $b = \varphi(a)$ in the general solution such that the integral surface obtained passes through the curve Γ. Let this one-parameter subfamily of surfaces be denoted by

$$\psi(x, y, z, a) = 0. \tag{4.128}$$

Suppose that a member of subfamily (4.128) passes through each point of the curve Γ, that is,

$$\psi\left(x_0(s), y_0(s), z_0(s), a\right) = 0 \tag{4.129}$$

where different points of the curve correspond to different values of the constant a.

Now, we show that the envelope of subfamily (4.128) contains the curve Γ. As at each point of the curve Γ, the tangent to the curve lies on the tangent plane to the surface, we have

$$\psi_x x_s + \psi_y y_s + \psi_z z_s = 0. \tag{4.130}$$

Differentiating partially Eq. (4.129) w.r.t. s and using chain rule, we get

$$\psi_x x_s + \psi_y y_s + \psi_z z_s + \psi_a a_s = 0. \tag{4.131}$$

From (4.130) and (4.131), we obtain $\psi_a a_s = 0$. But $a_s \neq 0$, we have $\psi_a = 0$. Thus, the components of the curve Γ satisfy $\psi = 0$ and $\psi_a = 0$ identically w.r.t. s. It follows that the envelope of subfamily

(4.128) contains curve Γ and hence this envelope forms the required integral surface of (4.126) passing through the curve Γ.

Our final attempt now is to find the function $b = \varphi(a)$ for which (4.129) and (4.130) are satisfied. As components of curve Γ satisfy the family (4.127), we have

$$f\left(x_0(s), y_0(s), z_0(s), a, b\right) = 0. \tag{4.132}$$

Since LHS of (4.130) represents the derivative w.r.t. s of the LHS of (4.129), we must write, therefore,

$$f_s\left(x_0(s), y_0(s), z_0(s), a, b\right) = 0. \tag{4.133}$$

On eliminating s between (4.132) and (4.133), we get the required functional relation between a and b of the form

$$\phi(a, b) = 0. \tag{4.134}$$

However, there may be many solutions of Eq. (4.134) as it can be factored into a set of alternative equations, say $b = \varphi_1(a)$, $b = \varphi_2(a)$, and so on. Each one of them defines a subfamily and the envelope of each of these subfamilies, if it exists, forms a particular solution of the problem.

Therefore, the Cauchy problem for non-linear equations is more complicated as compared to quasilinear equations. For the sake of simplicity, we furnish the following illustrative examples, which are solved in three steps.

Example 4.45.   *Find the integral surface of the PDE*

$$(p^2 + q^2)x = pz$$

containing the curve Γ: $x = 0$, $y = s^2$, $z = 2s$.

Solution: *Equation of the given PDE is*

$$F \equiv (p^2 + q^2)x - pz = 0.$$

Parametric equations of initial data curve Γ are

$$x_0(s) = 0, \ y_0(s) = s^2, \ z_0(s) = 2s.$$

Step-I (Complete Solution):   The Charpit's auxiliary equations are

$$\frac{dx}{-2px + z} = \frac{dy}{-2qx} = \frac{dz}{-pz} = \frac{dp}{q^2} = \frac{dq}{-pq}.$$

Taking the last two fractions, we get $p\,dp + q\,dq = 0$ so that

$$p^2 + q^2 = a^2.$$

Hence, we have

$$p = \frac{a^2 x}{z} \text{ and } q = \frac{a}{z}\sqrt{z^2 - a^2 x^2}.$$

Putting these values of p and q in $dz = pdx + qdy$, we get

$$dz = \frac{a^2 x}{z} dx + \frac{a}{z}\sqrt{z^2 - a^2 x^2} dy$$

or

$$\frac{zdz - a^2 xdx}{\sqrt{z^2 - a^2 x^2}} = ady.$$

On integrating it, we obtain

$$\sqrt{z^2 - a^2 x^2} = ay + b$$

or

$$z^2 = a^2 x^2 + (ay + b)^2 \tag{4.135}$$

which is a complete solution of the given PDE.

Step-II (One-Parameter Subfamily): Putting $x = 0$, $y = s^2$, $z = 2s$ in Eq. (4.135), we get

$$4s^2 = (as^2 + b)^2. \tag{4.136}$$

Differentiating (4.136) w.r.t. s, we obtain

$$2 = a(as^2 + b). \tag{4.137}$$

Eliminating s between (4.136) and (4.137), we obtain $ab = 1$. Putting $b = 1/a$ in (4.135), we obtain

$$z^2 = a^2 x^2 + \left(ay + \frac{1}{a}\right)^2$$

or

$$a^4(x^2 + y^2) + a^2(2y - z^2) + 1 = 0 \tag{4.138}$$

which is the one-parameter subfamily of the family of complete solution.

Step-III (Envelope): Differentiating partially w.r.t. a, we get

$$2a^2(x^2 + y^2) + (2y - z^2) = 0. \tag{4.139}$$

Eliminating a between (4.138) and (4.139), we obtain the envelope of subfamily (4.138) given by

$$(2y - z^2)^2 = 4(x^2 + y^2)$$

or

$$z^2 = 2(y \pm \sqrt{x^2 + y^2}).$$

But $y \leq \sqrt{x^2 + y^2}$ so that $y - \sqrt{x^2 + y^2} \leq 0$, the minus sign must be discarded. Thus, the required integral surface of the given PDE is

$$z^2 = 2(y + \sqrt{x^2 + y^2}).$$

Example 4.46. *Find the integral surface of the PDE*

$$p^2 x + qy = z$$

passing through the line $x + z = 0$, $y = 1$.

Solution: *Equation of the given PDE is*

$$F \equiv p^2 x + qy - z = 0.$$

Parametric equations of the given line (initial data curve) are

$$x_0(s) = s, \ y_0(s) = 1, \ z_0(s) = -s.$$

Step-I (Complete Solution): The Charpit's auxiliary equations are

$$\frac{dx}{-2px} = \frac{dy}{-y} = \frac{dz}{-2p^2x - qy} = \frac{dp}{p(1+p)} = \frac{dq}{0}.$$

Last fraction gives rise to $q = a$. Hence, we have $p = \sqrt{\dfrac{z - ay}{x}}$. Putting these values of p and q in $dz = p\,dx + q\,dy$, we get

$$dz = \sqrt{\frac{z - ay}{x}}\,dx + a\,dy$$

or

$$\frac{dz - a\,dy}{\sqrt{z - ay}} = \frac{dx}{\sqrt{x}}.$$

On integrating it, we get

$$\sqrt{z - ay} = \sqrt{x} + \sqrt{b}$$

or

$$(x + ay - z + b)^2 = 4bx \tag{4.140}$$

which is the complete solution of the given PDE.

Step-II (One-Parameter Subfamily): Making use of $x = s$, $y = 1$, $z = -s$, Eq. (4.140) reduces to

$$(a + b + 2s)^2 = 4bs. \tag{4.141}$$

Differentiating (4.141) w.r.t. s, we get

$$a + 2s = 0. \tag{4.142}$$

Eliminating s between (4.141) and (4.142), we get $b^2 + 2ab = 0$ yielding thereby $b = 0$ or $b = -2a$. For $b = 0$, one cannot find the particular solution. Putting $b = -2a$ in Eq. (4.140), we obtain

$$(ay - z + x - 2a)^2 = -8ax \tag{4.143}$$

which forms a one-parameter subfamily of the family of complete solution.

Step-III (Envelope): On differentiating partially w.r.t. a, the above equation becomes

$$(y - 2)(ay - z + x - 2a)^2 = -4x. \tag{4.144}$$

Eliminating a between (4.143) and (4.144), we get the envelope of subfamily (4.141) given by

$$xy = z(y - 2)$$

which is the integral surface of PDE as desired.

4.9 Cauchy's Method of Characteristics: A General Method for Solving Cauchy Problems

In Section 4.8, we have already discussed the classical technique to deduce the integral surface of a Cauchy problem for quasilinear and non-linear PDEs from their general and complete solutions. In this section, we describe a general method to determine a particular solution of the Cauchy problem without using the general/complete solution. This method is called **Cauchy's Method of Characteristics** and is based mainly on geometrical ideas. Historically, a French mathematician *Gaspard Monge* (1746–1818) initiated the geometric theory of PDEs by introducing the concepts of characteristic curves and Monge cones. *Augustin Louis Cauchy* (1789–1857) thus furthered the theory of characteristics originated by Monge. However, the method of characteristics was developed in a proper way in the middle of the 19th century by *Saint Venant* (1797–1886) and *William Rowan Hamilton* (1805–1865). Here it can be pointed out that neither Monge nor Cauchy used the term 'characteristics'. It was used first by Hamilton when he applied this method to solve the well-known *Eikonal equation*.

The most general PDE of order one has the form

$$F(x, y, z, p, q) = 0. \tag{4.145}$$

Here, F is a twice continuously differentiable function w.r.t. its arguments: x, y, z, p, q, and F is not degenerate so that $F_p^2 + F_q^2 \neq 0$. Without loss of generality, we shall assume that $F_q \neq 0$.

The chief feature of the method of characteristics is reducing the integral surface of Eq. (4.145) passing through a given (initial) curve Γ to a particular solution of the initial value problem for a system of ODEs associated with Eq. (4.145). The required integral surface thus may be visualised as the surface of the form $x = X(t, s)$, $y = Y(t, s)$, $z = Z(t, s)$ such that t changes along the characteristic curve C (that is, the curve determined by the system of ODEs), whereas s changes along the initial curve Γ.

Monge Cone: Suppose that $\mathbf{S} : z = z(x, y)$ is an integral surface of Eq. (4.145) and $A(\bar{x}, \bar{y}, \bar{z})$ is fixed point on the surface $\mathbf{S}$. The equation of the tangent plane to $\mathbf{S}$ through the point A is

$$p(x - \bar{x}) + q(y - \bar{y}) = (z - \bar{z}). \tag{4.146}$$

As A lies on $\mathbf{S}$, Eq. (4.145) imposes the relation

$$F(\bar{x}, \bar{y}, \bar{z}, p, q) = 0 \tag{4.147}$$

which yields that at the point A, the derivatives p and q are not independent. We can calculate the value of q for given values of p so that Eq. (4.147) can be expressed explicitly as

$$q = \varphi(p). \tag{4.148}$$

The pair of equations represented by (4.146) and (4.148) forms a one-parameter (p) family of tangent planes. The envelope of this family is a cone with vertex at A, which is called the *Monge cone* or *elementary cone* at A of Eq. (4.145).

Generator of Monge Cone: Differentiating (4.146) w.r.t. parameter p, we get

$$(x - \bar{x}) + \frac{dq}{dp}(y - \bar{y}) = 0. \tag{4.149}$$

Again, differentiating Eq. (4.147) w.r.t. p and using chain rule, we obtain

$$F_p + F_q \frac{dq}{dp} = 0. \tag{4.150}$$

Eliminating $\dfrac{dq}{dp}$ between (4.149) and (4.150), we get

$$\frac{x - \bar{x}}{F_p} = \frac{y - \bar{y}}{F_q} \tag{4.151}$$

which represents the equation of the projections of generators of Monge cone on xy-plane. Combining (4.146) and (4.151), we get

$$\frac{x - \bar{x}}{F_p} = \frac{y - \bar{y}}{F_q} = \frac{z - \bar{z}}{pF_p + qF_q} \tag{4.152}$$

which is the equation of the generators of Monge cone at the point A for a given p and the corresponding $q = \varphi(p)$.

A geometrical interpretation immediately reveals the following fact.

Theorem 4.8. *A surface* **S** $: z = z(x, y)$ *is an integral surface of Eq. (4.145) iff at each point on the surface* **S***, its tangent plane should touch the Monge cone along its generator.*

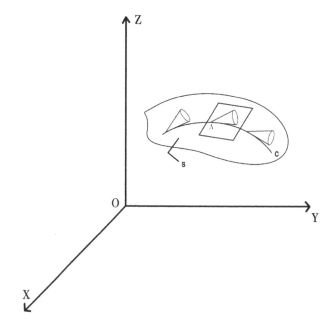

FIGURE 4.2 Monge Cone

Thus, the problem of solving Eq. (4.145) is equivalent to finding a surface, which touches the Monge cone at each point along a generator. In order to generate such a surface, we use the concept of characteristic curves that is defined in the followng.

Characteristic Curves: A space curve that at each point touches a generator of the Monge cone is called the *characteristic curve* or *Monge curve* of Eq. (4.145). In our subsequent discussions, a characteristic curve will be represented by parameterised form in parameter 't', where t belongs to an appropriate closed interval $[0, T]$, $T > 0$.

Characteristic Direction: At different points on the integral surface **S**, there exist different Monge cones of Eq. (4.145). According to Theorem 4.8, the generators of Monge cones of Eq. (4.145) are the lines of contact between the tangent planes of **S** and corresponding Monge cones. These generators, along which the surface **S** is touched, thus define a direction field on **S**, which is called *characteristic direction* or *Monge direction* of Eq. (4.145).

In terms of characteristic direction, we propose the following characterization of the characteristic curves.

Proposition 4.2. *A space curve C is the characteristic curve of Eq. (4.145) iff C is an integral curve of the characteristic direction of Eq. (4.145).*

Strip: A space curve $x = x(t), y = y(t), z = z(t)$ together with a pair of functions $p(t)$ and $q(t)$ is called a *strip* if at each point the tangent to the curve lies on a plane, whose normal has direction ratios $(p(t), q(t), -1)$. We denote such a strip by the ordered five-tuple $(x(t), y(t), z(t), p(t), q(t))$. For each $t = t_0$, the five-tuple of numbers $(x(t_0), y(t_0), z(t_0), p(t_0), q(t_0))$ is said to be an element of the strip. Further, a strip is called the *integral strip* of Eq. (4.145) if its components satisfy Eq. (4.145).

Proposition 4.3. *An ordered five-tuple of real differentiable functions $(x(t), y(t), z(t), p(t), q(t))$ forms a strip iff it satisfies the following condition (called strip condition):*

$$\frac{dz}{dt} = p(t)\frac{dx}{dt} + q(t)\frac{dy}{dt}.$$

Proof. We know that $\left(\dfrac{dx}{dt}, \dfrac{dy}{dt}, \dfrac{dz}{dt}\right)$ represent direction ratios of tangent line to the curve $x = x(t), y = y(t), z = z(t)$. This tangent line lies on a plane having normal $\mathbf{n} := (p(t), q(t), -1)$ if and only if the tangent line and the normal vector $\mathbf{n}$ are orthogonal, which is equivalent to

$$\left(\frac{dx}{dt}, \frac{dy}{dt}, \frac{dz}{dt}\right) \cdot (p(t), q(t), -1) = 0$$
$$\Leftrightarrow \quad p(t)\frac{dx}{dt} + q(t)\frac{dy}{dt} = \frac{dz}{dt}.$$

$\square$

Characteristic Strip: A strip $(x(t), y(t), z(t), p(t), q(t))$ is called the *characteristic strip* of Eq. (4.145) if the corresponding curve $x = x(t), y = y(t), z = z(t)$ is a characteristic curve. In other words, an ordered five-tuple of real differentiable functions $(x(t), y(t), z(t), p(t), q(t))$ is said to be a characteristic strip of Eq. (4.145) if the first three components, namely, $x = x(t), y = y(t), z = z(t)$ represent a characteristic curve C and at each point of C, $p(t)$ and $q(t)$ define a tangent plane with $(p, q, -1)$ as the normal vector.

Determination of Characteristic Curves and Strips: In view of Proposition 4.2, a characteristic curve C of Eq. (4.145) remains an integral curve of the characteristic direction of Eq. (4.145). Also from Eq. (4.152), it follows that the characteristic direction of Eq. (4.145) is $(F_p, F_q, pF_p + qF_q)$ (dropping the evaluation of arguments of each components at A). The slop of C therefore is

$$\frac{dx}{F_p} = \frac{dy}{F_q} = \frac{dz}{pF_p + qF_q}. \tag{4.153}$$

Introducing the parameter t and equating each fraction of the above system by dt, we obtain the following system of ODEs:

$$\left. \begin{aligned} \frac{dx}{dt} &= F_p \\[2mm] \frac{dy}{dt} &= F_q \\[2mm] \frac{dz}{dt} &= pF_p + qF_q. \end{aligned} \right\} \tag{4.154}$$

Clearly, the three equations of the system (4.154) are not sufficient to determine the characteristic curve C and we need to find equations for $p(t)$ and $q(t)$. By chain rule, along the characteristic curves on **S**, we have

$$\left. \begin{aligned} \frac{dp}{dt} &= p_x \frac{dx}{dt} + p_y \frac{dy}{dt} \\[2mm] \frac{dq}{dt} &= q_x \frac{dx}{dt} + q_y \frac{dy}{dt}. \end{aligned} \right\}$$

By virtue of $p_y = q_x$, $q_x = p_y$ and using the first two equations of (4.154), the above equations become, respectively,

$$\left. \begin{aligned} \frac{dp}{dt} &= p_x F_p + q_x F_q \\[2mm] \frac{dq}{dt} &= p_y F_p + q_y F_q. \end{aligned} \right\} \tag{4.155}$$

Differentiating Eq. (4.145) w.r.t. x as well as y, we get

$$\left. \begin{aligned} F_x + F_z p + F_p p_x + F_q q_x &= 0 \\[2mm] F_y + F_z q + F_p p_y + F_q q_y &= 0. \end{aligned} \right\} \tag{4.156}$$

Making use of (4.156), (4.155) reduce to

$$\left.\begin{aligned}
\frac{dp}{dt} &= -(F_x + pF_z) \\
\frac{dq}{dt} &= -(F_y + qF_z).
\end{aligned}\right\} \tag{4.157}$$

Thus, combining the five ODEs represented by (4.154) and (4.157), we get the following system

$$\left.\begin{aligned}
\frac{dx}{dt} &= F_p \\
\frac{dy}{dt} &= F_q \\
\frac{dz}{dt} &= pF_p + qF_q \\
\frac{dp}{dt} &= -(F_x + pF_z) \\
\frac{dq}{dt} &= -(F_y + qF_z)
\end{aligned}\right\} \tag{4.158}$$

These equations are called the *characteristic equations* of Eq. (4.145). In fact, these characteristic equations are parametric formulation of Charpit's auxiliary equations. The last three equations of the system (4.158) are also referred as the *compatibility conditions*. Thus, we conclude that the general solution of the system (4.158) of the form

$$x = x(t), y = y(t), z = z(t), p = p(t), q = q(t) \tag{4.159}$$

determines the characteristic curve $C : x = x(t), y = y(t), z = z(t)$. Further, from the first three equations of (4.158) and Proposition 4.3, it follows that the general solution (4.159) forms the characteristic strip of Eq. (4.145).

Proposition 4.4. *Along every characteristic strip of Eq. (4.145), the function $F(x, y, z, p, q)$ is a constant.*

Proof. Differentiating $F \equiv F\big(x(t), y(t), z(t), p(t), q(t)\big)$ w.r.t. t and using chain rule, we get

$$\frac{dF}{dt} = \frac{\partial F}{\partial x}\frac{dx}{dt} + \frac{\partial F}{\partial y}\frac{dy}{dt} + \frac{\partial F}{\partial z}\frac{dz}{dt} + \frac{\partial F}{\partial p}\frac{dp}{dt} + \frac{\partial F}{\partial q}\frac{dq}{dt}.$$

Using characteristic equations in RHS of the above equation, we obtain

$$\begin{aligned}
\frac{dF}{dt} &= F_x F_p + F_y F_q + F_z(pF_p + qF_q) - F_p(F_x + pF_z) - F_q(F_y + qF_z) \\
&= 0
\end{aligned}$$

implying thereby,

$$F(x, y, z, p, q) = c, \text{ a constant along characteristic strip.}$$

□

The following two lemmas are needed to prove our main result.

Lemma 4.1. *If a characteristic strip contains an element $(\overline{x}, \overline{y}, \overline{z}, \overline{p}, \overline{q}) \in \mathbb{R}^5$ such that*

$$F(\overline{x}, \overline{y}, \overline{z}, \overline{p}, \overline{q}) = 0$$

then the characteristic strip remains an integral strip of Eq. (4.145).

Proof. Let $\big(x(t), y(t), z(t), p(t), q(t)\big)$ be a characteristic strip of Eq. (4.145), which contains an element $(\overline{x}, \overline{y}, \overline{z}, \overline{p}, \overline{q})$. It follows that there exists $t = t_0$ such that $x(t_0) = \overline{x}$, $y(t_0) = \overline{y}$, $z(t_0) = \overline{z}$, $p(t_0) = \overline{p}$ and $q(t_0) = \overline{q}$. At initial point $t = t_0$, the characteristic strip $(x(t), y(t), z(t), p(t), q(t))$ thus remains a unique solution of the characteristic equations passing through this point. In addition, with a view of Proposition 4.4, we have

$$F\big(x(t), y(t), z(t), p(t), q(t)\big) = c, \quad \text{where } c = F(\overline{x}, \overline{y}, \overline{z}, \overline{p}, \overline{q}).$$

But by the hypothesis, we have $c = 0$. This implies that

$$F\big(x(t), y(t), z(t), p(t), q(t)\big) = 0 \quad \forall\, t \in [0, T].$$

Thus, $(x(t), y(t), z(t), p(t), q(t))$ is an integral strip of Eq. (4.145). □

Lemma 4.2. *Let $z = z(x, y)$ be an integral surface of Eq. (4.145). For a fixed element $(\overline{x}, \overline{y}) \in \mathbb{R}^2$, define $\overline{z} = z(\overline{x}, \overline{y})$, $\overline{p} = z_x(\overline{x}, \overline{y})$, $\overline{q} = z_y(\overline{x}, \overline{y})$. If $(x(t), y(t), z(t), p(t), q(t))$ is the particular solution of characteristic equations (4.158) with initial conditions*

$$x(t_0) = \overline{x}, \ y(t_0) = \overline{y}, \ z(t_0) = \overline{z}, \ p(t_0) = \overline{p}, \ q(t_0) = \overline{q}$$

at a specific point $t = t_0$, then

$$z(t) = z(x(t), y(t))$$
$$p(t) = z_x(x(t), y(t))$$
$$q(t) = z_y(x(t), y(t)).$$

In other words, if a characteristic strip has an element $(\overline{x}, \overline{y}, \overline{z}, \overline{p}, \overline{q}) \in \mathbb{R}^5$ in common with an integral surface, then it lies completely on the integral surface.

Proof. Consider the initial value problem

$$
\left.
\begin{aligned}
\frac{dx}{dt} &= F_p(x, y, z(x, y), z_x(x, y), z_y(x, y)) \\
\frac{dy}{dt} &= F_q(x, y, z(x, y), z_x(x, y), z_y(x, y))
\end{aligned}
\right\}
$$

with the initial conditions $x(t_0) = \bar{x}$, $y(t_0) = \bar{y}$. The above system therefore uniquely determines a curve $x = x(t)$ and $y = y(t)$ in the xy-plane.

Set $z(t) = z(x(t), y(t))$, $p(t) = z_x(x(t), y(t))$ and $q(t) = z_y(x(t), y(t))$, then the space curve $x = x(t), y = y(t), z = z(t)$ lies on the integral surface. Therefore, we have

$$
\begin{aligned}
z(t_0) &= z(\bar{x}, \bar{y}) = \bar{z} \\
p(t_0) &= z_x(\bar{x}, \bar{y}) = \bar{p} \\
q(t_0) &= z_y(\bar{x}, \bar{y}) = \bar{q}.
\end{aligned}
$$

Further, on the space curve $x = x(t), y = y(t), z = z(t)$, we have

$$
\frac{dz}{dt} = z_x \frac{dx}{dt} + z_y \frac{dy}{dt} = z_x F_p + z_y F_q
$$

$$
\frac{dp}{dt} = z_{xx} F_p + z_{xy} F_q \tag{4.160}
$$

$$
\frac{dq}{dt} = z_{yx} F_p + z_{yy} F_q. \tag{4.161}
$$

As $z = z(x, y)$ is an integral surface of Eq. (4.145), we have

$$
F(x, y, z(x, y), z_x(x, y), z_y(x, y)) = 0.
$$

Differentiating the above equation w.r.t. x and y, we get respectively,

$$
\begin{aligned}
F_x + F_z z_x + F_{z_x} z_{xx} + F_{z_y} z_{yx} &= 0 \\
F_y + F_z z_y + F_{z_x} z_{xy} + F_{z_y} z_{yy} &= 0.
\end{aligned}
$$

Making use of the above equations, (4.160) and (4.161) reduce to, respectively,

$$
\frac{dp}{dt} = -F_x - F_z z_x \tag{4.162}
$$

$$
\frac{dq}{dt} = -F_y - F_z z_y. \tag{4.163}
$$

It follows that the five functions $x = x(t)$, $y = y(t)$, $z = z(x(t), y(t))$, $p = z_x(x(t), y(t))$, and $q = z_y(x(t), y(t))$ being satisfy the characteristic equations determine a characteristic strip. Owing to uniqueness of the characteristic strip with the initial element $\bar{x}, \bar{y}, \bar{z}, \bar{p}, \bar{q}$, this coincides with the given strip. However, this characteristic strip lies on the integral surface by definition. Hence, our result follows. □

4.9.1 Solution of a Cauchy Problem for Non-linear PDE

In foregoing discussion, we provided the background materials and essential tools regarding the method of characteristics. We are now in a position to discuss the method of characteristics for solving the Cauchy problem governed by a first-order (non-linear) PDE. Consider the Cauchy problem for Eq. (4.145) with initial data curve

$$\Gamma : \quad x = x_0(s), y = y_0(s), z = z_0(s) \tag{4.164}$$

where x_0, y_0, z_0 are continuously differentiable functions on an interval $I \subset \mathbb{R}$.

In order to construct the integral surface of the Cauchy problem, we find the characteristic strips defined by the characteristic equations issuing from each point of the initial data curve. The characteristic curves associated with these characteristic strips generate the required integral surface. Naturally, the initial conditions on x, y, and z are given by the initial data curve. In order to solve the system of characteristic equations, we need to prescribe initial conditions on p and q also. Along the initial data curve Γ, if we specify the functions $p_0(s)$ and $q_0(s)$ such that $(x_0(s), y_0(s), z_0(s), p_0(s), q_0(s))$ becomes an integral strip of Eq. (4.145), then such a strip is called an **initial strip** for the given Cauchy problem. Notice that in this case, the **initial strip condition** is defined as

$$\frac{dz_0}{ds} = p_0(s)\frac{dx_0}{ds} + q_0(s)\frac{dy_0}{ds}.$$

In so doing, we replace our initial data curve Γ with an initial strip. Consequently, the statement of the Cauchy problem becomes:

To find the integral surface **S** $: z = z(x, y)$ *of* $F(x, y, z, p, q) = 0$ *such that* **S** *contains an initial strip* $x = x_0(s)$, $y = y_0(s)$, $z = z_0(s)$, $p = p_0(s)$, $q = q_0(s)$.

It is possible that there exist more than one pair $(p_0(s), q_0(s))$ satisfying the above-mentioned compatibility conditions for the initial strip. Therefore, there may be several initial strips associated with a single initial data curve. Consequently, there can be more than one integral surface passing through a given initial data curve. However, for each choice of the pair $(p_0(s), q_0(s))$, there is a unique solution of the Cauchy problem.

Finally, we are equipped to prove the following result regarding the existence and uniqueness of the solution of the Cauchy problem in a neighbourhood of the initial data curve, which actually justifies the method of characteristics.

Theorem 4.9. *Consider the Cauchy problem governed by Eq. (4.145) with initial data curve* Γ *given by (4.164). Let $p_0(s)$ and $q_0(s)$ be a pair of continuously differentiable functions of s such that $(x_0, y_0, z_0, p_0, q_0)$ forms an initial strip of Eq. (4.145). If this initial strip satisfies the following* *transversality condition:*

$$\frac{dx_0}{ds} F_q(x_0, y_0, z_0, p_0, q_0) - \frac{dy_0}{ds} F_p(x_0, y_0, z_0, p_0, q_0) \neq 0$$

then in a neighbourhood of initial curve Γ' (where Γ' is a projection of Γ on xy-plane), there exists a unique solution of Cauchy problem, which contains the initial strip.

Proof. For the sake of simplicity, we shall prove our result in several steps.

Step 1: Without loss of generality, we choose the parameter 't' such that the characteristic curve is located on Γ when $t = 0$. Hence, using the existence and uniqueness theorem for the initial value problem regarding a system of ODEs[2], there exists a unique solution

$$x = X(t, s), y = Y(t, s), z = Z(t, s), p = P(t, s), q = Q(t, s) \tag{4.165}$$

of system (4.158) satisfying the initial conditions

$$X(0, s) = x_0(s), Y(0, s) = y_0(s), Z(0, s) = z_0(s), P(0, s) = p_0(s), Q(0, s) = q_0(s).$$

Further, by the definition of initial strip, we have

$$F\left(X(0, s), Y(0, s), Z(0, s), P(0, s), Q(0, s)\right) = 0 \quad \forall \, s \in I.$$

Consequently using Lemma 4.1, we have

$$F\left(X(t, s), Y(t, s), Z(t, s), P(t, s), Q(t, s)\right) = 0 \quad \forall \, t \in [0, T], \, s \in I.$$

It follows that for each fixed s, $(X(t, s), Y(t, s), Z(t, s), P(t, s), Q(t, s))$ is an integral strip, in which first three equations, viz. $x = X(t, s), y = Y(t, s), z = Z(t, s)$ form a characteristic curve, varying s, we get certain parameterized surface.

Step 2: We show that the parameterized surface represented by $x = X(t, s), y = Y(t, s), z = Z(t, s)$ defines a smooth surface of the form $z = z(x, y)$. From the characteristic equations and transversality condition, it follows that the Jacobian

$$J|_{t=0} : \quad = \left. \frac{(\partial X, Y)}{(\partial s, t)} \right|_{t=0} = X_s(0, s)Y_t(0, s) - Y_s(0, s)X_t(0, s)$$

$$= \frac{dx_0}{ds} F_q - \frac{dy_0}{ds} F_p \neq 0.$$

Therefore, using inverse mapping theorem, there exists a neighbourhood $N(x_0, y_0)$ of a point $(x_0(s), y_0(s))$ of Γ' (corresponding to $t = 0$) such that the mapping $(X(t, s), Y(t, s))$ is inverted to a smooth mapping $(\tau(x, y), \sigma(x, y))$ in $N(x_0, y_0)$ so that $t = \tau(x, y)$ and $s = \sigma(x, y)$. Thus, we can find a smooth function $z = z(x, y)$ as follows:

$$z = Z(t, s) = Z(\tau(x, y), \sigma(x, y)) \equiv z(x, y).$$

---

[2]See Theorem 10.5 [6].

Step 3: We show that the smooth surface $z = z(x, y)$ so constructed remains an integral surface as desired. Set

$$p = P(t, s) = P(\tau(x, y), \sigma(x, y)) \equiv p(x, y)$$

and

$$q = Q(t, s) = Q(\tau(x, y), \sigma(x, y)) \equiv q(x, y).$$

From Lemma 4.1 and Lemma 4.2, we obtain

$$F(x, y, z(x, y), p(x, y), q(x, y)) = 0 \quad \forall (x, y) \in N(x_0, y_0).$$

Therefore, it suffices to show that $p(x, y) = z_x(x, y)$ and $q(x, y) = z_y(x, y)$. Define the auxiliary function

$$R(s, t) = Z_s - PX_s - QY_s. \tag{4.167}$$

Taking $t = 0$ in (4.167) and using initial strip condition, we obtain

$$R(s, 0) = \frac{dz_0}{ds} - p_0 \frac{dx_0}{ds} - q_0 \frac{dy_0}{ds} = 0. \tag{4.168}$$

We now show that $R = 0$ for all $t \in (0, T]$. Differentiating Eq. (4.167) w.r.t. t, we get

$$\frac{\partial R}{\partial t} = Z_{st} - P_t X_s - Q_t Y_s - PX_{st} - QY_{st}$$

$$= \frac{\partial}{\partial s}(Z_t - PX_t - QY_t) + P_s X_t + Q_s Y_t - Q_t Y_s - P_t X_s$$

which, by using characteristic equations, reduces to

$$\frac{\partial R}{\partial t} = P_s F_p + Q_s F_q + (F_x + F_z P)X_s + (F_y + F_z Q)Y_s.$$

Adding and subtracting $F_z Z_s$ and rearranging the terms, we get

$$\frac{\partial R}{\partial t} = (F_x X_s + F_y Y_s + F_z Z_s + F_p P_s + F_q Q_s) - F_z(Z_s - PX_s - QY_s)$$

$$= F_s - F_z R.$$

Differentiating (4.166) partially w.r.t. s, we get $F_s = 0$. Therefore, the above equation reduces to linear homogeneous differential equation of the form

$$\frac{\partial R}{\partial t} = -F_z R(s, t).$$

The solution of this equation is

$$R(s, t) = R(s, 0) \exp\left[-\int_0^t F_z(s, t)dt \right]$$

which, by using (4.168), reduces to $R(s, t) = 0$ for all $s \in I$ and $t \in [0, T]$ implying thereby

$$Z_s = PX_s + QY_s. \tag{4.169}$$

From the characteristics equations, we have

$$Z_t = PX_t + QY_t. \tag{4.170}$$

Differentiating the identity $Z(t, s) = Z(\tau(x, y), \sigma(x, y)) \equiv z(x, y)$ w.r.t. s and t, we get respectively,

$$Z_s = z_x X_s + z_y Y_s \tag{4.171}$$

and

$$Z_t = z_x X_t + z_y Y_t. \tag{4.172}$$

The four Eqs (4.169)-(4.172) reduce to

$$X_s(P - z_x) + Y_s(Q - z_y) = 0$$
$$X_t(P - z_x) + Y_t(Q - z_y) = 0.$$

which is a system of linear homogeneous equations in $(P - z_x)$ and $(Q - z_y)$. As $J|_{t=0} \neq 0$, the system has a trivial solution so that

$$z_x(x, y) = P(s, t) = p(x, y) \quad \text{and} \quad z_y(x, y) = Q(s, t) = q(x, y) \quad \forall \, (x, y) \in N(x_0, y_0).$$

Step 4: We show that integral surface $z = z(x, y)$ contains the initial strip. The conclusion is immediate due to the following relations:

$$z(x_0, y_0) = z(x(s, 0), y(s, 0)) = Z(s, 0) = z_0(s)$$
$$z_x(x_0, y_0) = p(x_0, y_0) = p(x(s, 0), y(s, 0)) = P(s, 0) = p_0(s)$$
$$z_y(x_0, y_0) = q(x_0, y_0) = q(x(s, 0), y(s, 0)) = Q(s, 0) = q_0(s).$$

Step 5: We show that the integral surface $z = z(x, y)$ so obtained is unique. Assume that $z = u(x, y)$ is another integral surface of Eq. (4.145) containing the initial strip. At an arbitrary point of initial data curve Γ corresponding to $s = \bar{s}$, both the surfaces $z = z(x, y)$ and $z = u(x, y)$ contain an element $(x_0(\bar{s}), y_0(\bar{s}), z_0(\bar{s}), p_0(\bar{s}), q_0(\bar{s}))$, and hence the characteristic strip issuing from this element through that point. Therefore, by Lemma 4.2, this characteristic strip and hence the associated characteristic curve must lie entirely on both surfaces. The same conclusion holds for each point on Γ, therefore both surfaces being generated by the same set of characteristic curves must be coincide, that is, $z(x, y) = u(x, y)$. This completes proof.

□

It can be easily seen that on initial data curve Γ, if $\dfrac{dx_0}{ds}F_q - \dfrac{dy_0}{ds}F_p = 0$ then Γ must be a characteristic curve. In this case, there may be infinitely many solutions or no solution of the Cauchy problem. Due to this fact, instead of saying that the transversality condition holds, some authors termed as Γ is non-characteristic.

Summary and Algorithm for Method of Characteristics: We have discussed the method of characteristics deeply. However, due to involving geometrical ideas, it is time taking for a teacher and difficult for students at the undergraduate level. In practice, all the involved geometrical objects and ideas need not be utilised. Hence, we present a simple analytic algorithm for the method of characteristics, which serves our purpose in solving Cauchy problems.

Cauchy Problem: *To find the integral surface $z = z(x, y)$ of PDE*

$$F(x, y, z, p, q) = 0$$

passing through the curve

$$\Gamma : \ x = x_0(s), y = y_0(s), z = z_0(s).$$

Step I (Initial Strip): Determine $p_0(s)$ and $q_0(s)$ to solve the system

$$\left.\begin{aligned} & F(x_0, y_0, z_0, p_0, q_0) = 0 \\ & \frac{dz_0}{ds} = p_0\frac{dx_0}{ds} + q_0\frac{dy_0}{ds}. \end{aligned}\right\}$$

The second equation of above system is called initial strip condition. The five-tuple of functions $(x_0(s), y_0(s), z_0(s), p_0(s), q_0(s))$ so obtained is called an initial strip.

Step II (Characteristic Equations): Determine the following system of five ODEs:

$$\left.\begin{aligned} & \frac{dx}{dt} = F_p \\ & \frac{dy}{dt} = F_q \\ & \frac{dz}{dt} = pF_p + qF_q \\ & \frac{dp}{dt} = -(F_x + pF_z) \\ & \frac{dq}{dt} = -(F_y + qF_z). \end{aligned}\right\}$$

These ODEs are called the characteristic equations of the given PDE.

Step III (Characteristic Strip): Find the general solution of the system of characteristic equations of the form

$$x = x(t), y = y(t), z = z(t), p = p(t), q = q(t).$$

This solution is called the characteristic strip of the given PDE. Naturally, it contains arbitrary constants of integration.

Step IV (Integral Surface): Obtain a unique solution of the form

$$x = X(t, s), y = Y(t, s), z = Z(t, s), p = P(t, s), q = Q(t, s)$$

of IVP for the characteristic equations with initial conditions

$$X(0, s) = x_0(s), Y(0, s) = y_0(s), Z(0, s) = z_0(s), P(0, s) = p_0(s), Q(0, s) = q_0(s).$$

To compute this solution, apply IC on the characteristic strip (that is, the general solution of characteristic equations) so that the values of arbitrary constants are determined in terms of s. Hence, the characteristic strips issuing from each point of initial data curves now constitute a unique solution of IVP, each argument of which is a function of t and s.

Finally, we solve the system $x = X(t, s)$, $y = Y(t, s)$ to obtain $t = \tau(x, y)$ and $s = \sigma(x, y)$ and to construct the required integral surface $z = z(x, y)$ as follows:

$$z = Z(t, s) = Z(\tau(x, y), \sigma(x, y)) \equiv z(x, y).$$

Now, we furnish several examples to demonstrate the method of characteristics.

Example 4.47. *Find the characteristic strip of the equation $pq = z$ and determine the integral surface which passes through the straight line $x = 1, z = y$.*

Solution. *The given PDE can be written as*

$$F \equiv pq - z = 0. \tag{4.173}$$

Parametric equations of initial data curve are

$$x_0(s) = 1, y_0(s) = s, z_0(s) = s. \tag{4.174}$$

Step I (Initial Strip): If $(x_0, y_0, z_0, p_0, q_0)$ is the initial strip, then it satisfies Eq. (4.173), that is,

$$p_0 q_0 - z_0 = 0$$

or

$$p_0 q_0 = s. \tag{4.175}$$

From Eq. (4.174), we get

$$\frac{dx_0}{ds} = 0, \frac{dy_0}{ds} = 1, \frac{dz_0}{ds} = 1.$$

Thus, the initial strip condition

$$\frac{dz_0}{ds} = p_0 \frac{dx_0}{ds} + q_0 \frac{dy_0}{ds}$$

becomes

$$1 = p_0 0 + q_0 1$$

or

$$q_0 = 1.$$

Putting this value of q_0 in Eq. (4.175), we get

$$p_0 = s.$$

Hence, the initial strip is

$$\left. \begin{array}{l} x_0 = 1 \\ y_0 = s \\ z_0 = s \\ p_0 = s \\ q_0 = 1. \end{array} \right\} \tag{4.176}$$

Step II (Characteristic Equations): From Eq. (4.173), we have

$$F_x = 0, F_y = 0, F_z = -1, F_p = q, F_q = p.$$

The characteristic equations of Eq. (4.173) are

$$\frac{dx}{dt} = F_p = q$$

$$\frac{dy}{dt} = F_q = p$$

$$\frac{dz}{dt} = pF_p + qF_q = 2pq$$

$$\frac{dp}{dt} = -(F_x + pF_z) = p$$

$$\frac{dq}{dt} = -(F_y + qF_z) = q$$

or

$$\left.\begin{aligned}
dx &= qdt \\
dy &= pdt \\
dz &= 2pqdt \\
dp &= pdt \\
dq &= qdt.
\end{aligned}\right\} \qquad (4.177)$$

Step III (Characteristic Strip): Integrating last two equations of the system (4.177), we get

$$p = c_1 e^t \text{ and } q = c_2 e^t.$$

Hence, the first three equations of the system (4.177) become, respectively,

$$dx = c_2 e^t dt, \ dy = c_1 e^t dt, \ dz = 2c_1 c_2 e^{2t} dt.$$

Integrating these equations, we get

$$x = c_2 e^t + c_3, \ y = c_1 e^t + c_4, \ z = c_1 c_2 e^{2t} + c_5.$$

Hence, the characteristic strip of Eq. (4.173) is

$$\left.\begin{aligned}
x &= c_2 e^t + c_3 \\
y &= c_1 e^t + c_4 \\
z &= c_1 c_2 e^{2t} + c_5 \\
p &= c_1 e^t \\
q &= c_2 e^t.
\end{aligned}\right\} \qquad (4.178)$$

Step IV (Integral Surface): At the initial value $t = 0$, the characteristic strip (4.178) coincides with initial strip (4.176). Hence, we have

$$c_2 + c_3 = 1, \; c_1 + c_4 = s, \; c_1 c_2 + c_5 = s, \; c_1 = s, \; c_2 = 1.$$

On simplifying above, we obtain

$$c_1 = s, \; c_2 = 1, \; c_3 = 0, \; c_4 = 0, \; c_5 = 0.$$

Using these values of constants, the first three equations of system (4.178) reduce to

$$x = e^t, \; y = se^t, \; z = se^{2t}.$$

Eliminating t and s between these equations, we get

$$z = se^{2t} = e^t . se^t = xy$$

or

$$z = xy$$

which is the required integral surface.

Example 4.48. *By the method of characteristics, find the solution of Eikonal equation*

$$p^2 + q^2 = 1$$

through the initial curve

$$x = \cos s, y = \sin s, z = 1, \quad 0 \le s \le 2\pi.$$

Solution. The given PDE can be written as

$$F \equiv p^2 + q^2 - 1 = 0. \tag{4.179}$$

Parametric equations of initial data curve are

$$x_0(s) = \cos s, \; y_0(s) = \sin s, \; z_0(s) = 1. \tag{4.180}$$

Step I (Initial Strip): If $(x_0, y_0, z_0, p_0, q_0)$ is the initial strip, then it satisfies Eq. (4.179), that is,

$$p_0^2 + q_0^2 - 1 = 0$$

or

$$p_0^2 + q_0^2 = 1. \tag{4.181}$$

From Eq. (4.180), we get

$$\frac{dx_0}{ds} = -\sin s, \frac{dy_0}{ds} = \cos s, \frac{dz_0}{ds} = 0.$$

Thus, the initial strip condition

$$\frac{dz_0}{ds} = p_0 \frac{dx_0}{ds} + q_0 \frac{dy_0}{ds}$$

becomes

$$0 = -p_0 \sin s + q_0 \cos s$$

or

$$q_0 = p_0 \tan s.$$

Putting this value of q_0 in Eq. (4.181), we get

$$p_0^2 + p_0^2 \tan^2 s = 1$$

or

$$p_0^2 \sec^2 s = 1$$

or

$$p_0 = \cos s.$$

Also, we get

$$q_0 = \sin s.$$

Hence, the initial strip is

$$\left. \begin{aligned} x_0 &= \cos\ s \\ y_0 &= \sin\ s \\ z_0 &= 1 \\ p_0 &= \cos\ s \\ q_0 &= \sin\ s. \end{aligned} \right\} \tag{4.182}$$

Step II (Characteristic Equations): From Eq. (4.179), we have

$$F_x = 0, F_y = 0, F_z = 0, F_p = 2p, F_q = 2q.$$

The characteristic equations of Eq. (4.179) are

$$\frac{dx}{dt} = F_p = 2p$$

$$\frac{dy}{dt} = F_q = 2q$$

$$\frac{dz}{dt} = pF_p + qF_q = 2p^2 + 2q^2 = 2 \quad \text{(from Eq. (4.179))}$$

$$\frac{dp}{dt} = -(F_x + pF_z) = 0$$

$$\frac{dq}{dt} = -(F_y + qF_z) = 0$$

or

$$\left.\begin{aligned}
dx &= 2pdt \\
dy &= 2qdt \\
dz &= 2dt \\
dp &= 0 \\
dq &= 0.
\end{aligned}\right\} \tag{4.183}$$

Step III (Characteristic Strips): Integrating the last three equations of the system (4.183), we get

$$z = 2t + c_1, \quad p = c_2 \text{ and } q = c_3.$$

Hence, the first two equations become, respectively,

$$dx = 2c_2 dt, \quad dy = 2c_3 dt.$$

Integrating these equations, we get

$$x = 2c_2 t + c_4, \quad y = 2c_3 t + c_5.$$

Hence, the characteristic strip of Eq. (4.179) is

$$\left.\begin{aligned} x &= 2c_2 t + c_4 \\ y &= 2c_3 t + c_5 \\ z &= 2t + c_1 \\ p &= c_2 \\ q &= c_3. \end{aligned}\right\} \tag{4.184}$$

Step IV (Integral Surface): At the initial value $t = 0$, the characteristic strip (4.184) coincides with initial strip (4.182) so that we get

$$c_1 = 1, \ c_2 = \cos s, \ c_3 = \sin s, \ c_4 = \cos s, \ c_5 = \sin s.$$

Using these values of constants, the first three equations of system (4.184) reduces to

$$\begin{aligned} x &= (2t+1)\cos s \\ y &= (2t+1)\sin s \\ z &= 2t+1. \end{aligned}$$

Eliminating s and t among these equations, we obtain

$$x^2 + y^2 = (2t+1)^2(\cos^2 + \sin^2) = (2t+1)^2 = z^2$$

or

$$x^2 + y^2 = z^2$$

which is the integral surface as desired.

4.9.2 Standard Strategy for Quasilinear PDE

Let us consider the Cauchy problem for a quasilinear equation

$$F(x,y,z,p,q) \equiv P(x,y,z)p + Q(x,y,z)q - R(x,y,z) = 0 \tag{4.185}$$

with initial data curve

$$\Gamma: \ x = x_0(s), y = y_0(s), z = z_0(s). \tag{4.186}$$

Differentiating Eq. (4.185) partially w.r.t. p and q, we get

$$F_p = P \quad \text{and} \quad F_q = Q.$$

Using the above and Eq. (4.185), we get

$$pF_q + qF_q = Pp + Qq = R.$$

The first three equations of the system (4.158) thus reduce to

$$\left.\begin{array}{l} \dfrac{dx}{dt} = P(x, y, z) \\[2mm] \dfrac{dy}{dt} = Q(x, y, z) \\[2mm] \dfrac{dz}{dt} = R(x, y, z). \end{array}\right\}$$

This system is indeed a parametric formulation of Lagrange's auxiliary equations. Since P, Q, R all are independent of p and q, therefore the system (4.187) is consistent to determine the characteristic curve. Consequently, unlike non-linear equations, it is not required to determine the initial strip as well as the characteristic strip in such a case. Only determination of characteristic curves serves the purpose to generate the integral surface as desired. Thus, the method of characteristics is simpler to apply for quasilinear equations as compared to non-linear equations.

In this case, for a fixed $A(\overline{x}, \overline{y}, \overline{z})$, the relation $F(\overline{x}, \overline{y}, \overline{z}, p, q) = 0$ represents a linear equation in p and q. Monge cone thus degenerates into the single straight line represented by

$$\frac{x - \overline{x}}{P} = \frac{y - \overline{y}}{Q} = \frac{z - \overline{z}}{R}$$

which is called **Monge axis**. Monge axis determines (P, Q, R) as the characteristic direction.

In quasilinear case, Theorem 4.9 reduces to the following local existence and uniqueness theorem of Cauchy problem.

Theorem 4.10. *Consider the Cauchy problem governed by Eq. (4.185) with initial data curve Γ given by (4.186). On the curve Γ, if the following transversality condition:*

$$\frac{dx_0}{ds} Q(x_0, y_0, z_0) - \frac{dy_0}{ds} P(x_0, y_0, z_0) \neq 0$$

holds then in a neighbourhood of initial curve Γ' (where Γ' is a projection of Γ on xy-plane), there exists a unique solution of Cauchy problem, which contains the initial data curve.

To demonstrate the method just discussed, let us furnish the following Cauchy problem.

Example 4.49. *Using the method of characteristics, solve Cauchy problem for quasilinear equation $zp + q = 1$, when the initial data curve is $x = s, y = s, z = \dfrac{1}{2}s, \ 0 \leq s \leq 1$.*

Solution. *Here $P = z, Q = 1, R = 1$. Thus, the characteristic equations are*

$$\frac{dx}{dt} = z, \frac{dy}{dt} = 1, \frac{dz}{dt} = 1$$

or

$$dx = z\,dt, dy = dt, dz = dt \tag{4.187}$$

Integrating last two equations of the system (4.187), we get

$$y = t + c_1 \text{ and } z = t + c_2.$$

Hence, the first equation of the system becomes

$$dx = (t + c_2)dt.$$

Integrating these equations, we get

$$x = \frac{1}{2}t^2 + c_2 t + c_3.$$

Hence, the characteristic curve of the given PDE is

$$\left.\begin{array}{l} x = \dfrac{1}{2}t^2 + c_2 t + c_3 \\[2mm] y = t + c_1 \\[2mm] z = t + c_2. \end{array}\right\} \tag{4.188}$$

Parametric equations of initial data curve are

$$x_0(s) = s, y_0(s) = s, z_0(s) = \frac{1}{2}s.$$

At the initial point $t = 0$, the characteristic curve (4.188) coincides with initial data curve. Therefore, we have

$$c_1 = s, \ c_2 = \frac{s}{2}, \ c_3 = s.$$

Putting these values of constants in system (4.187), we get

$$\left.\begin{array}{l} x = \dfrac{1}{2}t^2 + \dfrac{s}{2}t + s \\[2mm] y = t + s \\[2mm] z = t + \dfrac{s}{2}. \end{array}\right\} \tag{4.189}$$

Solving the first two equations of system (4.189), we get

$$s = \frac{2x - y^2}{2 - y}, t = \frac{2(y - x)}{2 - y}.$$

Putting these values of t and s in the third equation of the system (4.189), we get

$$z = \frac{2(y-x)}{2-y} + \frac{1}{2}\frac{2x-y^2}{2-y} = \frac{2y-x-y^2/2}{2-y}.$$

Hence, the required integral surface is

$$z(y-2) = x - 2y + \frac{1}{2}y^2.$$

4.10 Simultaneous PDEs

Two first-order PDEs

$$\left.\begin{array}{c} F(x,y,z,p,q) = 0 \\ G(x,y,z,p,q) = 0 \end{array}\right\} \tag{4.190}$$

satisfying $J := \dfrac{\partial(F,G)}{\partial(p,q)} \neq 0$[3] are said to be compatible if both admit a common solution. If the equations are compatible, then their common solution represents a one-parameter family of surfaces of the form $h(x,y,z,c) = 0$.

The following result provides the necessary and sufficient condition for a system of PDEs to be compatible.

Theorem 4.11. *The PDEs represented by (4.190) are compatible iff*

$$[F,G] := \frac{\partial(F,G)}{\partial(x,p)} + \frac{\partial(F,G)}{\partial(y,q)} + p\frac{\partial(F,G)}{\partial(z,p)} + q\frac{\partial(F,G)}{\partial(z,q)} = 0.$$

Proof. As $J \neq 0$, the system (4.190) is solvable for p and q so that

$$p = \phi(x,y,z) \text{ and } q = \psi(x,y,z).$$

Thus, we have

$$dz = pdx + qdx = \phi dx + \psi dy$$

or

$$\phi dx + \psi dy - dz = 0. \tag{4.191}$$

The compatibility condition of the given equations then reduces to the condition that the total differential equation represented by (4.191) should be integrable. The necessary and sufficient condition for the integrability of Eq. (4.191) is (see Theorem 1.2)

$$\phi\left[\frac{\partial\psi}{\partial z} - 0\right] + \psi\left[0 - \frac{\partial\phi}{\partial z}\right] + (-1)\left[\frac{\partial\phi}{\partial y} - \frac{\partial\psi}{\partial x}\right] = 0$$

---

[3] J cannot be vanished identically. Otherwise, F and G regarding as functions of p and q must be dependent, which contradicts the compatibility of the given system.

which is equivalent to

$$\psi_x + \phi\psi_z = \phi_y + \psi\phi_z. \tag{4.192}$$

Putting ϕ and ψ for p and q, respectively, in the first equation of the system (4.190) and differentiating it w.r.t. x and z, we obtain

$$F_x + F_p\phi_x + F_q\psi_x = 0$$

and

$$F_z + F_p\phi_z + F_q\psi_z = 0.$$

Multiplying the last equation by ϕ and adding it to the first one, we obtain

$$F_x + \phi F_z + F_p(\phi_x + \phi\phi_z) + F_q(\psi_x + \phi\psi_z) = 0.$$

Similarly, from the second equation of the system (4.190), we may deduce that

$$G_x + \phi G_z + G_p(\phi_x + \phi\phi_z) + G_q(\psi_x + \phi\psi_z) = 0.$$

Solving these equations for $\psi_x + \phi\psi_z$, we get

$$\psi_x + \phi\psi_z = \frac{1}{J}\left[\frac{\partial(F,G)}{\partial(x,p)} + \phi\frac{\partial(F,G)}{\partial(z,p)}\right]. \tag{4.193}$$

Now, differentiating each equation of the system (4.190) w.r.t. y and z and proceeding as earlier, we obtain

$$\phi_y + \psi\psi_z = -\frac{1}{J}\left[\frac{\partial(F,G)}{\partial(y,q)} + \psi\frac{\partial(F,G)}{\partial(z,q)}\right]. \tag{4.194}$$

Using (4.193) and (4.194), Eq. (4.192) becomes

$$\frac{\partial(F,G)}{\partial(x,p)} + \phi\frac{\partial(F,G)}{\partial(z,p)} = -\left[\frac{\partial(F,G)}{\partial(y,q)} + \psi\frac{\partial(F,G)}{\partial(z,q)}\right].$$

Replacing ϕ and ψ by p and q, respectively, in the above equation, we get

$$\frac{\partial(F,G)}{\partial(x,p)} + \frac{\partial(F,G)}{\partial(y,q)} + p\frac{\partial(F,G)}{\partial(z,p)} + q\frac{\partial(F,G)}{\partial(z,q)} = 0$$

which is the compatibility condition as desired. $\qquad\square$

With a view to illustrate the foregoing description, let us adopt the following examples.

Example 4.50. *Show that the system of equations*

$$xp = yq, \quad z(xp + yq) = 2xy$$

is compatible and hence solve the system.

Solution. *The given equations can be written as*

$$F \equiv xp - yq = 0 \tag{4.195}$$

and

$$G \equiv z(xp + yq) - 2xy = 0. \tag{4.196}$$

Thus, we have

$$\frac{\partial(F, G)}{\partial(x, p)} = \begin{vmatrix} p & x \\ zp - 2y & zx \end{vmatrix} = 2xy$$

$$\frac{\partial(F, G)}{\partial(y, q)} = \begin{vmatrix} -q & -y \\ zq - 2x & zy \end{vmatrix} = -2xy$$

$$\frac{\partial(F, G)}{\partial(z, p)} = \begin{vmatrix} 0 & x \\ xp + yq & zx \end{vmatrix} = -(x^2 p + xyq)$$

$$\frac{\partial(F, G)}{\partial(z, q)} = \begin{vmatrix} 0 & -y \\ xp + yq & zy \end{vmatrix} = xyp + y^2 q.$$

Now, we obtain

$$\begin{aligned}
[F, G] &= \frac{\partial(F, G)}{\partial(x, p)} + \frac{\partial(F, G)}{\partial(y, q)} + p\frac{\partial(F, G)}{\partial(z, p)} + q\frac{\partial(F, G)}{\partial(z, q)} \\
&= 2xy - 2xy - (x^2 p + xyq)p + (xyp + y^2 q)q \\
&= y^2 q^2 - x^2 p^2 \\
&= (yq + xp)(yq - xp) \\
&= 0 \quad \left(\text{using Eq. (4.195)}\right).
\end{aligned}$$

It follows that the given system is compatible. Solving (4.195) and (4.196) for p and q, we obtain

$$p = \frac{y}{z} \text{ and } q = \frac{x}{z}.$$

Putting the values of p and q in relation dz = pdx + qdy, we get

$$dz = \frac{ydx + xdy}{z}$$

or

$$zdz = d(xy).$$

Integrating, we get

$$z^2 = 2xy + c$$

which is the solution of the given system as desired.

Example 4.51. *Show that the equations*

$$xp - yq = x, \quad x^2 p + q = xz$$

are compatible. Also, find the one-parameter family of their common solutions.

Solution. *The given equations can be written as*

$$F \equiv xp - yq - x = 0 \tag{4.197}$$

and

$$G \equiv x^2 p + q - xz = 0. \tag{4.198}$$

Thus, we have

$$\frac{\partial(F,G)}{\partial(x,p)} = \begin{vmatrix} p-1 & x \\ 2xp - z & x^2 \end{vmatrix} = xz - x^2 p - x^2$$

$$\frac{\partial(F,G)}{\partial(y,q)} = \begin{vmatrix} -q & -y \\ 0 & 1 \end{vmatrix} = -q$$

$$\frac{\partial(F,G)}{\partial(z,p)} = \begin{vmatrix} 0 & x \\ -x & -x^2 \end{vmatrix} = x^2$$

$$\frac{\partial(F,G)}{\partial(z,q)} = \begin{vmatrix} 0 & -y \\ -x & 1 \end{vmatrix} = -xy.$$

Now, we obtain

$$\begin{aligned}
[F,G] &= \frac{\partial(F,G)}{\partial(x,p)} + \frac{\partial(F,G)}{\partial(y,q)} + p\frac{\partial(F,G)}{\partial(z,p)} + q\frac{\partial(F,G)}{\partial(z,q)} \\
&= xz - x^2 p - x^2 - q + px^2 - qxy \\
&= xz - x^2 - q - xyq \\
&= (xz - q) - x^2 - xyq \\
&= x^2 p - x^2 - xyq \quad \left(using\ Eq.\ (4.198)\right) \\
&= x(xp - x - yq) \\
&= x \cdot 0 = 0 \quad \left(using\ Eq.\ (4.197)\right)
\end{aligned}$$

which implies that the given PDEs are compatible. Solving (4.197) and (4.198) for p and q, we obtain

$$p = \frac{1 + yz}{1 + xy} \quad and \quad q = \frac{x(z - x)}{1 + xy}.$$

Putting the values of p and q in relation dz = pdx + qdy, we get

$$dz = \frac{(1 + yz)dx + x(z - x)dy}{1 + xy}$$

or

$$dz - dx = \frac{(1 + yz)dx + x(z - x)dy}{1 + xy} - dx$$
$$= \frac{y(z - x)dx + x(z - x)dy}{1 + xy}$$

or

$$\frac{dz - dx}{z - x} = \frac{ydx + xdy}{1 + xy}$$

or

$$\frac{d(z - x)}{z - x} = \frac{d(1 + xy)}{1 + xy}.$$

Integrating the above, we get

$$\log(z - x) = \log(1 + xy) + \log c$$

or

$$z = x + c(1 + xy)$$

which is the required solution of the given system.

4.11 PDEs in More Than Two Independent Variables: Jacobi's Method

Consider the PDE of the form

$$F(x, y, z, u_x, u_y, u_z) = 0. \tag{4.199}$$

Here, x, y, z are independent variables and u is a dependent variable, which does not occur explicitly in the equation. A complete solution of Eq. (4.199) contains three arbitrary constants (parameters). The method to determine the complete solution of such an equation called **Jacobi's method** is due to the German mathematician *Karl Gustav Jacob Jacobi of Potsdam* (1804–1851). The fundamental idea of Jacobi's method is very similar to Charpit's method.

Definition 4.6. *The simultaneous total differential equations*

$$\frac{dx}{-F_{u_x}} = \frac{dy}{-F_{u_y}} = \frac{dz}{-F_{u_z}} = \frac{du_x}{F_x} = \frac{du_y}{F_y} = \frac{du_z}{F_z}$$

are called Jacobi's auxiliary equations of Eq. (4.199).

Theorem 4.12. *A complete solution of Eq. (4.199) is the integral of total differential equation*

$$du = u_x dx + u_y dy + u_z dz$$

where u_x, u_y and u_z are determined from Eq. (4.199) and two independent integrals of its Jacobi's auxiliary equations.

Proof. As u is a function of x, y, z, we have

$$du = u_x dx + u_y dy + u_z dz. \tag{4.200}$$

Now, we have to find two additional PDEs of the form

$$\Phi_1(x, y, z, u_x, u_y, u_z) = 0 \tag{4.201}$$

$$\Phi_2(x, y, z, u_x, u_y, u_z) = 0 \tag{4.202}$$

such that u_x, u_y, and u_z can be determined from Eqs (4.199), (4.201), and (4.202) as functions of x, y, z that make Eq. (4.200) integrable. In lieu of Theorem 4.11, each of (4.201) and (4.202) is compatible with Eq. (4.199) iff

$$\frac{\partial(F, \Phi)}{\partial(x, u_x)} + \frac{\partial(F, \Phi)}{\partial(y, u_y)} + \frac{\partial(F, \Phi)}{\partial(z, u_z)} = 0$$

where $\Phi = \Phi_i$ ($i = 1, 2$). The above equation after rearranging can be rewritten as

$$-F_{u_x}\frac{\partial\Phi}{\partial x} - F_{u_y}\frac{\partial\Phi}{\partial y} - F_{u_z}\frac{\partial\Phi}{\partial z} + F_x\frac{\partial\Phi}{\partial u_x} + F_y\frac{\partial\Phi}{\partial u_y} + F_z\frac{\partial\Phi}{\partial u_z} = 0 \tag{4.203}$$

which is a semilinear PDE of order one in Φ as dependent variable and x, y, z, u_x, u_y, u_z as independent variables. The Lagrange's auxiliary equations for Eq. (4.203) are

$$\frac{dx}{-F_{u_x}} = \frac{dy}{-F_{u_y}} = \frac{dz}{-F_{u_z}} = \frac{du_x}{F_x} = \frac{du_y}{F_y} = \frac{du_z}{F_z} = \frac{d\Phi}{0} \tag{4.204}$$

which are indeed Jacobi's auxiliary equations. Any integral of (4.204) will also satisfy (4.203). Thus, with the help of (4.204), we find two independent integrals of the forms (4.201) and (4.202) containing arbitrary constants, say a and b, respectively. These integrals along with Eq. (4.204) provide the values of u_x, u_y, and u_z, which after putting in relation (4.200) and on integrating gives rise to the solution of given PDE of the form

$$f(x, y, z, a, b, c) = 0.$$

Here c remains constant of integration obtained from integration of (4.200). $\qquad\square$

The natural inherent property of Jacobi's method lies in the fact that it can readily be generalised for any number of variables. Thus, for an equation of the form

$$F(x_1, x_2, \ldots, x_n, u_1, u_2, \ldots, u_n) = 0 \qquad (4.205)$$

where $u_i := \dfrac{\partial u}{\partial x_i}$ $(i = 1, 2, \ldots, n)$, the complete solution is determined by integrating the total differential equation

$$du = \sum_{i=1}^{n} u_i dx_i \qquad (4.206)$$

so that $u_1, u_2, \ldots, u_n$ are obtained from Eq. (4.205) as well as $n - 1$ independent integrals of Jacobi's auxiliary equations given by

$$\frac{dx_1}{-F_{u_1}} = \frac{dx_2}{-F_{u_2}} = \cdots = \frac{dx_n}{-F_{u_n}} = \frac{du_1}{F_{x_1}} = \frac{du_2}{F_{x_2}} = \cdots = \frac{du_n}{F_{x_n}}.$$

The complete solution of Eq. (4.205) contains n arbitrary constants. Quite contrary to this, it is interesting to observe that such type of direct generalisation is not possible in the context of Charpit's method. But the main difficulty in Jacobi's method is that it cannot be applicable for those classes of PDEs, which contain u explicitly.

To illustrate Jacobi's method, let us consider the following examples.

Example 4.52. *Find the complete solution of PDE*

$$z^2 + zu_z - u_x^2 - u_y^2 = 0.$$

Solution. *The given PDE can be written as*

$$F \equiv z^2 + zu_z - u_x^2 - u_y^2 = 0. \qquad (4.207)$$

The Jacobi's auxiliary equations are

$$\frac{dx}{2u_x} = \frac{dy}{2u_y} = \frac{dz}{-z} = \frac{du_x}{0} = \frac{du_y}{0} = \frac{du_z}{2z + u_z}.$$

From the fourth and fifth fractions, we get

$$du_x = 0 \text{ and } du_y = 0$$

which on integrating give rise to

$$u_x = a \text{ and } u_y = b.$$

Putting these values in Eq. (4.207), we obtain

$$z^2 + zu_z - a^2 - b^2 = 0$$

or

$$u_z = \frac{1}{z}(a^2 + b^2) - z.$$

Now, we have

$$du = u_x dx + u_y dy + u_z dz$$
$$= adx + bdy + (a^2 + b^2)\frac{dz}{z} - zdz.$$

Integrating the above, we get

$$u = ax + by + (a^2 + b^2)\log z - \frac{1}{2}z^2 + c$$

which is the required complete solution.

Example 4.53. *Find the complete solution of PDE*

$$u_x u_y u_z = u^3 xyz.$$

Solution. *The given PDE can be written as*

$$\frac{u_x}{u}\frac{u_y}{u}\frac{u_z}{u} = xyz.$$

Put $v = \log u$. Then the above equation reduces to

$$v_x v_y v_z = xyz \qquad\qquad (4.208)$$

or

$$F \equiv xyz - v_x v_y v_z = 0.$$

The Jacobi's auxiliary equations are

$$\frac{dx}{v_y v_z} = \frac{dy}{v_x v_z} = \frac{dz}{v_x v_y} = \frac{dv_x}{yz} = \frac{dv_y}{xz} = \frac{dv_z}{xy}.$$

Taking the first and fourth fractions, we get

$$\frac{dx}{v_y v_z} = \frac{dv_x}{yz}$$

yielding thereby

$$\frac{v_x dx}{v_x v_y v_z} = \frac{xdv_x}{xyz}.$$

Using (4.208), the above equation reduces to

$$v_x dx = x dv_x$$

or

$$\frac{dv_x}{v_x} = \frac{dx}{x}.$$

Integrating, we obtain

$$v_x = ax.$$

Similarly, from the second and fifth fractions, we obtain

$$v_y = by.$$

Putting the values of v_x and v_y in (4.208), we get

$$v_z = \frac{1}{ab}z.$$

Now, we have

$$dv = v_x dx + v_y dy + v_z dz$$
$$= ax dx + by dy + \frac{1}{ab} z dz.$$

Integrating the above, we get

$$v = \frac{ax^2}{2} + \frac{by^2}{2} + \frac{1}{ab}\frac{z^2}{2} + \frac{c}{2}$$

or

$$2 \log u = ax^2 + by^2 + \frac{z^2}{ab} + c$$

which is the complete solution as desired.

Example 4.54. *Find the complete solution of PDE*

$$u_x u_y u_z + u_t^3 xyzt^3 = 0.$$

Solution. *The given PDE can be written as*

$$F \equiv u_x u_y u_z + u_t^3 xyzt^3 = 0. \tag{4.209}$$

The Jacobi's auxiliary equations are

$$\frac{dx}{-u_y u_z} = \frac{dy}{-u_x u_z} = \frac{dz}{-u_x u_y} = \frac{dt}{-3u_t^2 xyzt^3} = \frac{du_x}{u_t^3 yzt^3} = \frac{du_y}{u_t^3 xzt^3} = \frac{du_z}{u_t^3 xyt^3} = \frac{du_t}{3u_t^3 xyt^2}.$$

Taking the first and fifth fractions, we get

$$\frac{dx}{-u_y u_z} = \frac{du_x}{u_t^3 yzt^3}$$

which implies that

$$\frac{u_x dx}{-u_x u_y u_z} = \frac{x du_x}{u_t^3 xyzt^3}.$$

Using (4.209), the above equation reduces to

$$u_x dx = x du_x$$

or

$$\frac{du_x}{u_x} = \frac{dx}{x}.$$

Integrating, we obtain

$$u_x = ax.$$

Similarly, we can obtain

$$u_y = by \text{ and } u_z = cz.$$

Putting the values of u_x, u_y, and u_z in Eq. (4.209), we get

$$u_t = -(abc)^{1/3} \frac{1}{t}.$$

Now, we have

$$du = u_x dx + u_y dy + u_z dz + u_t dt$$
$$= axdx + bydy + czdz - (abc)^{1/3} \frac{dt}{t}.$$

Integrating the above, we get

$$u = \frac{1}{2}(ax^2 + by^2 + cz^2) - (abc)^{1/3} \log t + d$$

which is the required complete solution.

▶ Exercises

Find the general solutions of the linear PDEs described in problems 1–7 using coordinate method.

1. $p - q = z$.   Ans. $z = \phi(x + y)e^{-y}$.
2. $p + yq = 0$.   Ans. $z = \phi(ye^{-x})$.
3. $(1 + x^2)p + q = 0$.   Ans. $z = \phi(y - \tan^{-1}x)$.
4. $xp + yq = x^n$.   Ans. $z = \dfrac{x^n}{n} + \phi\left(\dfrac{y}{x}\right)$.
5. $\alpha p + \beta q + \gamma z = \lambda$, where $\alpha, \beta, \gamma, \lambda$ are constants and $\alpha^2 + \beta^2 \neq 0$.
   Ans. $z = e^{-\gamma x/\alpha}\phi(\alpha y - \beta x) + \dfrac{\lambda}{\gamma}$.
6. $x\dfrac{\partial z}{\partial x} + y\dfrac{\partial z}{\partial y} = nz$ (Euler's relation).   Ans. $z = x^n\phi\left(\dfrac{y}{x}\right)$.
7. $y^2p - xyq = x(z - 2y)$.   Ans. $z = y + \dfrac{1}{y}\phi(x^2 + y^2)$.

Solve the semilinear PDEs described in problems 8–14.

8. $xp + yq = xyz^2$.   Ans. $\dfrac{1}{2}xy + \dfrac{1}{z} = \phi(x/y)$.
9. $xp - yq = y^2(1 - z)$, $y \neq 0$.   Ans. $z = 1 + e^{y^2/2}\phi(xy)$
10. $-2p + 4q = e^{x+3y} - 5z$.   Ans. $z = \dfrac{1}{15}e^{x+3y} + \phi(y + 2x)e^{5x/2}$.
11. $p + q = z^2$, $z(x, 0) = 1$.   Ans. $\dfrac{1}{z} = 1 - y$.
12. $yp - xq = 0$, (a) : $z(x,\ 0) = x$,(b) : $z(0, y) = \cos y^2$.   Ans. (a) : $z^2 = x^2 + y^2$,
    (b) : $z = \cos(x^2 + y^2)$.
13. $p + 2xq = 2xz$, (a) : $z(x,\ 0) = x^2$,(b) : $z(0, y) = y^2$.   Ans. (a) : $z = (x^2 - y)e^y$,
    (b) : $z = (y - x^2)e^{x^2}$.
14. $yp + xq = x^2 + y^2$, (a) : $z(x,0) = 1 + x^2$,(b) : $z(0,y) = 1 + y^2$.
    Ans. (a) : $\dfrac{z - 1}{x^2 + y^2} = 1 - \dfrac{\pi}{2} + \sin^{-1}\left(\dfrac{x}{x^2 + y^2}\right)$, (b) : $\dfrac{z - 1}{x^2 + y^2} = 1 + \sin^{-1}\left(\dfrac{x}{x^2 + y^2}\right)$.

Use Lagrange's method to obtain the general solutions of the PDEs described in problems 15–25.

15. $zp + q = 0$.   Ans. $z = \phi(x - yz)$.
16. $xp + yq = z$. $\phi\left[\dfrac{x}{y}, \dfrac{x}{z}\right] = 0$.
17. $xp - yq = xy$. $xe^{-z/xy} = \phi(xy)$.
18. $p\tan x + q\tan y = \tan z$. $\phi\left[\dfrac{\sin x}{\sin y}, \dfrac{\sin y}{\sin z}\right] = 0$.
19. $(x^2 + 2y^2)p - xyq = xz$. $y^2x^2 + y^2 = \phi(yz)$.

20. $(3x + y - z)p + (x + y - z)q = 2(z - y)$.

Ans.

$$\phi\left[-x + 3y + z, \frac{(x - y + z)^2}{x + y - z}\right] = 0.$$

21. $(y + zx)p - (x + yz)q = x^2 - y^2$.

Ans. $xy + z = \phi(x^2 + y^2 - z^2)$.

22. $(x^2 - yz)p + (y^2 - zx)q = z^2 - xy$.

Ans. $\phi\left[\frac{x - y}{y - z}, \frac{y - z}{z - x}\right] = 0$.

23. $z(xy + z^2)(xp - yq) = x^4$.

Ans. $xy = \phi(x^4 - 2xyz^2 - z^2)$.

24. $x\dfrac{\partial u}{\partial x} + y\dfrac{\partial u}{\partial y} + z\dfrac{\partial u}{\partial z} = xyz$.

Ans. $\phi\left[\dfrac{x}{y}, \dfrac{z}{y}, xyz - 3u\right] = 0$.

25. $(y + z + u)\dfrac{\partial u}{\partial x} + (z + u + x)\dfrac{\partial u}{\partial y} + (u + x + y)\dfrac{\partial u}{\partial z} = x + y + z$.

Ans. $\phi[(x - u)w, (y - u)w, (z - u)w] = 0$ where $w = (x + y + z + u)^{1/3}$.

26. Obtain the general equation of surfaces that cuts orthogonally the cones of the system $x^2 + y^2 + z^2 = cxy$.
Ans. $x^2 - y^2 = z^2\phi(x^2 + y^2 + z^2)$.

27. Find the equation of the surface that cuts orthogonally the surfaces of the family $2xz + 3yz = c(z + 2)$ and passes through the circle $x^2 + y^2 = 9, z = 0$. Ans. $3(x^2 + y^2 - z^2) - z^3 = 27$.

28. Find the surface that is orthogonal to the one-parameter system $z = cxy(x^2 + y^2)$ that passes through the hyperbola $x^2 - y^2 = a^2, z = 0$.
Ans. $(x^2 + y^2 + 4z^2)^2(x^2 - y^2)^2 = a^4(x^2 + y^2)$.

Find the complete solutions of the PDEs described in problems 29–64 by standard methods.

29. $q = 3p^2$.

Ans. $z = ax + 3a^2y + c$.

30. $p = e^q$.

Ans. $z = ax + y\log a + c$.

31. $\sqrt{p} + \sqrt{q} = 1$.

Ans. $z = ax + (1 - \sqrt{a})^2 y + c$.

32. $p^2 - q^2 = 4$.

Ans. $z = ax + \sqrt{a^2 - 4}\, y + c$.

33. $p = 2q^2 + 1$.

Ans. $z = ax + \dfrac{1}{\sqrt{2}}\sqrt{a - 1}\, y + c$.

34. $p + q = pq$.

Ans. $z = ax + \dfrac{ay}{a - 1} + c$.

35. $p^2 + 6p + 2q + 4 = 0$.

Ans. $z = ax - \left(2 + 3a + \dfrac{a^2}{2}\right)y + c$.

36. $z = pq$.

Ans. $4az = (x + ay + z)^2$.

37. $1 + p^2 = qz$.

Ans.
$a^2z^2 \pm [az\sqrt{a^2z^2 - 4} - 4\log(az + \sqrt{a^2z^2 - 4})] = 4(x + ay) + b$.

38. $z^2(p^2 + q^2 + 1) = c^2$.

Ans. $(a^2 + 1)(c^2 - z^2) = (x + ay + b)^2$,

39. $z^2p^2 + q^2 = 1$.

Ans.
$z\sqrt{z^2 + a^2} + a^2\log[z + \sqrt{z^2 + a^2}] = 2(x + ay + b)$.

40. $p(1 - q^2) = q(1 - z)$.

Ans. $4(1 - a + az) = (x + ay + b)^2$.

41. $p(1 + q^2) = q(z - c)$.

Ans. $4[a(z - c) - 1] = (x + ay + b)^2$.

42. $p^2 = z^2(1 - pq)$.

Ans. $\pm(x + ay + b) =$
$$- \sinh\left(\frac{1}{z\sqrt{a}}\right) + \sqrt{1 + az^2}.$$

43. $q^2 = z^2 p^2(1 - p^2)$.

Ans. $(x + ay + b)^2 = z^2 - a^2$.

44. $9(p^2 z + q^2) = 4$.

Ans. $(x + ay + b)^2 = (z + a^2)^3$.

45. $zpq - p - q = 0$.

Ans. $z^2 = \dfrac{2(1 + a)}{a}(ax + y) + b$.

46. $4(1 + z^3) = 9z^4 pq$.

Ans. $a(1 + z^3) = (x + ay + b)^2$.

47. $p^3 + q^3 = 3pqz$.

Ans. $(1 + a^3) \log z = 3a(x + ay) + b$.

48. $q - p + x - y = 0$.

Ans. $2z = (x + a)^2 + (y + a)^2 + b$.

49. $p = 2qx$.

Ans. $z = ax^2 + ay + b$.

50. $p + q = \sin x + \sin y$.

Ans.
$z = a(x - y) - (\cos x + \cos y) + b$.

51. $p - 3x^2 = q^2 - y$.

Ans. $z = x^3 + ax \pm \dfrac{2}{3}(y + a)^{3/2} + b$.

52. $p^2 + q^2 = x + y$.

Ans.
$z = \dfrac{2}{3}(x + a)^{3/2} + \dfrac{2}{3}(y - a)^{3/2} + b$.

53. $p^2 + q^2 = x^2 + y^2$.

Ans. $2z = x\sqrt{x^2 + a^2} + y\sqrt{y^2 - a^2} +$
$a^2 \sinh^{-1}\left(\dfrac{x}{a}\right) - a^2 \cosh^{-1}\left(\dfrac{y}{a}\right) + b$

54. $p - q = x^2 + y^2$.

Ans. $z = a(x + y) + \dfrac{1}{3}(x^3 - y^3) + b$.

55. $\sqrt{p} + \sqrt{q} = 2x$.

Ans. $z = \dfrac{1}{6}(a + 2x)^3 + a^2 y + b$.

56. $2\sqrt{p} + 3\sqrt{q} = 6x + 2y$.

Ans.
$z = \dfrac{1}{72}(6x + a)^3 + \dfrac{1}{54}(2y - a)^3 + b$.

57. $p^2 y(1 + x^2) = qx^2$.

Ans. $z = a\sqrt{1 + x^2} + \dfrac{1}{2}a^2 y^2 + b$.

58. $p^2 q(x^2 + y^2) = p^2 + q$.

Ans. $z = \log(x + \sqrt{x^2 + a}) +$
$\dfrac{1}{2\sqrt{a}} \log\left(\dfrac{y - \sqrt{a}}{y + \sqrt{a}}\right) + b$.

59. $p^2 q^2 = 9p^2 y^2(x^2 + y^2) - 9x^2 y^2$.

Ans. $z =$
$\dfrac{1}{3}(9x^2 - a^2)^{1/2} + \dfrac{1}{27}(9y^2 + a^2)^{3/2} + b$.

60. $z = xp + yq + pq$.

Ans. $z = ax + by + ab$,

61. $z = xp + yq + 2\sqrt{pq}$.

Ans. $z = ax + by + 2\sqrt{ab}$.

62. $z = xp + yq + \log pq$.

Ans. $z = ax + by + \log ab$.

63. $z = xp + yq + \dfrac{p}{q} - p$.

Ans. $z = ax + by + \dfrac{a}{b} - a$.

64. $pqz = p^2(xq + p^2) + q^2(yp + q^2)$.

Ans. $abz = a^2(bx + a^2) + b^2(ay + b^2)$.

Find the complete solutions of the PDEs described in problems 65–82 by reducing them into standard forms.

65. $z^2 = pqxy$.   Ans. $z = bx^a y^{1/a}$.

66. $x^4 p^2 + y^2 zq = 2z^2$.   Ans. $\log z =$
$$\frac{1}{2}(a \pm \sqrt{a^2 + 8}\,)\left(\frac{1}{x} + \frac{a}{y}\right) + b.$$

67. $\dfrac{x^2}{p} + \dfrac{y^2}{q} = z$.   Ans. $\left(\dfrac{a+1}{a}\right)(x^3 + ay^3) = \dfrac{z^2}{2} + b.$

68. $x^4 p^2 - yzq = z^2$.   Ans. $\log z =$
$$\frac{1}{2}(a \pm \sqrt{a^2 + 4}\,)\left(\frac{1}{x} + a\log y\right) + b.$$

69. $z = 2(xp + yq) + \sqrt{xp^2 + yq^2}$.   Ans. $z = a\sqrt{x} + b\sqrt{y} + \dfrac{1}{2}\sqrt{a^2 + b^2}.$

70. $4xyz = pq + 2x^2 yp + 2xy^2 q$.   Ans. $z = ax^2 + by^2 + ab.$

71. $p^2 + q^2 = z^2(x^2 + y^2)\, k = -1$.   Ans.
$$2\log z = x\sqrt{x^2 + a^2} + y\sqrt{y^2 - a^2}$$
$$+ a^2 \sinh^{-1}\left(\frac{x}{a}\right) - a^2 \cosh^{-1}\left(\frac{y}{a}\right) + b.$$

72. $2x(z^2 q^2 + 1) = pz$.   Ans. $\dfrac{1}{2}z^2 = (1 + a^2)x + ay + b.$

73. $z(p^2 - q^2) = x - y$.   Ans. $z^{3/2} = (x + a)^{3/2} + (y + a)^{3/2} + b.$

74. $yzp^2 = q$.   Ans. $z^2 = 2ax + a^2 y + b.$

75. $z(xp - yq) = y^2 - x^2$.   Ans. $z^2 = 2a\log xy - x^2 - y^2 + b.$

76. $z^4 q^2 - z^2 p = 1$.   Ans. $z^3 = ax \pm \sqrt{3a + 9}\, y + b.$

77. $2x^4 p^2 - yzq - 3z^2 = 0$.   Ans. $\log z = \dfrac{a}{x} + (2a^2 - 3)\log y + b.$

78. $x^2 p^2 + y^2 q^2 = z$.   Ans.
$$yz = \left(a\log x + \sqrt{1 - a^2}\log y + b\right)^2.$$

79. $\dfrac{p^2}{x} - \dfrac{q^2}{y} = \dfrac{1}{z}\left(\dfrac{1}{x} + \dfrac{1}{y}\right)$.   Ans.
$$az^{3/2} = (ax + 1)^{3/2} + (ay - 1)^{3/2} + b.$$

80. $(1 - x^2)yp^2 + x^2 q = 0$.   Ans. $z = a\sqrt{1 - x^2} - \dfrac{1}{2}a^2 y^2 + c.$ (Put
$$X = \sqrt{1 - x^2},\ Y = \frac{y^2}{2}).$$

81. $z^2(p^2 + q^2 + 1) = 1$.   Ans. $(x - a)^2 + (y - b)^2 + z^2 = 1.$

82. $x^2 p^2 + y^2 q^2 = 4$.   Ans. $z = a\log x + \sqrt{4 - a^2}\log y + b.$

Use Charpit's method to obtain the complete solutions of the PDEs described in problems 83–95.

83. $xpq + yq^2 = 1$.   Ans. $(z + b)^2 = 4(ax + b).$

84. $xp + yq + z = xq^2$.   Ans. $xz = -\dfrac{a^2}{2x^2} + \dfrac{ay}{x} + b.$

85. $pq + qx = y$.   Ans. $2az = 2a^2 x - ax^2 + y^2 + b.$

86. $pxy + pq + qy = yz$. Ans. $(z - ax)(y + a)^a = be^y$.

87. $2z + p^2 + qy + 2y^2 = 0$. Ans. $2y^2z + y^2(x - a)^2 + y^4 = b$.

88. $p(q^2 + 1) + (b - z)q = 0$. Ans. $(x + cy + a)^2 = 4[c(z - b) - 1]$.

89. $(p^2 + q^2)x = pz$. Ans. $z^2 = a^2x^2 + (ay + b)^2$.

90. $q = -xp + p^2$. Ans. $z = axe^{-y} - \dfrac{1}{2}a^2e^{-2y} + b$.

91. $q = xp + p^2$. Ans. $z = -\dfrac{x^2}{4} \pm \dfrac{1}{4}[x\sqrt{x^2 + 4a} +$

$4a \log(x + \sqrt{x^2 + 4a}\,)] + ay + b$.

92. $xp + yq + pq = 0$. Ans. $2az + (ax + y)^2 = b$.

93. $xp + yq = z\sqrt{1 + pq}$.

Ans. $-\log z + \dfrac{v^2}{2} + \dfrac{1}{2}v\sqrt{v^2 - 1} - \dfrac{1}{2}\log(v + \sqrt{v^2 - 1}\,) = b$, where $v = \dfrac{ax + y}{z\sqrt{a}}$.

94. $2xz - px^2 - 2qxy + pq = 0$ Ans. $z = ay + b(x^2 - a)$.

95. $2(pq + yp + xq) + x^2 + y^2 = 0$.

Ans. $z =$

$-\dfrac{1}{2}(x^2 + y^2) + \dfrac{1}{2}a(x + y) + \dfrac{1}{2\sqrt{2}}\left[(x - y)\sqrt{\dfrac{a^2}{2} + (x - y)^2} + \dfrac{a^2}{2}\log\left(x - y + \sqrt{\dfrac{a^2}{2} + (x - y)^2}\right)\right] + b$.

96. Solve the four standard forms given in Section 4.4 using Charpit's method.

97. Find the general and singular solutions of the PDE described in problem 34.
Ans. General Solution: $z = ax + \dfrac{ay}{a - 1} + \phi(a)$, $0 = x - \dfrac{y}{(a - 1)^2} + \phi'(a)$, Singular Solution: None.

98. Find the singular solution of the PDE described in problem 38. Also, obtain a particular solution from the general solution of the PDE.
Ans. Singular Solution: $z^2 = c^2$, General Solution: $(x - h)^2 + (y - k)^2 + z^2 = c^2$ for $b = -ak - h$.

99. Find the singular solution of the PDE described in problem 43.
Ans. $z = 0$.

100. Determine the singular solution of the PDE described in problem 44 if exists.
Ans. does not exist.

101. Find the singular solution of the PDE described in problem 46.
Ans. $1 + z^3 = 0$.

102. Find the general solution of the PDE described in problem 47.
Ans. $(1 + a^3)\log z = 3a(x + ay) +$
$\phi(a), 3a^2\log z = 3x + 6ay + \phi'(a)$.

103. Find the general and singular solutions of the PDE described in problem 60.
Ans. General Solution: $z = ax + \phi(a)y + a\phi(a)$, $0 = x + \phi'(a)y + \phi(a) + a\phi'(a)$, Singular Solution: $z + xy = 0$.

104. Find the general and singular solutions of the PDE described in problem 62.
Ans. General Solution: $z = ax + \phi(a)y + \log a\phi(a)$, $0 = x + \phi'(a)y + \dfrac{1}{a} + \dfrac{\phi'(a)}{\phi(a)}$, Singular Solution:
$z + \log xy + 2 = 0$.

105. Find the singular solution of the PDE described in problem 70.
Ans. $z + x^2y^2 = 0$.

106. Find the singular solution of the PDE described in problem 78.
Ans. $z = 0$.

107. Find the singular solution of the PDE described in problem 81. Also, obtain a particular solution from the general solution of the PDE.
    Ans. Singular Solution: $z = \pm 1$, General Solution: $(x - y)^2 + 2z^2 = 2$ for $b = a$.

108. Find the singular solution of the PDE described in problem 94.
    Ans. $z = x^2 y$.

109. Find the integral surface of the equation $x^2 p + y^2 q + z^2 = 0$, passing through the hyperbola $xy = x + y, z = 1$.
    Ans. $xz + yz + 2xy = 3xyz$.

110. Find the integral surface of the equation $(2xy - 1)p + (z - 2x^2)q = 2(x - yz)$, passing through the line $x(t) = 1, y(t) = 0, z(t) = t$.
    Ans. $x^2 + y^2 - xz - y + z = 1$.

111. Find the integral surface of the equation $x^3 p + y(3x^2 + y)q = z(2x^2 + y)$, passing through the curve $x(t) = 1, y(t) = t, z(t) = t(1 + t)$.
    Ans. $yz = (x^2 + y)(xz - y)$.

112. Find the integral surface of the equation $z = \dfrac{1}{2}(p^2 + q^2) + (p - x)(q - y)$, which passes through the x-axis.
    Ans. $2z = 4xy - 3y^2$.

113. Show that the integral surface of the equation $z(1 - q^2) - 2(px + qy) = 0$, which passes through the line $x = 1$, $y = hz + k$ has the equation $z = \dfrac{y - kx}{\sqrt{(1 + h^2)x - 1}}$.

114. By using the method of characteristics, find the integral surface of $pq = xy$, which passes through the straight line $z = x, y = 0$.
    Ans. $z^2 = x^2(1 + y^2)$.

115. Apply method of characteristics to determine the integral surface of $p^2 + q^2 = 2$, which passes through the straight line $x = 0, z = y$.
    Ans. $z = y \pm x$.

116. Use the method of characteristics to obtain the integral surface of $pq = z$, which passes through the parabola $x = 0, y^2 = z$.
    Ans. $16z = (x + 4y)^2$.

117. Using the method of characteristics, find the solution of $xp + yq - pq = 0$ passing through the curve $z = \dfrac{x}{2}, y = 0$.
    Ans. $x^2 - 4z^2 + 8xyz = 0$.

118. By the method of characteristics, determine the integral surface of $z = p^2 - q^2$ passing through the parabola $4z + x^2 = 0, y = 0$.
    Ans. $4z + (x + \sqrt{2}y)^2 = 0$.

119. Use the method of characteristics to obtain the integral surface of the quasilinear equation $2p + yq = z$, which passes through the curve $x = s, y = s^2, z = s$, $1 \le s \le 2$.
    Ans. $z^2 = y \exp\left(\dfrac{xz - y}{2z}\right)$.

120. Applying the method of characteristics, find the solution of Cauchy problem for the quasilinear equation $p - zq + z = 0$ for $x > 0$ with initial data curve Γ : $x = 0, y = s, z = -2s, \infty < s < \infty$.
    Ans. $z = \dfrac{2y}{2 - 3e^x}$.

121. Show that the equations $p^2 + q^2 = 1$ and $(p^2 + q^2)x = pz$ are compatible and find their common solution.
    Ans. $z^2 = x^2 + (y + c)^2$.

122. Show that the equation $z = xp + yq$ is compatible with any equation $H(x, y, z, p, q) = 0$ that is homogeneous in x, y, and z.

123. Solve completely the simultaneous equations $z = xp + yq$ and $2xy(p^2 + q^2) = z(yp + xq)$.
    Ans. $z^2 = axy$, $z^2 = b(x^2 + y^2)$.

124. Show that the equations $p = f(x, y)$ and $q = g(x, y)$ are compatible if

$$\frac{\partial f}{\partial y} = \frac{\partial g}{\partial x}.$$

125. Verify that the system of equations $p = x^2 - my$ and $q = y^2 - mx$ is compatible. Also solve the system. Ans. $z = \dfrac{1}{3}(x^3 + y^3) - mxy + c$.

126. Show that the equations $F(x, y, z, u_x, u_y, u_z) = 0$ and $G(x, y, z, u_x, u_y, u_z) = 0$ are compatible if

$$\frac{\partial(F, G)}{\partial(x, u_x)} + \frac{\partial(F, G)}{\partial(y, u_y)} + \frac{\partial(F, G)}{\partial(z, u_z)} = 0.$$

127. Reconstruct the proof of Theorem 4.5 by using Theorem 4.11.

Using Jacobi's method, find the complete solutions of the PDEs described in problems 128–132.

128. $z + 2u_z - (u_x + u_y)^2 = 0$

Ans.
$u = ax + by + \dfrac{1}{2}(a + b)^2 z - \dfrac{1}{4}z^2 + c$.

129. $u_x^2 + u_y^2 + u_z = 1$.

Ans. $u = ax + by + (1 - a^2 - b^2)z + c$.

130. $(y + z)(u_x + u_z)^2 + zu_x = 0$.

Ans. $\log u + ax + b(y - z) \pm \sqrt{a(y + z)} + c = 0$.

131. $u_x^2 + u_y u_z - z(u_y + u_z) = 0$.

Ans. $(1 + ab)\log u = (a + b)(x + ay + bz + c)$.

132. $2xzu_x + 3z^2 u_y + u_y^2 u_z = 0$.

Ans.
$u = a \log x + by - \dfrac{1}{b^2}(az^2 + bz^3) + c$.

5

Second-Order Partial Differential Equations

In this chapter, we discuss various methods for solving the partial differential equations (PDEs) of order two. Second-order PDEs play a fascinating role in applied sciences describing a variety of physical problems such as transverse vibrations of a string as well as a membrane, longitudinal vibrations in a bar, sound waves, electromagnetic waves, electric signals in cables, gravitational potential, electrostatic potential, magnetostatic potential, irrotational motion of a fluid, steady currents, steady flow of heat, surface waves on a fluid, conduction of heat in a bar, diffusion in isotropic substances, slowing down of neutrons in a matter, transmission line, nuclear reactors, *etc*.

We shall begin this chapter with two standard forms of linear PDEs, whose solutions are relatively more natural. In the next section, first we classify the semilinear PDEs and then for each class, we reduce the equation to its canonical form in order to obtain the general solution. Thereafter, a well-known general method, namely, Monge's method, has been discussed for solving quasilinear and non-linear equations. We also describe the Fourier method to solve homogeneous linear PDEs associated with boundary conditions (BCs). Finally, we present the derivations and solutions of some well-known fundamental PDEs of mathematical physics.

5.1 Linear PDEs: Standard Forms

As discussed in Chapter 2, the most general linear PDE of order two in two independent variables x and y is of the form

$$R(x, y)r + S(x, y)s + T(x, y)t + P(x, y)p + Q(x, y)q + Z(x, y)z = V(x, y). \tag{5.1}$$

Here, the coefficients R, S, and T do not vanish simultaneously.

In this section, by particularising the coefficients suitably in Eq. (5.1), we obtain some standard forms, which on integrating reduce easily to the first-order PDEs. Then, solving these equations for p or q, we obtain the general solution as desired. In the following lines, we discuss two types of such equations.

Type I (Equations Reducible to First-Order Linear Equations): In this type, we adopt the following forms:

$$Rr + Pp = V(x, y), \quad \text{here } S = T = Q = Z = 0, R \neq 0, P \neq 0$$
$$Ss + Pp = V(x, y), \quad \text{here } R = T = Q = Z = 0, S \neq 0, P \neq 0$$
$$Ss + Qq = V(x, y), \quad \text{here } R = T = P = Z = 0, S \neq 0, Q \neq 0$$
$$Tt + Qq = V(x, y), \quad \text{here } R = S = P = Z = 0, T \neq 0, Q \neq 0.$$

Such forms can be rewritten, respectively, as

$$R\frac{\partial p}{\partial x} + Pp = V$$

$$S\frac{\partial p}{\partial y} + Pp = V$$

$$S\frac{\partial q}{\partial x} + Qq = V$$

$$T\frac{\partial q}{\partial y} + Qq = V.$$

These are linear PDEs of order one, in which either p or q is the dependent variable and hence can be solved by the method discussed in Section 3.2.

Now, we adopt some examples of these forms.

Example 5.1. *Solve $xr + p = 9x^2y^2$.*

Solution. *The given PDE can be written as*

$$\frac{\partial p}{\partial x} + \frac{1}{x}p = 9xy^2.$$

Its integrating factor (IF) is

$$\text{IF} = e^{\int \frac{\partial x}{x}} = e^{\log x} = x.$$

Now, the solution of the PDE is

$$px = \int 9x^2y^2 \partial x + \phi(y) = 3x^3y^2 + \phi(y)$$

which yields that

$$\frac{\partial z}{\partial x} = 3x^2y^2 + \frac{1}{x}\phi(y).$$

Integrating partially w.r.t. x, we obtain

$$z = x^3y^2 + \log x \cdot \phi(y) + \psi(y)$$

which is the required solution of the given equation.

Example 5.2. *Solve $xys - yq = x^2$.*

Solution.   The given PDE can be written as

$$\frac{\partial q}{\partial x} - \frac{1}{x}q = \frac{x}{y}.$$

Its IF is

$$IF = e^{\int -\frac{\partial x}{x}} = e^{-\log x} = \frac{1}{x}.$$

Now, the solution of the PDE is

$$q \cdot \frac{1}{x} = \int \frac{1}{x}\frac{x}{y}\partial x = \int \frac{1}{y}\partial x = \frac{x}{y} + \phi(y)$$

yielding thereby

$$\frac{\partial z}{\partial y} = \frac{x^2}{y} + x\phi(y).$$

Integrating partially w.r.t. y, we obtain

$$z = x^2 \log y + x \int \phi(y)dy + \psi(x).$$

Hence, the general solution of the given equation is

$$z = x^2 \log y + x\theta(y) + \psi(x)$$

where $\theta(y) = \int \phi(y)dy$.

Example 5.3. *Solve $t - xq = -\sin y - x\cos y$.*

Solution.   The given PDE can be written as

$$\frac{\partial q}{\partial y} - xq = -\sin y - x\cos y.$$

Its IF is

$$IF = e^{\int -x\partial y} = e^{-xy}.$$

Now, the solution of the PDE is

$$qe^{-xy} = \int e^{-xy}(-\sin y - x\cos y)\partial y + \phi(x)$$

$$= -\int e^{-xy}\sin y\,\partial y - x\int e^{-xy}\cos y\,\partial y + \phi(x)$$

$$= e^{-xy}\cos y + x\int e^{-xy}\cos y\,\partial y - x\int e^{-xy}\cos y\,\partial y + \phi(x)$$

$$= e^{-xy}\cos y + \phi(x)$$

implying thereby

$$\frac{\partial z}{\partial y} = \cos y + e^{xy}\phi(x).$$

Integrating partially w.r.t. y, we obtain

$$z = \sin y + \frac{1}{x}e^{xy}\phi(x) + \psi(x)$$

or

$$z = \sin y + e^{xy}\theta(x) + \psi(x), \quad \text{where } \theta(x) = \frac{1}{x}\phi(x)$$

which is the required solution of the given equation.

Example 5.4. *Solve $ys - p = xy^2\cos xy$.*

Solution. *The given PDE can be written as*

$$\frac{\partial p}{\partial y} - \frac{1}{y}p = xy\cos xy.$$

Hence,

$$\text{IF} = e^{\int -\frac{1}{y}\partial y} = e^{-\log y} = \frac{1}{y}.$$

Now, the solution of the PDE is

$$p\cdot\frac{1}{y} = \int x\cos xy\,\partial y + \phi(x) = \sin xy + \phi(x)$$

implying thereby

$$\frac{\partial z}{\partial x} = y\sin xy + y\phi(x).$$

Integrating partially w.r.t. x, we obtain

$$z = -\cos xy + y \int \phi(x)dx + \psi(y).$$

Therefore, the general solution of the given equation is

$$z = -\cos xy + y\theta(x) + \psi(y)$$

where $\theta(x) = \int \phi(x)dx.$

Example 5.5. *Solve* $xs + q - xp - z = (1 - y)(1 + \log x).$

Solution. *The given PDE can be written as*

$$\frac{\partial p}{\partial y} + \frac{1}{x}\frac{\partial z}{\partial y} - p - \frac{1}{x}z = \frac{(1-y)}{x}(1 + \log x)$$

or

$$\frac{\partial}{\partial y}\left(p + \frac{z}{x}\right) - \left(p + \frac{z}{x}\right) = \frac{(1-y)}{x}(1 + \log x).$$

Set $p + \dfrac{z}{x} = u,$ *then the above equation becomes*

$$\frac{\partial u}{\partial y} - u = \frac{(1-y)}{x}(1 + \log x).$$

Its IF is

$$\text{IF} = e^{\int(-1)\partial y} = e^{-y}.$$

Now, we have

$$ue^{-y} = \int \frac{(1-y)}{x}(1 + \log x)e^{-y}\partial y + \phi(x)$$

$$= \frac{(1 + \log x)}{x}\int(1 - y)e^{-y}\partial y + \phi(x)$$

$$= \frac{(1 + \log x)}{x}\left[(1 - y)(-e^{-y}) - \int e^{-y}\partial y\right] + \phi(x)$$

$$= \frac{(1 + \log x)}{x}ye^{-y} + \phi(x)$$

yielding thereby

$$u = \frac{(1 + \log x)}{x}y + e^{y}\phi(x)$$

which, on substituting the value of u, becomes

$$\frac{\partial z}{\partial x} + \frac{1}{x}z = \frac{(1 + \log x)}{x}y + e^y \phi(x).$$

The IF of the above equation is

$$IF = e^{\int \frac{1}{x} \partial x} = e^{\log x} = x.$$

Therefore, the solution of the given PDE is

$$zx = \int \left[\frac{y}{x}(1 + \log x) + e^y \phi(x) \right] x \partial x + \psi(y)$$

$$= y \int (1 + \log x)dx + e^y \int x\phi(x)dx + \psi(y)$$

$$= y \left[(1 + \log x)x - \int \frac{1}{x}x dx \right] + e^y \theta(x) + \psi(y)$$

or

$$zx = xy \log x + e^y \theta(x) + \psi(y)$$

where $\theta(x) = \int x\phi(x)dx.$

Type II (Equations Reducible to First-Order Semilinear Equations): In this type, we undertake the following two forms of Eq. (5.1):

$$Rr + Ss + Pp = V(x, y), \quad \text{here } T = Q = Z = 0, R \neq 0, S \neq 0, P \neq 0$$
$$Ss + Tt + Qq = V(x, y), \quad \text{here } R = P = Z = 0, S \neq 0, T \neq 0, Q \neq 0.$$

These forms can be rewritten, respectively, as

$$R\frac{\partial p}{\partial x} + S\frac{\partial p}{\partial y} = V - Pp$$

$$S\frac{\partial q}{\partial x} + T\frac{\partial q}{\partial y} = V - Qq$$

which are semilinear (and hence, quasilinear) PDEs of order one with p (or q) as the dependent variable. Therefore, these equations can be solved for p or q (as the case may be) by Lagrange's method.

Let us consider several examples of these forms.

Example 5.6. *Solve $r + s + p = 1$.*

Solution.   The given PDE can be written as

$$\frac{\partial p}{\partial x} + \frac{\partial p}{\partial y} = 1 - p \tag{5.2}$$

which is a quasilinear equation in p. Hence, the Lagrange's auxiliary equations are

$$\frac{dx}{1} = \frac{dy}{1} = \frac{dp}{1 - p}.$$

Taking the first two fractions, we get $dx - dy = 0$, which on integrating gives rise to

$$x - y = c_1. \tag{5.3}$$

The last two fractions give $dy = \dfrac{dp}{1 - p}$. Integrating, we obtain

$$y = -\log(1 - p) + \log c_2$$

or

$$1 - p = c_2 e^{-y}. \tag{5.4}$$

Using (5.3) and (5.4), we obtain the general solution of Eq. (5.2) given by

$$1 - p = e^{-y}\phi(x - y)$$

which implies that

$$\frac{\partial z}{\partial x} = 1 - e^{-y}\phi(x - y).$$

Integrating the above equation partially w.r.t. x, we get

$$z = x - e^{-y} \int \phi(x - y)\partial x + \psi(y).$$

Hence, the general solution of the given equation is

$$z = x + e^{-y}\theta(x - y) + \psi(y)$$

where $\theta(x - y) = -\int \phi(x - y)\partial x$.

Example 5.7. *Solve* $s - t = \dfrac{x}{y^2}$.

Solution. *The given PDE can be written as*

$$\frac{\partial q}{\partial x} - \frac{\partial q}{\partial y} = \frac{x}{y^2} \tag{5.5}$$

which is a quasilinear equation in q. Hence, the Lagrange's auxiliary equations are

$$\frac{dx}{1} = \frac{dy}{-1} = \frac{dq}{x/y^2}.$$

From the first two fractions, we get $dx + dy = 0$, which on integrating gives rise to

$$x + y = c_1. \tag{5.6}$$

Taking the last two fractions, we get

$$dq = -\frac{x}{y^2} dy$$

which, on using (5.6), reduces to

$$dq = \frac{y - c_1}{y^2} dy = \frac{1}{y} dy - \frac{c_1}{y^2} dy.$$

Integrating, we obtain

$$q = \log y + \frac{c_1}{y} + c_2$$

which, by using (5.6), becomes

$$q - \log y - \frac{x + y}{y} = c_2. \tag{5.7}$$

Using (5.6) and (5.7), we obtain the general solution of Eq. (5.5) given by

$$q - \log y - \frac{x + y}{y} = \phi(x + y)$$

implying thereby

$$\frac{\partial z}{\partial y} = 1 + \log y + \frac{x}{y} + \phi(x + y).$$

Integrating the above equation partially w.r.t. y, we get

$$z = y + \int \log y \, \partial y + x \int \frac{1}{y} \, \partial y + \int \phi(x + y) \partial y + \psi(x)$$

$$= y + y(\log y - 1) + x \log y + \theta(x + y) + \psi(x)$$

where $\theta(x + y) = \int \phi(x + y) \partial y$. Thus, the general solution of the given equation is

$$z = (x + y) \log y + \psi(x) + \theta(x + y).$$

Example 5.8. Solve $yt + xs + q = 8yx^2 + 9y^2$.

Solution. The given PDE can be written as

$$x\frac{\partial q}{\partial x} + y\frac{\partial q}{\partial y} = -q + 8yx^2 + 9y^2 \qquad (5.8)$$

which is a quasilinear equation in dependent variable q. Hence, the Lagrange's auxiliary equations are

$$\frac{dx}{x} = \frac{dy}{y} = \frac{dq}{-q + 8yx^2 + 9y^2}.$$

Taking the first two fractions, we get $x\,dy - y\,dx = 0$, which on integrating gives rise to

$$\frac{y}{x} = c_1. \qquad (5.9)$$

From the last two fractions, we get

$$\frac{dq}{dy} + \frac{1}{y}q = 8x^2 + 9y$$

which, by using (5.9), reduces to

$$\frac{dq}{dy} + \frac{1}{y}q = \frac{8}{c_1^2}y^2 + 9y. \qquad (5.10)$$

Its IF is

$$\text{IF} = e^{\int \frac{dy}{y}} = e^{\log y} = y.$$

Hence, the general solution of Eq. (5.10) is

$$qy = \int \left(\frac{8}{c_1^2}y^2 + 9y\right) y\,dy + c_2$$

$$= \frac{8}{c_1^2} \int y^3 dy + 9 \int y^2 dy + c_2$$

$$= \frac{2}{c_1^2}y^4 + 3y^3 + c_2$$

which, by using (5.9), becomes

$$qy - 2x^2y^2 - 3y^3 = c_2. \tag{5.11}$$

Using (5.9) and (5.11), we obtain the general solution of Eq. (5.8) given by

$$qy - 2x^2y^2 - 3y^3 = \phi\left(\frac{y}{x}\right)$$

which yields that

$$\frac{\partial z}{\partial y} = 2x^2y + 3y^2 + \frac{1}{y}\phi\left(\frac{y}{x}\right).$$

Integrating the above equation partially w.r.t. y, we get

$$z = x^2y^2 + y^3 + \int \frac{1}{y}\phi\left(\frac{y}{x}\right)dy + \psi(x).$$

$$= x^2y^2 + y^3 + \int \frac{1}{u}\phi(u)\partial u + \psi(x) \quad \left(\text{put } \frac{y}{x} = u \text{ so that } y = xu \text{ and } \partial y = x\partial u\right)$$

$$= x^2y^2 + y^3 + \theta(u) + \psi(x) \quad \left(\text{set } \int \frac{1}{u}\phi(u)\partial u = \theta(u)\right).$$

Hence, the general solution of the given equation is

$$z = x^2y^2 + y^3 + \theta\left(\frac{y}{x}\right) + \psi(x).$$

Example 5.9. *Solve $xyr + x^2s - yp = x^3e^y$.*

Solution. *The given PDE can be written as*

$$xy\frac{\partial p}{\partial x} + x^2\frac{\partial p}{\partial y} = yp + x^3e^y \tag{5.12}$$

which is a quasilinear equation in dependent variable p. Hence, the Lagrange's auxiliary equations are

$$\frac{dx}{xy} = \frac{dy}{x^2} = \frac{dp}{yp + x^3e^y}.$$

Taking the first two fractions, we get $xdx - ydy = 0$, which on integrating gives rise to

$$x^2 - y^2 = c_1. \tag{5.13}$$

From the first and third fractions, we get

$$\frac{dp}{dx} - \frac{p}{x} = \frac{x^2e^y}{y}$$

which, on using (5.13), reduces to

$$\frac{dp}{dx} - \frac{p}{x} = \frac{x^2 e^{\sqrt{x^2 - c_1}}}{\sqrt{x^2 - c_1}} \qquad (5.14)$$

$$\text{IF} = e^{\int -\frac{dx}{x}} = e^{-\log x} = \frac{1}{x}.$$

Hence, the general solution of Eq. (5.14) is

$$p \cdot \frac{1}{x} = \int \frac{1}{x} \frac{x^2 e^{\sqrt{x^2 - c_1}}}{\sqrt{x^2 - c_1}} dx + c_2 = \int \frac{x e^{\sqrt{x^2 - c_1}}}{\sqrt{x^2 - c_1}} dx + c_2$$

$$= \int e^u du + c_2 \quad \left(\text{put } \sqrt{x^2 - c_1} = u \text{ so that } \frac{x}{\sqrt{x^2 - c_1}} dx = du \right)$$

$$= e^u + c_2 = e^{\sqrt{x^2 - c_1}} + c_2$$

which, by using (5.13), becomes

$$\frac{p}{x} - e^y = c_2. \qquad (5.15)$$

Using (5.13) and (5.15), we obtain the general solution of Eq. (5.12) given by

$$\frac{p}{x} - e^y = \phi(x^2 - y^2).$$

yielding thereby

$$\frac{\partial z}{\partial x} = xe^y + x\phi(x^2 - y^2).$$

Integrating the above equation partially w.r.t. x, we get

$$z = \frac{1}{2}x^2 e^y + \int x\phi(x^2 - y^2)\partial x + \psi(y).$$

Hence, the general solution of the given equation is

$$z = \frac{1}{2}x^2 e^y + \theta(x^2 - y^2) + \psi(y)$$

where $\theta(x^2 - y^2) = \int x\phi(x^2 - y^2)\partial x$.

5.2 Semilinear PDEs: Reduction to Canonical Forms

Recall that a semilinear (or almost-linear) PDE of second order in two variables is of the form

$$R(x, y)r + S(x, y)s + T(x, y)t = V(x, y, z, p, q). \tag{5.16}$$

Here, the coefficients R, S, T do not vanish simultaneously. We also assume that the dependent variable z and the coefficients R, S, T are twice continuously differentiable functions.

LHS of Eq. (5.16) is often referred to as the principal part of the equation, which is instrumental in determining the properties of the solution. First of all, we present some relevant notions, which will be utilised in the forthcoming discussion.

Definition 5.1. *The function $\triangle$ defined by*

$$\triangle(x_0, y_0) = S^2(x_0, y_0) - 4R(x_0, y_0)T(x_0, y_0)$$

is called the discriminant of Eq. (5.16) at a point (x_0, y_0).

We classify semilinear equations according to the sign of the discriminant $\triangle$.

Definition 5.2. *Equation (5.16) is called*

- *hyperbolic at (x_0, y_0) if $\triangle(x_0, y_0) > 0$*
- *parabolic at (x_0, y_0) if $\triangle(x_0, y_0) = 0$*
- *elliptic at (x_0, y_0) if $\triangle(x_0, y_0) < 0$.*

Moreover, Eq. (5.16) is called hyperbolic (respectively, parabolic or elliptic) in a domain $D \subseteq \mathbb{R}^2$ if it is hyperbolic (respectively, parabolic or elliptic) at each $(x, y) \in D$.

For illustration, consider the equation $xr - t + q = \sin x$. The discriminant is $\triangle = S^2 - 4RT = 4x$ and hence, the given equation is hyperbolic, parabolic, and elliptic accordingly $x <, =, > 0$.

Definition 5.3. *The two ordinary differential equations (ODEs)*

$$\frac{dy}{dx} = \frac{S \pm \sqrt{\triangle}}{2R}$$

*are called the **characteristic equations** of Eq. (5.16). The solutions of these characteristic equations are called the **characteristic curves** or the **characteristic projections** or simply the **characteristics** of Eq. (5.16).*

Clearly, a hyperbolic equation admits two distinct real characteristic curves, whereas a parabolic equation has a single, real characteristic curve. As a vital change, an elliptic equation admits two complex conjugate characteristic curves but no real characteristic curve.

Example 5.10. *Classify the following PDEs. Also, find their corresponding characteristic equations and characteristic curves.*

(i) $\dfrac{\partial^2 z}{\partial x^2} = \dfrac{\partial z}{\partial y}$.

(ii) $\dfrac{\partial^2 z}{\partial x^2} = \dfrac{\partial^2 z}{\partial y^2}$.

(iii) $\dfrac{\partial^2 z}{\partial x^2} + \dfrac{\partial^2 z}{\partial y^2} = 0$.

Solution. *(i) Rewrite the given PDE as*

$$r - q = 0.$$

Comparing with general Eq. (5.16), we get $R = 1, S = T = 0$. Hence, the discriminant of the equation is

$$\triangle = S^2 - 4RT = 0.$$

It follows that the given equation is parabolic. The characteristic equation is

$$\frac{dy}{dx} = \frac{S \pm \sqrt{\triangle}}{2R} = 0.$$

Its solution is $y = c$, which remains the single characteristic curve of the given PDE.
(ii) Rewrite the given PDE as

$$r - t = 0.$$

Comparing with general Eq. (5.16), we get $R = 1, S = 0, T = -1$. Hence, the discriminant of the equation is

$$\triangle = S^2 - 4RT = 4 > 0.$$

It follows that the given equation is hyperbolic. The characteristic equations are

$$\frac{dy}{dx} = \frac{S \pm \sqrt{\triangle}}{2R} = \frac{\pm 2}{2} = \pm 1$$

which become

$$dy \pm dx = 0.$$

On integrating these equations, we get

$$y + x = c_1 \ \text{ and } \ y - x = c_2$$

which are the two real and distinct characteristic curves of the given PDE.
(iii) Rewrite the given PDE as

$$r + t = 0.$$

Comparing with general Eq. (5.16), we get $R = 1, S = 0, T = 1$. Hence, the discriminant of the equation is

$$\triangle = S^2 - 4RT = -4 < 0.$$

It follows that the given equation is elliptic. The characteristic equations are

$$\frac{dy}{dx} = \frac{S \pm \sqrt{\triangle}}{2R} = \frac{\pm 2i}{2} = \pm i$$

which become

$$dy \pm i\,dx = 0.$$

On integrating these equations, we get

$$y + ix = c_1 \quad \text{and} \quad y - ix = c_2$$

which are the two complex conjugate characteristic curves of the given PDE.

Coordinate Transformation: By changing the independent variables, a semilinear PDE reduces to a new semilinear equation. Consider the general transformation of independent variables

$$\xi = \xi(x, y) \text{ and } \eta = \eta(x, y). \tag{5.17}$$

With a view to use the chain rule, we compute the partial derivatives of $z = z(\xi, \eta)$ as follows:

$$p = \frac{\partial z}{\partial x} = \frac{\partial z}{\partial \xi}\frac{\partial \xi}{\partial x} + \frac{\partial z}{\partial \eta}\frac{\partial \eta}{\partial x} = \xi_x z_\xi + \eta_x z_\eta \tag{5.18}$$

$$q = \frac{\partial z}{\partial y} = \frac{\partial z}{\partial \xi}\frac{\partial \xi}{\partial y} + \frac{\partial z}{\partial \eta}\frac{\partial \eta}{\partial y} = \xi_y z_\xi + \eta_y z_\eta. \tag{5.19}$$

From (5.18) and (5.19), we get, respectively,

$$\frac{\partial}{\partial x} = \xi_x \frac{\partial}{\partial \xi} + \eta_x \frac{\partial}{\partial \eta} \tag{5.20}$$

$$\frac{\partial}{\partial y} = \xi_y \frac{\partial}{\partial \xi} + \eta_y \frac{\partial}{\partial \eta}. \tag{5.21}$$

Using (5.18) and (5.20), we get

$$r = \frac{\partial p}{\partial x} = \frac{\partial}{\partial x}(\xi_x z_\xi + \eta_x z_\eta)$$

$$= \xi_{xx} z_\xi + \xi_x \frac{\partial}{\partial x}(z_\xi) + \eta_{xx} z_\eta + \eta_x \frac{\partial}{\partial x}(z_\eta)$$

$$= \xi_{xx} z_\xi + \xi_x \left(\xi_x \frac{\partial}{\partial \xi} + \eta_x \frac{\partial}{\partial \eta} \right) z_\xi + \eta_{xx} z_\eta + \eta_x \left(\xi_x \frac{\partial}{\partial \xi} + \eta_x \frac{\partial}{\partial \eta} \right) z_\eta$$

$$= \xi_x^2 z_{\xi\xi} + 2\xi_x \eta_x z_{\xi\eta} + \eta_x^2 z_{\eta\eta} + \xi_{xx} z_\xi + \eta_{xx} z_\eta.$$

Using (5.19) and (5.21), we get

$$t = \frac{\partial q}{\partial y} = \frac{\partial}{\partial y}(\xi_y z_\xi + \eta_y z_\eta)$$

$$= \xi_{yy} z_\xi + \xi_y \frac{\partial}{\partial y}(z_\xi) + \eta_{yy} z_\eta + \eta_y \frac{\partial}{\partial y}(z_\eta)$$

$$= \xi_{yy} z_\xi + \xi_y \left(\xi_y \frac{\partial}{\partial \xi} + \eta_y \frac{\partial}{\partial \eta} \right) z_\xi + \eta_{yy} z_\eta + \eta_y \left(\xi_y \frac{\partial}{\partial \xi} + \eta_y \frac{\partial}{\partial \eta} \right) z_\eta$$

$$= \xi_y^2 z_{\xi\xi} + 2\xi_y \eta_y z_{\xi\eta} + \eta_y^2 z_{\eta\eta} + \xi_{yy} z_\xi + \eta_{yy} z_\eta.$$

Using (5.19) and (5.20), we get

$$s = \frac{\partial q}{\partial x} = \frac{\partial}{\partial x}(\xi_y z_\xi + \eta_y z_\eta)$$

$$= \xi_{xy} z_\xi + \xi_y \frac{\partial}{\partial x}(z_\xi) + \eta_{xy} z_\eta + \eta_y \frac{\partial}{\partial x}(z_\eta)$$

$$= \xi_{xy} z_\xi + \xi_y \left(\xi_x \frac{\partial}{\partial \xi} + \eta_x \frac{\partial}{\partial \eta} \right) z_\xi + \eta_{xy} z_\eta + \eta_y \left(\xi_x \frac{\partial}{\partial \xi} + \eta_x \frac{\partial}{\partial \eta} \right) z_\eta$$

$$= \xi_x \xi_y z_{\xi\xi} + (\xi_x \eta_y + \xi_y \eta_x) z_{\xi\eta} + \eta_x \eta_y z_{\eta\eta} + \xi_{xy} z_\xi + \eta_{xy} z_\eta.$$

Collecting all these derivatives, we have

$$p = \xi_x z_\xi + \eta_x z_\eta$$

$$q = \xi_y z_\xi + \eta_y z_\eta$$

$$r = \xi_x^2 z_{\xi\xi} + 2\xi_x \eta_x z_{\xi\eta} + \eta_x^2 z_{\eta\eta} + \xi_{xx} z_\xi + \eta_{xx} z_\eta$$

$$t = \xi_y^2 z_{\xi\xi} + 2\xi_y \eta_y z_{\xi\eta} + \eta_y^2 z_{\eta\eta} + \xi_{yy} z_\xi + \eta_{yy} z_\eta$$

$$s = \xi_x \xi_y z_{\xi\xi} + (\xi_x \eta_y + \xi_y \eta_x) z_{\xi\eta} + \eta_x \eta_y z_{\eta\eta} + \xi_{xy} z_\xi + \eta_{xy} z_\eta.$$

Substituting the values of these partial derivatives in Eq. (5.16) and simplifying, we get

$$A z_{\xi\xi} + B z_{\xi\eta} + C z_{\eta\eta} = V^*(\xi, \eta, z, z_\xi, z_\eta) \tag{5.23}$$

where

$$A = R\xi_x^2 + S\xi_x\xi_y + T\xi_y^2 \tag{5.24}$$

$$B = 2R\xi_x\eta_x + S(\xi_x\eta_y + \xi_y\eta_x) + 2T\xi_y\eta_y \tag{5.25}$$

$$C = R\eta_x^2 + S\eta_x\eta_y + T\eta_y^2 \tag{5.26}$$

and

$$V^* = V - (R\xi_{xx} + S\xi_{xy} + T\xi_{yy})z_\xi - (R\eta_{xx} + S\eta_{xy} + T\eta_{yy})z_\eta.$$

Hence, under the transformation (5.17), Eq. (5.16) reduces to a new semilinear PDE of the form (5.23), in which z is a dependent variable, whereas ξ and η are independent variables.

Keeping in view of future need of invertibility of ξ and η, we assume that the Jacobian of the transformation (5.17) does not vanish at (x, y), *that is,*

$$J := \frac{\partial(\xi, \eta)}{\partial(x, y)} = \xi_x\eta_y - \xi_y\eta_x \neq 0$$

which enables us to determine x and y uniquely.

Proposition 5.1. *The type of a second-order semilinear PDE in two variables is invariant under an invertible transformation of independent variables.*

Proof. Consider the semilinear equation (5.16) with discriminant $\triangle = S^2 - 4RT$. We have shown that under an invertible transformation (5.17), Eq. (5.16) reduces to (5.23). If the discriminant of Eq. (5.23) is $\triangle'$, then using (5.24-5.26), we obtain

$$\begin{aligned}
\triangle' &= B^2 - 4AC \\
&= \left[2R\xi_x\eta_x + S(\xi_x\eta_y + \xi_y\eta_x) + 2T\xi_y\eta_y \right]^2 \\
&\quad -4\left[R\xi_x^2 + S\xi_x\xi_y + T\xi_y^2 \right]\left[R\eta_x^2 + S\eta_x\eta_y + T\eta_y^2 \right]
\end{aligned}$$

which, on simplification, becomes

$$\triangle' = (\xi_x\eta_y - \xi_y\eta_x)^2(S^2 - 4RT)$$

or

$$\triangle' = J^2 \triangle.$$

As $J \neq 0$, the above relation indicates that the sign of the discriminant does not alter under the transformation (5.17), which yields that the type of Eq. (5.16) remains invariant. In other words, Eq. (5.23) is hyperbolic (respectively, parabolic or elliptic) if Eq. (5.16) is hyperbolic (respectively, parabolic or elliptic). □

5.2.1 Laplace Transformation and Canonical Forms

By a suitable choice of variables (ξ, η), Eq. (5.16) can be transformed into a simpler form, which is often called the **canonical form** or **normal form** of Eq. (5.16). Such new coordinates (ξ, η), under which the original equation reduces to its canonical form, are called the **canonical coordinates**, whereas such a transformation is called **Laplace transformation**. The choice of canonical coordinates depends upon whether the original PDE (5.16) is hyperbolic, parabolic, or elliptic. First, we present the following result concerning the canonical form of a hyperbolic equation.

Theorem 5.1. *Let Eq. (5.16) be a hyperbolic equation, whose characteristic curves are*

$$u(x, y) = c_1 \text{ and } v(x, y) = c_2.$$

Then, the transformation

$$\xi = u(x, y) \text{ and } \eta = v(x, y)$$

is invertible. Moreover, under this transformation, Eq. (5.16) reduces to its canonical form given by

$$z_{\xi\eta} = \Phi(\xi, \eta, z, z_\xi, z_\eta).$$

Furthermore, under a new transformation

$$\alpha = \xi + \eta \text{ and } \beta = \xi - \eta \tag{5.27}$$

Eq. (5.16) reduces to its second canonical form given by

$$z_{\alpha\alpha} + z_{\beta\beta} = \Psi(\alpha, \beta, z, z_\alpha, z_\beta).$$

Proof. Denote $\lambda_1 = \dfrac{-S + \sqrt{\triangle}}{2R}$ and $\lambda_2 = \dfrac{-S - \sqrt{\triangle}}{2R}$, then λ_1 and λ_2 become the roots of λ-quadratic equation

$$R\lambda^2 + S\lambda + T = 0. \tag{5.28}$$

Using the above notations, the two characteristic equations of Eq. (5.16) can be written as

$$\frac{dy}{dx} + \lambda_1 = 0 \tag{5.29}$$

and

$$\frac{dy}{dx} + \lambda_2 = 0. \tag{5.30}$$

Suppose that the characteristic curves $u(x, y) = c_1$ and $v(x, y) = c_2$ are solutions of Eqs (5.29) and (5.30), respectively.

Along the curve $\xi \equiv u(x, y) = c_1$, we have

$$d\xi = 0$$

which gives rise to

$$\xi_x dx + \xi_y dy = 0$$

implying thereby

$$\frac{dy}{dx} = -\frac{\xi_x}{\xi_y}.$$

Comparing the above equation with Eq. (5.29), we get

$$\lambda_1 = \frac{\xi_x}{\xi_y}. \tag{5.31}$$

Similarly, along the curve $\eta \equiv v(x, y) = c_2$, we have $\dfrac{dy}{dx} = -\dfrac{\eta_x}{\eta_y}$ and hence comparing with Eq. (5.30), we get

$$\lambda_2 = \frac{\eta_x}{\eta_y}. \tag{5.32}$$

From (5.31) and (5.32), we get $\xi_x = \lambda_1 \xi_y$ and $\eta_x = \lambda_2 \eta_y$, respectively. With the help of these relations, we have

$$J = \xi_x \eta_y - \xi_y \eta_x = \lambda_1 \xi_y \eta_y - \lambda_2 \xi_y \eta_y = (\lambda_1 - \lambda_2)\xi_y \eta_y = \frac{\sqrt{\triangle}}{R}\xi_y \eta_y \neq 0$$

which yields that the given transformation is invertible.

Now, both λ_1 and λ_2 are roots of Eq. (5.28). Therefore, from Eqs (5.31), (5.32), and (5.28), we have

$$R\left(\frac{\xi_x}{\xi_y}\right)^2 + S\left(\frac{\xi_x}{\xi_y}\right) + T = 0$$

$$R\left(\frac{\eta_x}{\eta_y}\right)^2 + S\left(\frac{\eta_x}{\eta_y}\right) + T = 0.$$

Using the above equations, Eqs (5.24) and (5.26) reduce to

$$A = R\xi_x^2 + S\xi_x \xi_y + T\xi_y^2 = 0$$

$$C = R\eta_x^2 + S\eta_x \eta_y + T\eta_y^2 = 0.$$

Thus, putting $A = C = 0$ in Eq. (5.23), we get

$$Bz_{\xi\eta} = V^*(\xi, \eta, z, z_\xi, z_\eta).$$

Since the given equation is hyperbolic, in view of Proposition 5.1, we have $B^2 > 0$ yielding thereby $B \neq 0$. So on dividing by B, the above equation reduces to

$$z_{\xi\eta} = \Phi(\xi, \eta, z, z_\xi, z_\eta) \tag{5.33}$$

which is the required first canonical form of hyperbolic equation.

Now, we derive the second canonical form. Using the new transformation (5.27) and applying the chain rule of partial derivatives, we get

$$\left.\begin{aligned}
\frac{\partial z}{\partial \xi} &= \frac{\partial z}{\partial \alpha}\frac{\partial \alpha}{\partial \xi} + \frac{\partial z}{\partial \beta}\frac{\partial \beta}{\partial \xi} = \frac{\partial z}{\partial \alpha} + \frac{\partial z}{\partial \beta} = z_\alpha + z_\beta \\
\frac{\partial z}{\partial \eta} &= \frac{\partial z}{\partial \alpha}\frac{\partial \alpha}{\partial \eta} + \frac{\partial z}{\partial \beta}\frac{\partial \beta}{\partial \eta} = \frac{\partial z}{\partial \alpha} - \frac{\partial z}{\partial \beta} = z_\alpha - z_\beta.
\end{aligned}\right\} \tag{5.34}$$

With the help of (5.34), we obtain

$$z_{\xi\eta} = \frac{\partial}{\partial \xi}\left(\frac{\partial z}{\partial \eta}\right) = \left(\frac{\partial}{\partial \alpha} + \frac{\partial}{\partial \beta}\right)(z_\alpha - z_\beta) = z_{\alpha\alpha} - z_{\beta\beta}. \tag{5.35}$$

Using (5.34), (5.35), and inverse transformation of (5.27), Eq. (5.33) takes the form

$$z_{\alpha\alpha} - z_{\beta\beta} = \Psi(\alpha, \beta, z, z_\alpha, z_\beta)$$

which is the second canonical form of hyperbolic equation as desired. $\qquad\square$

Next, we investigate the canonical form of a parabolic equation.

Theorem 5.2. *Let Eq. (5.16) be a parabolic equation, whose characteristic curve is*

$$u(x, y) = c.$$

Assume that $v(x, y)$ is arbitrarily chosen such that the Jacobian $\dfrac{\partial(u, v)}{\partial(x, y)} \neq 0$. Then, under the invertible transformation

$$\xi = u(x, y) \text{ and } \eta = v(x, y)$$

Eq. (5.16) reduces to its canonical form given by

$$z_{\eta\eta} = \Phi(\xi, \eta, z, z_\xi, z_\eta).$$

Proof. In this case, $\triangle = 0$ so that the characteristic equation reduces to

$$\frac{dy}{dx} = \frac{S}{2R}.$$

Also, $\lambda = -S/2R$ remains a repeated root of λ-quadratic equation

$$R\lambda^2 + S\lambda + T = 0.$$

Given that $u(x, y) = c$ is characteristic curve and $\xi = u(x, y)$. Now, applying the procedure similar to Theorem 5.1, we can obtain

$$A = R\xi_x^2 + S\xi_x\xi_y + T\xi_y^2 = 0.$$

Since the given equation is parabolic, we have $S = 2\sqrt{RT}$. Therefore, the above equation reduces to

$$(\sqrt{R}\xi_x + \sqrt{T}\xi_y)^2 = 0$$

which yields that

$$\sqrt{R}\xi_x + \sqrt{T}\xi_y = 0. \tag{5.36}$$

Now, due to $S = 2\sqrt{RT}$, Eq. (5.25) becomes

$$B = 2(\sqrt{R}\xi_x + \sqrt{T}\xi_y)(\sqrt{R}\eta_x + \sqrt{T}\eta_y)$$

which, making use of (5.36), gives rise to

$$B = 0.$$

Hence, putting $A = B = 0$ in Eq. (5.23), we get

$$Cz_{\eta\eta} = V^*(\xi, \eta, z, z_\xi, z_\eta). \tag{5.37}$$

We claim that $C \neq 0$. To prove this, we suppose on contrary that $C = 0$, *that is,*

$$R\eta_x^2 + S\eta_x\eta_y + T\eta_y^2 = 0$$

which can be rewritten as

$$R\left(\frac{\eta_x}{\eta_y}\right)^2 + S\left(\frac{\eta_x}{\eta_y}\right) + T = 0.$$

Solving the above equation for η_x/η_y, we get

$$\frac{\eta_x}{\eta_y} = -\frac{S}{2R} = \frac{\xi_x}{\xi_y}$$

which contradicts the hypothesis $\dfrac{\partial(\xi, \eta)}{\partial(x, y)} \neq 0$. Hence, $C \neq 0$. Dividing Eq. (5.37) by C, we get

$$z_{\eta\eta} = \Phi(\xi, \eta, z, z_\xi, z_\eta)$$

which is the required canonical form of parabolic equation. □

In practice, for parabolic PDEs, we select the new variable $v(x, y)$ in such a way that the functions $u(x, y)$ and $v(x, y)$ must be independent so that $\dfrac{\partial(u, v)}{\partial(x, y)} \neq 0$.

Finally, we establish the result regarding the canonical form of an elliptic equation.

Theorem 5.3. *Let Eq. (5.16) be an elliptic equation, whose characteristic curves are*

$$u(x, y) + iv(x, y) = c_1 \quad \text{and} \quad u(x, y) - iv(x, y) = c_2.$$

Then, the transformation

$$\xi = u(x, y) \text{ and } \eta = v(x, y)$$

is invertible. Moreover, under this transformation, Eq. (5.16) reduces to its canonical form given by

$$z_{\xi\xi} + z_{\eta\eta} = \Phi(\xi, \eta, z, z_\xi, z_\eta).$$

Proof. Introduce the two conjugate complex variables (α, β) such that

$$\left.\begin{aligned} \alpha(x, y) &= u(x, y) + iv(x, y) = \xi + i\eta \\ \beta(x, y) &= u(x, y) - iv(x, y) = \xi - i\eta. \end{aligned}\right\} \tag{5.38}$$

Clearly, in this case, the complex characteristics $\alpha(x, y) = c_1$ and $\beta(x, y) = c_2$ are distinct. Therefore, proceeding as Theorem 5.1, we can show that the transformation $\alpha = \alpha(x, y), \beta = \beta(x, y)$ is invertible, that is, $\dfrac{\partial(\alpha, \beta)}{\partial(x, y)} \neq 0$ and it transforms Eq. (5.16) to the complex canonical form

$$z_{\alpha\beta} = \Theta(\alpha, \beta, z, z_\alpha, z_\beta). \tag{5.39}$$

To make it real canonical form, consider the inverse transformation from the system (5.38)

$$\xi = \frac{\alpha + \beta}{2} \tag{5.40}$$

$$\eta = \frac{\alpha - \beta}{2i}. \tag{5.41}$$

Clearly, $\dfrac{\partial(\xi, \eta)}{\partial(\alpha, \beta)} \neq 0$ and hence, we have

$$J = \frac{\partial(\xi, \eta)}{\partial(x, y)} = \frac{\partial(\xi, \eta)}{\partial(\alpha, \beta)} \cdot \frac{\partial(\alpha, \beta)}{\partial(x, y)} \neq 0$$

which yields that the given transformation is invertible.

Now, using (5.40), (5.41) and applying the chain rule of partial derivatives, we get

$$
\left.\begin{aligned}
\frac{\partial z}{\partial \alpha} &= \frac{\partial z}{\partial \xi}\frac{\partial \xi}{\partial \alpha} + \frac{\partial z}{\partial \eta}\frac{\partial \eta}{\partial \alpha} = \frac{1}{2}\left(\frac{\partial z}{\partial \xi} - i\frac{\partial z}{\partial \eta}\right) = \frac{1}{2}(z_\xi - iz_\eta) \\
\frac{\partial z}{\partial \beta} &= \frac{\partial z}{\partial \xi}\frac{\partial \xi}{\partial \beta} + \frac{\partial z}{\partial \eta}\frac{\partial \eta}{\partial \beta} = \frac{1}{2}\left(\frac{\partial z}{\partial \xi} + i\frac{\partial z}{\partial \eta}\right) = \frac{1}{2}(z_\xi + iz_\eta).
\end{aligned}\right\}
\tag{5.42}
$$

With the help of (5.42), we obtain

$$
z_{\alpha\beta} = \frac{\partial}{\partial \alpha}\left(\frac{\partial z}{\partial \beta}\right) = \frac{1}{4}\left(\frac{\partial}{\partial \xi} - i\frac{\partial}{\partial \eta}\right)(z_\xi + iz_\eta) = \frac{1}{4}(z_{\xi\xi} + z_{\eta\eta}).
\tag{5.43}
$$

Using (5.38), (5.42), and (5.43), Eq. (5.39) takes the form

$$
z_{\xi\xi} + z_{\eta\eta} = \Phi(\xi, \eta, z, z_\xi, z_\eta)
$$

which is the required (real) canonical form of elliptic equation. $\square$

The following examples are adopted to explain the canonical forms.

Example 5.11. *Reduce the PDE*

$$
r + 4s - 5t - p^2 + 10q^2 - 9z = 0
$$

to a canonical form.

Solution.   *Here*

$$
R = 1, \quad S = 4, \quad T = -5.
$$

Step I (Classification): The discriminant of the given equation is

$$
\triangle = S^2 - 4RT = 16 + 20 = 36 > 0.
$$

Therefore, the given equation is hyperbolic.

Step II (Characteristics): The characteristic equations are

$$
\frac{dy}{dx} = \frac{S \pm \sqrt{\triangle}}{2R} = \frac{4 \pm 6}{2}
$$

which become

$$
\frac{dy}{dx} = 5 \quad \text{and} \quad \frac{dy}{dx} = -1
$$

or

$$dy - 5dx = 0 \ \text{ and } \ dy + dx = 0.$$

Integrating the above equations, we get

$$y - 5x = c_1 \ \text{ and } \ y + x = c_2$$

which are the characteristic curves of the given PDE.

Step III (Canonical Form): The canonical coordinates are

$$\xi = y - 5x \ \text{ and } \ \eta = y + x.$$

Hence, we have

$$\xi_x = -5, \ \xi_y = 1, \ \xi_{xx} = \xi_{xy} = \xi_{yy} = 0$$

$$\eta_x = 1, \ \eta_y = 1, \ \eta_{xx} = \eta_{xy} = \eta_{yy} = 0.$$

Using the above, Eq. (5.22) becomes

$$p = -5z_\xi + z_\eta$$
$$q = z_\xi + z_\eta$$
$$r = 25z_{\xi\xi} - 10z_{\xi\eta} + z_{\eta\eta}$$
$$t = z_{\xi\xi} + 2z_{\xi\eta} + z_{\eta\eta}$$
$$s = -5z_{\xi\xi} - 4z_{\xi\eta} + z_{\eta\eta}.$$

Putting these values in the given PDE, we get

$$25z_{\xi\xi} - 10z_{\xi\eta} + z_{\eta\eta} + 4(-5z_{\xi\xi} - 4z_{\xi\eta} + z_{\eta\eta}) - 5(z_{\xi\xi} + 2z_{\xi\eta} + z_{\eta\eta}) - (-5z_\xi + z_\eta)^2 + 10(z_\xi + z_\eta)^2 - 9z = 0$$

which, on rearranging, gives rise to

$$-36z_{\xi\eta} - 15z_\xi^2 + 30z_\xi z_\eta + 9z_\eta^2 - 9z = 0$$

or

$$z_{\xi\eta} = -\frac{5}{12}z_\xi^2 + \frac{5}{6}z_\xi z_\eta + \frac{1}{4}z_\eta^2 - \frac{1}{4}z \tag{5.44}$$

which is the first canonical form of the given equation as desired. To obtain the second canonical form, we consider the transformation

$$\alpha = \xi + \eta, \ \beta = \xi - \eta.$$

Using (5.34) and (5.35), we obtain

$$z_\xi = z_\alpha + z_\beta, z_\eta = z_\alpha - z_\beta, z_{\xi\eta} = z_{\alpha\alpha} - z_{\beta\beta}.$$

Putting these values in (5.44) and simplifying, we get the second canonical form given by

$$z_{\alpha\alpha} - z_{\beta\beta} = \frac{2}{3}z_\alpha^2 - \frac{4}{3}z_\alpha z_\beta - z_\beta^2 - \frac{1}{4}z.$$

Example 5.12. *Reduce the PDE*

$$r - 6s + 9t - (6p + 9q)z^2 = 0$$

to a canonical form.

Solution. *Here*

$$R = 1, \quad S = -6, \quad T = 9.$$

Step I (Classification): The discriminant of the given equation is

$$\triangle = S^2 - 4RT = 36 - 36 = 0.$$

Therefore, the given equation is parabolic.

Step II (Characteristics): The characteristic equations are

$$\frac{dy}{dx} = \frac{S}{2R} = \frac{-6}{2}$$

$$\text{or} \quad dy + 3dx = 0.$$

Solving it, we get

$$y + 3x = c$$

which is the characteristic curve of the given PDE.

Step III (Canonical Form): The canonical coordinates are

$$\xi = y + 3x \quad \text{and} \quad \eta = y.$$

Notice that we have chosen $\eta = y$ in such a way that ξ and η are independent. Now, we have

$$\xi_x = 3, \ \xi_y = 1, \ \xi_{xx} = \xi_{xy} = \xi_{yy} = 0$$

$$\eta_x = 0, \ \eta_y = 1, \ \eta_{xx} = \eta_{xy} = \eta_{yy} = 0.$$

Using the above, Eq. (5.22) becomes

$$p = 3z_\xi$$
$$q = z_\xi + z_\eta$$
$$r = 9z_{\xi\xi}$$
$$t = z_{\xi\xi} + 2z_{\xi\eta} + z_{\eta\eta}$$
$$s = 3z_{\xi\xi} + 3z_{\xi\eta}.$$

Putting these values in the given PDE, we get

$$9z_{\xi\xi} - 6(3z_{\xi\xi} + 3z_{\xi\eta}) + 9(z_{\xi\xi} + 2z_{\xi\eta} + z_{\eta\eta}) - \{18z_\xi + 9(z_\xi + z_\eta)\}z^2 = 0$$

which, on simplifying, gives rise to

$$9z_{\eta\eta} - 9(3z_\xi + z_\eta)z^2 = 0$$

or

$$z_{\eta\eta} = (3z_\xi + z_\eta)z^2$$

which is the required canonical form of the given equation.

Example 5.13. *Reduce the PDE*

$$r + 2s + 5t + pq = 0$$

to a canonical form.

Solution.   *Here*

$$R = 1, \quad S = 2, \quad T = 5.$$

Step I (Classification): The discriminant of the given equation is

$$\triangle = S^2 - 4RT = 4 - 20 = -16 < 0.$$

Therefore, the given equation is elliptic.

Step II (Characteristics): The characteristic equations are

$$\frac{dy}{dx} = \frac{S \pm \sqrt{\triangle}}{2R} = \frac{2 \pm 4i}{2} = 1 \pm 2i$$

or

$$dy - dx \pm 2i\,dx = 0.$$

Integrating the above equations, we get

$$(y - x) + 2xi = c_1 \text{ and } (y - x) - 2xi = c_2$$

which are the characteristic curves of the given PDE.

Step III (Canonical Form): The canonical coordinates are

$$\xi = y - x \text{ and } \eta = 2x.$$

Hence, we have

$$\xi_x = -1, \ \xi_y = 1, \ \xi_{xx} = \xi_{xy} = \xi_{yy} = 0$$

$$\eta_x = 2, \ \eta_y = 0, \ \eta_{xx} = \eta_{xy} = \eta_{yy} = 0.$$

Using the above, Eq. (5.22) becomes

$$p = -z_\xi + 2z_\eta$$
$$q = z_\xi$$
$$r = z_{\xi\xi} - 4z_{\xi\eta} + 4z_{\eta\eta}$$
$$t = z_{\xi\xi}$$
$$s = -z_{\xi\xi} + 2z_{\xi\eta}.$$

Putting these values in the given PDE, we get

$$z_{\xi\xi} - 4z_{\xi\eta} + 4z_{\eta\eta} + 2(-z_{\xi\xi} + 2z_{\xi\eta}) + 5z_{\xi\xi} + (-z_\xi + 2z_\eta)z_\xi = 0$$

which, on simplifying, gives rise to

$$4z_{\xi\xi} + 4z_{\eta\eta} - z_\xi^2 + 2z_\xi z_\eta = 0$$

or

$$z_{\xi\xi} + z_{\eta\eta} = \frac{1}{4}z_\xi^2 - \frac{1}{2}z_\xi z_\eta$$

which is the required canonical form of the given equation.

Example 5.14. *Reduce the PDE*

$$r = x^2 t$$

to a canonical form.

Solution. *Here*

$$R = 1, \ S = 0, \ T = -x^2.$$

Step I (Classification): The discriminant of the given equation is

$$\triangle = S^2 - 4RT = 4x^2 > 0.$$

Therefore, the given equation is hyperbolic for $x \neq 0$. Notice that for $x = 0$, the given equation is parabolic and reduces to $r = 0$, which is already in canonical form.

Step II (Characteristics): The characteristic equations are

$$\frac{dy}{dx} = \frac{S \pm \sqrt{\triangle}}{2R} = \frac{\pm 2x}{2} = \pm x$$

which become

$$dy \pm xdx = 0.$$

On integrating these equations, we get

$$y + \frac{1}{2}x^2 = c_1 \text{ and } y - \frac{1}{2}x^2 = c_2$$

which are the characteristic curves of the given PDE.

Step III (Canonical Form): The canonical coordinates are

$$\xi = y + \frac{1}{2}x^2 \text{ and } \eta = y - \frac{1}{2}x^2. \tag{5.45}$$

Thus, we have

$$\xi_x = x, \ \xi_y = 1, \ \xi_{xx} = 1, \ \xi_{xy} = 0, \ \xi_{yy} = 0$$

$$\eta_x = -x, \ \eta_y = 1, \ \eta_{xx} = -1, \ \eta_{xy} = 0, \ \eta_{yy} = 0.$$

Using the above, Eq. (5.22) becomes

$$r = x^2 z_{\xi\xi} - 2x^2 z_{\xi\eta} + x^2 z_{\eta\eta} + z_\xi - z_\eta$$
$$t = z_{\xi\xi} + 2z_{\xi\eta} + z_{\eta\eta}.$$

Putting these values in the given PDE, we get

$$x^2 z_{\xi\xi} - 2x^2 z_{\xi\eta} + x^2 z_{\eta\eta} + z_\xi - z_\eta = x^2(z_{\xi\xi} + 2z_{\xi\eta} + z_{\eta\eta})$$

or

$$4x^2 z_{\xi\eta} = z_\xi - z_\eta.$$

Using (5.45), we have $x^2 = \xi - \eta$ and hence the above equation reduces to

$$z_{\xi\eta} = \frac{1}{4(\xi - \eta)}(z_\xi - z_\eta)$$

which is the required canonical form of the given equation.

Example 5.15. *Reduce the PDE*

$$r + x^2 t = 0$$

to a canonical form.

Solution. Here

$$R = 1, \quad S = 0, \quad T = x^2.$$

Step I (Classification): The discriminant of the given equation is

$$\triangle = S^2 - 4RT = -4x^2 < 0.$$

Therefore, the given equation is elliptic for $x \neq 0$. It may be noted that for $x = 0$, the given equation is parabolic and reduces to $r = 0$, which is already in canonical form.

Step II (Characteristics): The characteristic equations are

$$\frac{dy}{dx} = \frac{S \pm \sqrt{\triangle}}{2R} = \frac{\pm 2ix}{2} = \pm ix$$

which become

$$dy \pm ixdx = 0.$$

On integrating these equations, we get

$$y + i\frac{x^2}{2} = c_1 \quad \text{and} \quad y - i\frac{x^2}{2} = c_2$$

which are the characteristic curves of the given PDE.

Step III (Canonical Form): The canonical coordinates are

$$\xi = y \quad \text{and} \quad \eta = \frac{x^2}{2}. \tag{5.46}$$

Thus, we have

$$\xi_x = 0, \; \xi_y = 1, \; \xi_{xx} = \xi_{xy} = \xi_{yy} = 0$$

$$\eta_x = x, \ \eta_y = 0, \ \eta_{xx} = 1, \ \eta_{xy} = \eta_{yy} = 0.$$

Using the above, Eq. (5.22) becomes

$$r = x^2 z_{\eta\eta} + z_\eta$$
$$t = z_{\xi\xi}.$$

Putting these values in the given PDE, we get

$$x^2 z_{\eta\eta} + z_\eta + x^2 z_{\xi\xi} = 0$$

or

$$z_{\xi\xi} + z_{\eta\eta} = -\frac{1}{x^2} z_\eta.$$

Using (5.46), we have $x^2 = 2\eta$ and hence the above equation reduces to

$$z_{\xi\xi} + z_{\eta\eta} = -\frac{1}{2\eta} z_\eta$$

which is the required canonical form of the given equation.

Example 5.16. *Reduce the PDE*

$$x^2 r - 2xys + y^2 t + xp + yq = 0$$

to a canonical form.

Solution. *Here*

$$R = x^2, \ S = -2xy, \ T = y^2.$$

Step I (Classification): The discriminant of the given equation is

$$\triangle = S^2 - 4RT = 4x^2y^2 - 4x^2y^2 = 0.$$

Therefore, the given equation is parabolic.

Step II (Characteristics): The characteristic equation is

$$\frac{dy}{dx} = \frac{S}{2R} = -\frac{y}{x}$$

or

$$x\,dy + y\,dx = 0.$$

Integrating the above equation, we get

$$xy = c$$

which is the characteristic curve of the given PDE.

Step III (Canonical Form): The canonical coordinates are

$$\xi = xy \ \text{ and } \ \eta = x. \tag{5.47}$$

Notice that we have chosen $\eta = x$ in such a way that ξ and η are independent. Now, we have

$$\xi_x = y, \ \xi_y = x, \ \xi_{xx} = 0, \ \xi_{xy} = 1, \ \xi_{yy} = 0$$

$$\eta_x = 1, \ \eta_y = 0, \ \eta_{xx} = \eta_{xy} = \eta_{yy} = 0.$$

Using the above, Eq. (5.22) becomes

$$p = yz_\xi + z_\eta$$
$$q = xz_\xi$$
$$r = y^2 z_{\xi\xi} + 2yz_{\xi\eta} + z_{\eta\eta}$$
$$t = x^2 z_{\xi\xi}$$
$$s = xyz_{\xi\xi} + xz_{\xi\eta} + z_\xi.$$

Putting these values in the given PDE, we get

$$x^2(y^2 z_{\xi\xi} + 2yz_{\xi\eta} + z_{\eta\eta}) - 2xy(xyz_{\xi\xi} + xz_{\xi\eta} + z_\xi) + y^2 x^2 z_{\xi\xi} + x(yz_\xi + z_\eta) + yxz_\xi = 0$$

or

$$xz_{\eta\eta} + z_\eta = 0.$$

Using $x = \eta$ (due to (5.47)), the above equation reduces to

$$z_{\eta\eta} = -\frac{1}{\eta} z_\eta$$

which is the required canonical form of the given equation.

Example 5.17. *Reduce the PDE*

$$r + xt = 0$$

to a canonical form.

Solution.   Here

$$R = 1, \quad S = 0, \quad T = x.$$

The discriminant of the given equation is

$$\triangle = S^2 - 4RT = -4x.$$

Therefore, the given equation is parabolic for $x = 0$, hyperbolic for $x < 0$, and elliptic for $x > 0$.

Case I: If $x = 0$, then the given equation reduces to $r = 0$, which is already in canonical form.

Case II: If $x < 0$, then characteristic equations become

$$\frac{dy}{dx} = \frac{S \pm \sqrt{\triangle}}{2R} = \frac{\pm 2\sqrt{-x}}{2} = \pm\sqrt{-x}.$$

On integrating, we obtain

$$\frac{3}{2}y - (-x)^{3/2} = c_1 \quad \text{and} \quad \frac{3}{2}y + (-x)^{3/2} = c_2.$$

Thus, the canonical coordinates are

$$\xi = \frac{3}{2}y - (-x)^{3/2} \quad \text{and} \quad \eta = \frac{3}{2}y + (-x)^{3/2}.$$

Using these coordinates, we can obtain the canonical form given by

$$z_{\xi\eta} = \frac{1}{6(\xi - \eta)}(z_\xi - z_\eta).$$

Case III: If $x > 0$, then characteristic equations become

$$\frac{dy}{dx} = \frac{S \pm \sqrt{\triangle}}{2R} = \frac{\pm 2i\sqrt{x}}{2} = \pm i\sqrt{x}.$$

On integrating, we obtain

$$\frac{3}{2}y - ix^{3/2} = c_1 \quad \text{and} \quad \frac{3}{2}y + ix^{3/2} = c_2.$$

Thus, the canonical coordinates are

$$\xi = \frac{3}{2}y \ \text{ and } \ \eta = x^{3/2}.$$

Using these coordinates, we can obtain the canonical form given by

$$z_{\xi\xi} + z_{\eta\eta} = -\frac{1}{3\eta}z_\eta.$$

5.2.2 General Solution

In general, it is not easy to obtain the general solution of a semilinear PDE. However, sometimes the canonical form of an equation is simple enough to be solved easily. In fact, the solution of a semilinear PDE and that of the corresponding canonical form share many exclusive qualitative properties. In this subsection, we explain the above-mentioned fact using several examples.

Example 5.18. *Find the general solution of the PDE*

$$y^2 r - 2xys + x^2 t = \frac{y^2}{x}p + \frac{x^2}{y}q.$$

Solution. *Here*

$$R = y^2, \ \ S = -2xy, \ \ T = x^2.$$

Step I (Classification): The discriminant of the given equation is

$$\triangle = S^2 - 4RT = 4x^2y^2 - 4x^2y^2 = 0.$$

Therefore, the given equation is parabolic.

Step II (Characteristics): The characteristic equation is

$$\frac{dy}{dx} = \frac{S}{2R} = -\frac{x}{y}$$

or

$$x dx + y dy = 0.$$

On integrating, we get

$$x^2 + y^2 = c$$

which is the characteristic curve of the given PDE.

Step III (Canonical Form): The canonical coordinates are

$$\xi = x^2 + y^2 \quad \text{and} \quad \eta = y^2.$$

Using these coordinates, the given equation can be reduced to its canonical form given by

$$z_{\eta\eta} = 0.$$

Step IV (General Solution): Integrating twice partially w.r.t. η, we obtain

$$z = \eta\phi(\xi) + \psi(\xi).$$

Putting the values of ξ and η in the above equation, we get

$$z = y^2\phi(x^2 + y^2) + \psi(x^2 + y^2)$$

which is the required general solution.

Example 5.19. *Find the general solution of the PDE*

$$xyr - (x^2 - y^2)s - xyt + py - qx = 2(x^2 - y^2), \ x \neq 0.$$

Solution. Here

$$R = xy, \quad S = -(x^2 - y^2), \quad T = -xy.$$

Step I (Classification): The discriminant of the given equation is

$$\triangle = S^2 - 4RT = (x^2 - y^2)^2 + 4x^2y^2 = (x^2 + y^2)^2 > 0 \quad \text{as } x \neq 0$$

Therefore, the given equation is hyperbolic.

Step II (Characteristics): The characteristic equations are

$$\frac{dy}{dx} = \frac{S \pm \sqrt{\triangle}}{2R} = \frac{-(x^2 - y^2) \pm (x^2 + y^2)}{2xy}$$

which become

$$\frac{dy}{dx} = \frac{y}{x} \quad \text{and} \quad \frac{dy}{dx} = -\frac{x}{y}$$

or

$$xdy - ydx = 0 \quad \text{and} \quad xdx + ydy = 0.$$

Integrating these equations, we get

$$\frac{y}{x} = c_1 \quad \text{and} \quad x^2 + y^2 = c_2$$

which are the characteristic curves of the given PDE.

Step III (Canonical Form): The canonical coordinates are

$$\xi = \frac{y}{x} \quad \text{and} \quad \eta = x^2 + y^2.$$

Using these coordinates, the given equation can be reduced to its canonical form given by

$$z_{\xi\eta} = \frac{\xi^2 - 1}{(\xi^2 + 1)^2}.$$

Step IV (General Solution): Integrating the above equation partially w.r.t. ξ, we obtain

$$z_\eta = \int \frac{\xi^2 - 1}{(\xi^2 + 1)^2} d\xi + \phi(\eta)$$

$$= \int \frac{1 - \frac{1}{\xi^2}}{\left(\xi + \frac{1}{\xi}\right)^2} d\xi + \phi(\eta)$$

$$= \int \frac{d\tau}{\tau^2} + \phi(\eta), \quad \left(\text{Putting } \xi + \frac{1}{\xi} = \tau\right)$$

$$= -\frac{1}{\tau} + \phi(\eta) = -\frac{\xi}{\xi^2 + 1} + \phi(\eta).$$

Integrating now partially w.r.t. η, we get

$$z = -\frac{\xi\eta}{\xi^2 + 1} + \int \phi(\eta) d\eta + \psi(\xi) = -\frac{\xi\eta}{\xi^2 + 1} + \theta(\eta) + \psi(\xi)$$

where $\theta(\eta) = \int \phi(\eta) d\eta$. Putting the values of ξ and η in the above equation, we get

$$z = \theta(x^2 + y^2) + \psi\left(\frac{y}{x}\right) - xy$$

which is the required general solution.

Example 5.20. *Find the general solution of the PDE*

$$yr + 3ys + 3p = 0, \ y \neq 0.$$

Solution.   *Here*

$$R = y, \quad S = 3y, \quad T = 0.$$

Step I (Classification): The discriminant of the given equation is

$$\triangle = S^2 - 4RT = 9y^2 > 0 \quad \text{as } y \neq 0.$$

Therefore, the given equation is hyperbolic.

Step II (Characteristics): The characteristic equation is

$$\frac{dy}{dx} = \frac{S \pm \sqrt{\triangle}}{2R} = \frac{3y \pm 3y}{2y} = 3, 0$$

which becomes

$$dy = 0 \quad \text{and} \quad dy - 3dx = 0.$$

Integrating these equations, we get

$$y = c_1 \quad \text{and} \quad y - 3x = c_2$$

which are the characteristic curves of the given PDE.

Step III (Canonical Form): The canonical coordinates are

$$\xi = y \quad \text{and} \quad \eta = y - 3x.$$

Using these coordinates, the given equation can be reduced to its canonical form given by

$$z_{\xi\eta} + \frac{1}{\xi} z_\eta = 0.$$

Step IV (General Solution): Setting $z_\eta = w$, the above equation reduces to

$$\frac{\partial w}{\partial \xi} + \frac{w}{\xi} = 0$$

or

$$\frac{\partial w}{w} + \frac{\partial \xi}{\xi} = 0.$$

Integrating partially, we obtain

$$w\xi = \phi(\eta)$$

or

$$z_\eta = \frac{1}{\xi}\phi(\eta)$$

which, on integrating w.r.t. η, gives rise to

$$z = \frac{1}{\xi}\int \phi(\eta)d\eta + \psi(\xi)$$

or

$$z = \frac{1}{\xi}\theta(\eta) + \psi(\xi)$$

where $\theta(\eta) = \int \phi(\eta)d\eta$. Putting the values of ξ and η in the above equation, we get

$$z = \frac{1}{y}\theta(y - 3x) + \psi(y)$$

which is the required general solution.

Example 5.21. *Find the general solution of the PDE*

$$x^2 r - 2xys + y^2 t + xp + yq = 0.$$

Solution. Here

$$R = x^2, \quad S = -2xy, \quad T = y^2.$$

Step I (Classification): The discriminant of the given equation is

$$\triangle = S^2 - 4RT = 4x^2 y^2 - 4x^2 y^2 = 0$$

Therefore, the given equation is parabolic.

Step II (Characteristics): The characteristic equation is

$$\frac{dy}{dx} = \frac{S}{2R} = -\frac{y}{x}$$

or

$$xdy + ydx = 0$$

which, on integrating, gives rise to

$$xy = c$$

which is the characteristic curve of the given PDE.

Step III (Canonical Form): The canonical coordinates are

$$\xi = xy \text{ and } \eta = x.$$

Using these coordinates, the given equation can be reduced to its canonical form given by

$$z_{\eta\eta} + \frac{1}{\eta} z_\eta = 0.$$

Step IV (General Solution): Setting $z_\eta = w$, the above equation reduces to

$$\frac{\partial w}{\partial \eta} + \frac{w}{\eta} = 0$$

or

$$\frac{\partial w}{w} + \frac{\partial \eta}{\eta} = 0.$$

Integrating partially, we obtain

$$w\eta = \phi(\xi)$$

or

$$z_\eta = \frac{1}{\eta} \phi(\xi)$$

which, on integrating w.r.t. η, gives rise to

$$z = \phi(\xi) \log \eta + \psi(\xi).$$

Putting the values of ξ and η in the above equation, we get

$$z = \phi(xy) \log x + \psi(xy)$$

which is the required general solution.

5.2.3 Cauchy Problem

In Chapter 4, we have discussed Cauchy problems for first-order PDEs. Now, let us visit Cauchy problems for second-order semilinear PDEs. Recall that the normal derivative of a function along a given curve or surface is the directional derivative taken in the direction of the normal to the curve or surface. The normal derivative of a function $z = z(x, y)$ is usually denoted by $\partial z / \partial \mathbf{n}$, whereas $\mathbf{n}$ is the normal vector to the given curve or surface. Thus far, we have $\dfrac{\partial z}{\partial \mathbf{n}} = \nabla z \cdot \mathbf{n}$.

Let Γ be a space curve represented by its parametric equations

$$x_0 = x_0(s), \ y_0 = y_0(s), \ z_0 = z_0(s)$$

where x_0, y_0, z_0 are continuously differentiable functions on an interval $I \subset \mathbb{R}$. Then the Cauchy problem for a semilinear equation (5.16) is the problem to determine in a certain neighbourhood N of the plane curve $\Gamma' \ : \ x = x_0(s), y = y_0(s)$ (that is, projection of Γ to xy-plane), an integral surface $z = z(x, y)$ of Eq. (5.16) containing the curve Γ and having a prescribed value $z^\perp(s)$ of the normal derivative along the curve Γ', *that is,*

$$z(x_0(s), y_0(s)) = z_0(s) \quad \forall \, s \in I \tag{5.48}$$

$$\frac{\partial z}{\partial \mathbf{n}}(x_0(s), y_0(s)) = z^\perp(s) \quad \forall \, s \in I. \tag{5.49}$$

The conditions (5.48) and (5.49) are called the **initial conditions** or **Cauchy conditions**. The functions $z_0(s)$ and $z^\perp(s)$ are called the **initial data** or **Cauchy data**. The curve Γ is called **initial data curve**, whereas Γ' is called **initial curve**. Geometrically, the Cauchy condition (5.49) represents a tangent plane to the integral surface along the curve Γ.

To appreciate the Cauchy problem better, we utilise the following example, in which the initial curve is x-axis. Consequently, y-axis being the normal to the initial curve provides $\partial z / \partial y$ as normal derivative.

Example 5.22. *Solve the Cauchy problem*

$$\frac{\partial^2 z}{\partial x^2} = \frac{\partial^2 z}{\partial y^2}$$

$$z(x, 0) = f(x) \tag{5.50}$$

$$\frac{\partial z}{\partial y}(x, 0) = g(x). \tag{5.51}$$

Solution. *Rewrite the given equation*

$$r - t = 0.$$

$$Here \ \ R = 1, \ \ S = 0, \ \ T = -1.$$

Hence, we have

$$\triangle = S^2 - 4RT = 4 > 0.$$

Therefore, the given equation is hyperbolic. The characteristic equation is

$$\frac{dy}{dx} = \frac{S \pm \sqrt{\triangle}}{2R} = \frac{\pm 2}{2} = \pm 1$$

or

$$dx \pm dy = 0$$

On integrating, we get the characteristic curves given by

$$x + y = c_1 \quad \text{and} \quad x - y = c_2.$$

Hence, the canonical coordinates are

$$\xi = x + y \quad \text{and} \quad \eta = x - y.$$

Using these coordinates, the given equation can be reduced to its canonical form given by

$$z_{\xi\eta} = 0.$$

Solution of the above equation $z = \phi(\xi) + \psi(\eta)$, *which in terms of original variables, becomes*

$$z(x, y) = \phi(x + y) + \psi(x - y). \tag{5.52}$$

Using IC (5.50) and IC (5.51) in (5.52), we get

$$\phi(x) + \psi(x) = f(x) \tag{5.53}$$

$$\phi'(x) - \psi'(x) = g(x). \tag{5.54}$$

On integrating (5.54), we get

$$\phi(x) - \psi(x) = \int_{x_0}^{x} g(t)dt \tag{5.55}$$

where x_0 *is an arbitrary constant. From (5.53) and (5.55), we obtain*

$$\phi(x) = \frac{1}{2}f(x) + \frac{1}{2}\int_{x_0}^{x} g(t)dt \tag{5.56}$$

$$\psi(x) = \frac{1}{2}f(x) - \frac{1}{2}\int_{x_0}^{x} g(t)dt. \tag{5.57}$$

Using (5.56) and (5.57), Eq. (5.52) becomes

$$z(x, y) = \frac{1}{2}f(x + y) + \frac{1}{2}\int_{x_0}^{x+y} g(t)dt + \frac{1}{2}f(x - y) - \frac{1}{2}\int_{x_0}^{x-y} g(t)dt$$

or

$$z(x, y) = \frac{1}{2}[f(x + y) + f(x - y)] + \frac{1}{2}\int_{x-y}^{x+y} g(t)dt$$

which is the required particular solution of the given equation.

5.2.4 Semilinear Equations in n Independent Variables

As usual, we write $x = (x_1, x_2, \ldots, x_n) \in \mathbb{R}^n$. A second-order semilinear PDE in n independent variables $x_1, x_2, \ldots, x_n$ can be written as

$$\sum_{i,j=1}^{n} a_{ij}(x) \frac{\partial^2 u}{\partial x_i \partial x_j} = F(x, u, \nabla u) \tag{5.58}$$

where $a_{ij}(x) = a_{ji}(x)$, u is the dependent variable and $\nabla u = (u_{x_1}, u_{x_2}, \ldots, u_{x_n})$ is the gradient of u.

The classification of Eq. (5.58) is based on the symmetric matrix determined by the coefficients a_{ij} given by

$$A(x) = \begin{bmatrix} a_{11}(x) & a_{12}(x) & \cdots & a_{1n}(x) \\ a_{21}(x) & a_{22}(x) & \cdots & a_{2n}(x) \\ \vdots & \vdots & \cdots & \vdots \\ a_{n1}(x) & a_{n2}(x) & \cdots & a_{nn}(x) \end{bmatrix}$$

The eigenvalues of the matrix A are the roots of the equation

$$|A - \lambda I| = 0.$$

Eq. (5.58), at a point $\bar{x} \in \mathbb{R}^n$, is said to be

(i) elliptic if all the eigenvalues of $A(\bar{x})$ are non-zero and have the same sign.

(ii) hyperbolic if all the eigenvalues of $A(\bar{x})$ are non-zero and all except one have the same sign.

(iii) parabolic if any one of the eigenvalues of $A(\bar{x})$ is zero (or equivalently, the matrix $A(\bar{x})$ is singular, *that is,* $|A(\bar{x})| = 0$).

As earlier, we have briefly discussed the classification of semilinear equations in n variables. We skip deep discussions regarding the PDE of the form (5.58) including the characteristics and canonical forms, as these topics involve the concepts of linear algebra and hence are beyond the scope of the present book. For a detailed discussion, we refer the relevant portions contained in the books of [15, 18, 26, 33].

Example 5.23. *Classify the PDE*

$$u_{xx} + u_{yy} = u_{zz}.$$

Solution: Here

$$a_{11} = a_{22} = 1, a_{33} = -1, a_{12} = a_{21} = a_{23} = a_{32} = 0.$$

Hence, we have

$$A = \begin{bmatrix} 1 & 0 & 0 \\ 0 & 1 & 0 \\ 0 & 0 & -1 \end{bmatrix}.$$

Consider the equation

$$|A - \lambda I| = 0$$

or

$$\begin{vmatrix} 1-\lambda & 0 & 0 \\ 0 & 1-\lambda & 0 \\ 0 & 0 & -1-\lambda \end{vmatrix} = 0$$

or

$$(1-\lambda)^2(1+\lambda) = 0.$$

Thus, the eigenvalues of A are

$$\lambda_1 = \lambda_2 = 1, \lambda_3 = -1.$$

It follows that the given equation is hyperbolic.

5.3 Quasilinear PDEs: Monge's Method

In Chapter 4, we have discussed Lagrange's method to solve a quasilinear PDE of order one. In what follows, we shall discuss the general method for solving the quasilinear PDEs of order two, which remains a very important method. This method was initiated by a French mathematician *Gaspard Monge* (1746–1818), and hence it is known as **Monge's method**. Consider the most general quasilinear PDE of order two

$$Rr + Ss + Tt = V. \tag{5.59}$$

Here, the coefficients R, S, T, and V are functions of x, y, z, p, and q. Also R, S, and T do not vanish simultaneously.

In Chapter 2, we have discussed that a PDE of the form (5.59) can be derived by eliminating the arbitrary function ϕ from the relation

$$v = \phi(u) \tag{5.60}$$

where u is a known function of x, y, z, whereas v is a known function of x, y, z, p, q. The chief feature of Monge's method is to obtain one or two first integrals of the form (5.60).

Definition 5.4. *The two simultaneous equations*

$$Rdy^2 - Sdxdy + Tdx^2 = 0$$
$$Rdpdy + Tdqdx - Vdxdy = 0$$

are known as Monge's subsidiary equations of Eq. (5.59).

Definition 5.5. *A relation between x, y, z, p, q involving an arbitrary function, which satisfy Monge's subsidiary equations, is called an intermediate integral.*

The following result establishes the equivalence between a quasilinear PDE and its Monge's subsidiary equations.

Theorem 5.4. *Each intermediate integral corresponding to Monge's subsidiary equations of the quasilinear Eq. (5.59) also satisfies Eq. (5.59).*

Proof. Since p is a function of x and y, we have

$$dp = \frac{\partial p}{\partial x}dx + \frac{\partial p}{\partial y}dy = rdx + sdy$$

so that

$$r = \frac{dp - sdy}{dx}.$$

Similarly, as q is a function of x and y, we have

$$dq = \frac{\partial q}{\partial x}dx + \frac{\partial q}{\partial y}dy = sdx + tdy$$

so that

$$t = \frac{dq - sdx}{dy}.$$

Substituting these values of r and t in Eq. (5.59), we get

$$R\left[\frac{dp - sdy}{dx}\right] + Ss + T\left[\frac{dq - sdx}{dy}\right] = V$$

which, on rearranging, becomes

$$(Rdpdy + Tdqdx - Vdxdy) - s(Rdy^2 - Sdxdy + Tdx^2) = 0. \tag{5.61}$$

Since Eq. (5.61) holds for arbitrary values of s, we set the expressions in each bracket to zero so that we obtain Monge's subsidiary equations. It follows that any relation between x, y, z, p, q satisfying Monge's subsidiary equations must also satisfy Eq. (5.61) and hence Eq. (5.59). □

In the following lines, we present the procedure for finding the intermediate integrals and the general solution of Eq. (5.59).

Algorithm for Monge's Method: The first Monge's subsidiary equation $Rdy^2 - Sdxdy + Tdx^2 = 0$ being quadratic can be resolved into two equations

$$A_1 dy + B_1 dx = 0$$
$$A_2 dy + B_2 dx = 0$$

so that $Rdy^2 - Sdxdy + Tdx^2 \equiv (A_1 dy + B_1 dx)(A_2 dy + B_2 dx) = 0$. Combining each of the equations with second Monge's equation, we have two systems

$$\begin{cases} A_1 dy + B_1 dx = 0 \\ Rdpdy + Tdqdx - Vdxdy = 0 \end{cases} \text{ and } \begin{cases} A_2 dy + B_2 dx = 0 \\ Rdpdy + Tdqdx - Vdxdy = 0. \end{cases}$$

From the first system (combined if necessary with $dz = pdx + qdy$), suppose that we obtain two integrals $u(x, y, z) = c_1$ and $v(x, y, z, p, q) = c_2$, then a relation of the form $v = \phi(u)$ forms an intermediate integral. Similarly, the second system provides another intermediate integral $v' = \psi(u')$. Solving these two intermediate integrals for p and q in terms of x, y, z. Finally, we integrate the relation

$$dz = pdx + qdy$$

to obtain the required solution of Eq. (5.59).

Alternately, we may use only one intermediate integral. As this intermediate integral is a quasilinear PDE of order one, we use the Lagrange's method to obtain the required solution of Eq. (5.59), especially, if either of the following special cases arise:

(**i**) only one of the systems is integrable
(**ii**) the expression $Rdy^2 - Sdxdy + Tdx^2$ remains perfect square so that we have only one system
(**iii**) either $R = 0$ or $T = 0$

then we can obtain only one intermediate integral and hence we have to use the latter technique.

We adopt the following examples to demonstrate Monge's method.

Example 5.24. *Solve* $(r - s)x = (t - s)y$.

Solution. *The given equation can be written as* $xr - (x - y)s - yt = 0$. *Here*

$$R = x, S = -(x - y), T = -y, V = 0.$$

Step I (Monge's Subsidiary Equations): The Monge's subsidiary equations are

$$xdy^2 + (x - y)dxdy - ydx^2 = 0 \tag{5.62}$$

$$xdpdy - ydqdx = 0. \tag{5.63}$$

Equation (5.62) can be factored as

$$(xdy - ydx)(dy + dx) = 0$$

which resolves into two equations

$$xdy - ydx = 0 \qquad (5.64)$$

$$dy + dx = 0. \qquad (5.65)$$

Step II (Intermediate Integrals): On integrating Eq. (5.64), we get

$$\frac{y}{x} = c_1.$$

From Eq. (5.64), we have $xdy = ydx$. Using this, Eq. (5.63) reduces to

$$ydx(dp - dq) = 0$$

yielding thereby

$$dp - dq = 0 \text{ (as } y \neq 0 \text{ and } dx \neq 0).$$

Integrating, we get $p - q = c_2$. Therefore, the intermediate integral corresponding to Eq. (5.64) is

$$p - q = \phi\left(\frac{y}{x}\right). \qquad (5.66)$$

Now, on integrating Eq. (5.65), we get

$$x + y = c_3.$$

From Eq. (5.65), we have $dy = -dx$. Using this, Eq. (5.63) reduces to

$$-dx(xdp + ydq) = 0$$

or

$$xdp + ydq = 0 \text{ (as } dx \neq 0)$$

or

$$(xdp + pdx + ydq + qdy) - (pdx + qdy) = 0$$

or

$$d(xp) + d(yq) - dz = 0.$$

Integrating, we get $xp + yq - z = c_4$. Therefore, the intermediate integral corresponding to (5.65) is

$$xp + yq = z + \psi(x + y). \qquad (5.67)$$

Step III (General Solution): Solving (5.66) and (5.67) for p and q, we obtain

$$p = \frac{1}{x+y}\left[z + y\phi\left(\frac{y}{x}\right) + \psi(x+y)\right]$$

and

$$q = \frac{1}{x+y}\left[z - x\phi\left(\frac{y}{x}\right) + \psi(x+y)\right].$$

Putting these values in relation to $dz = p\,dx + q\,dy$, we get

$$dz = \frac{1}{x+y}\left[z + y\phi\left(\frac{y}{x}\right) + \psi(x+y)\right]dx + \frac{1}{x+y}\left[z - x\phi\left(\frac{y}{x}\right) + \psi(x+y)\right]dy$$

or

$$(x+y)dz - z(dx+dy) = \phi\left(\frac{y}{x}\right)(y\,dx - x\,dy) + \psi(x+y)(dx+dy).$$

Now, dividing the above equation by $(x+y)^2$, we get

$$\frac{(x+y)dz - z\,d(x+y)}{(x+y)^2} = \frac{x^2}{(x+y)^2}\phi\left(\frac{y}{x}\right)\frac{y\,dx - x\,dy}{x^2} + \frac{\psi(x+y)}{(x+y)^2}d(x+y)$$

or

$$d\left(\frac{z}{x+y}\right) = -\frac{\phi(y/x)}{(1+y/x)^2}d\left(\frac{y}{x}\right) + \frac{\psi(x+y)}{(x+y)^2}d(x+y).$$

On integrating, we obtain

$$\frac{z}{x+y} = \theta(y/x) + \eta(x+y)$$

or

$$z = (x+y)\left[\theta(y/x) + \eta(x+y)\right]$$

which is the required solution of the given equation.

Example 5.25. *Solve* $q(yq+z)r - p(2yq+z)s + yp^2t + p^2q = 0.$

Solution. *Here*

$$R = q(yq+z), S = -p(2yq+z), T = yp^2, V = -p^2q.$$

Step I (Monge's Subsidiary Equations): The Monge's subsidiary equations are

$$q(yq + z)dy^2 + p(2yq + z)dxdy + yp^2dx^2 = 0 \qquad (5.68)$$

$$q(yq + z)dpdy + yp^2dqdx + p^2qdxdy = 0. \qquad (5.69)$$

Equation (5.68) can be factored as

$$(qdy + pdx)[(yq + z)dy + ypdx] = 0$$

which resolves into two equations

$$qdy + pdx = 0 \qquad (5.70)$$

$$(yq + z)dy + ypdx = 0 \qquad (5.71)$$

Step II (Intermediate Integrals): Equation (5.70) is equivalent to $dz = 0$ which, on integration, gives rise to

$$z = c_1.$$

From Eq. (5.70), we have $qdy = -pdx$. Using this, Eq. (5.69) reduces to

$$(yq + z)dp - ypdq - pqdy = 0$$

or

$$(yq + z)dp - p(ydq + qdy) = 0$$

or

$$(yq + z)dp - pd(yq) = 0$$

or

$$(yq + z)dp - pd(yq + z) = 0 \quad (\text{as } dz = 0)$$

or

$$\frac{d(yq + z)}{yq + z} = \frac{dp}{p}.$$

Integrating, we get $yq + z = pc_2$. Therefore, the intermediate integral corresponding to Eq. (5.70) is

$$yq + z = p\phi(z). \qquad (5.72)$$

Equation (5.71) is equivalent to $ydz + zdy = 0$, we get

$$yz = c_3.$$

From Eq. (5.71), we have $(yq + z)dy = -ypdx$. Using this, Eq. (5.69) reduces to

$$-yqdp + ypdq + pqdy = 0$$

or

$$\frac{dq}{q} + \frac{dy}{y} = \frac{dp}{p}.$$

Integrating, we get $yq = c_4 p$. Therefore, the intermediate integral corresponding to (5.71) is

$$yq = p\psi(yz). \tag{5.73}$$

Step III (General Solution): Solving (5.72) and (5.73) for p and q, we obtain

$$p = \frac{z}{\phi(z) - \psi(yz)}$$

and

$$q = \frac{z\psi(yz)}{y\left(\phi(z) - \psi(yz)\right)}.$$

Putting these values in relation to $dz = pdx + qdy$, we get

$$dz = \frac{zdx}{\phi(z) - \psi(yz)} + \frac{z\psi(yz)dy}{y\left(\phi(z) - \psi(yz)\right)}$$

or

$$\left(\phi(z) - \psi(yz)\right) dz = zdx + \frac{z\psi(yz)dy}{y}$$

or

$$zdx = \phi(z)dz - \psi(yz)\left[\frac{ydz + zdy}{y}\right]$$

or

$$dx = \frac{\phi(z)}{z}dz - \frac{\psi(yz)}{yz}d(yz).$$

On integrating, we obtain

$$x = \theta(z) + \eta(yz)$$

which is the required solution of the given equation.

Example 5.26. *Determine the general solution of the PDE given in Example 5.25 by using only one intermediate integral.*

Solution. *From Step II of the solution of Example 5.25, an intermediate integral is*

$$yq + z = p\phi(z).$$

Rewrite the above equation as

$$p\phi(z) - yq = z$$

which is a quasilinear equation of order one. The Lagrange's subsidiary equations are

$$\frac{dx}{\phi(z)} = \frac{dy}{-y} = \frac{dz}{z}.$$

From the last two fractions, we get $\dfrac{dy}{y} + \dfrac{dz}{z} = 0.$ *On integrating, we get*

$$yz = c_1.$$

Also, using the first and third fractions, we have $dx = \dfrac{\phi(z)}{z}dz$ *which, on integrating, gives rise to*

$$x = \int \frac{\phi(z)}{z}dz + c_2 = \theta(z) + c_2.$$

Hence, the general solution of the given equation is

$$x = \theta(z) + \eta(yz).$$

The solution may also be obtained by using the second intermediate integral $yq = p\psi(yz)$. *The Lagrange's subsidiary equations are*

$$\frac{dx}{\psi(yz)} = \frac{dy}{-y} = \frac{dz}{0}.$$

The last fraction leads to $z = c_1$. *From the first two fractions, we may obtain* $x = -\int \dfrac{\psi(c_1 y)}{y}dy + c_2 = \eta(yz) + c_2.$ *Thus, we get the solution*

$$x = \eta(yz) + \theta(z)$$

as desired.

Example 5.27. *Solve* $q(1 + q)r - (p + q + 2pq)s + p(1 + p)t = 0.$

Solution.   *Here*
$$R = q(1 + q), S = -(p + q + 2pq), T = p(1 + p), V = 0.$$

Step I (Monge's Subsidiary Equations): The Monge's subsidiary equations are

$$q(1 + q)dy^2 + (p + q + 2pq)dxdy + p(1 + p)dx^2 = 0 \tag{5.74}$$

$$q(1 + q)dpdy + p(1 + p)dqdx = 0. \tag{5.75}$$

Equation (5.74) can be factored as

$$(pdx + qdy)[(1 + p)dx + (1 + q)dy] = 0$$

which resolves into two equations

$$pdx + qdy = 0 \tag{5.76}$$

$$(1 + p)dx + (1 + q)dy = 0 \tag{5.77}$$

Step II (Intermediate Integral): Equation (5.76) is equivalent to $dz = 0$ which, on integration, gives rise to

$$z = c_1.$$

From Eq. (5.76), we have $qdy = -pdx$. Using this, Eq. (5.75) reduces to

$$(-pdx)(1 + q)dp + p(1 + p)dqdx = 0$$

or

$$(1 + q)dp - (1 + p)dq = 0$$

or

$$\frac{dp}{1 + p} = \frac{dq}{1 + q}.$$

Integrating, we get $1 + p = c_2(1 + q)$. Therefore, one intermediate integral is

$$1 + p = (1 + q)\phi(z). \tag{5.78}$$

Step III (General Solution): Rewrite Eq. (5.78) as $p - \phi(z)q = \phi(z) - 1$, which is a quasilinear equation of first order. Its Lagrange's equations are

$$\frac{dx}{1} = \frac{dy}{-\phi(z)} = \frac{dz}{\phi(z) - 1}.$$

Using multipliers $1, 1, 1$, we get

$$\text{eachfraction} = \frac{dx + dy + dz}{0}$$

so that

$$dx + dy + dz = 0.$$

Integrating, we get

$$x + y + z = c_3.$$

Now, the last two fractions give

$$dy + \frac{\phi(z)}{\phi(z) - 1} dz = 0.$$

On integrating, we obtain

$$y + \int \frac{\phi(z)}{\phi(z) - 1} dz = c_4$$

or

$$y + \theta(z) = c_4.$$

Therefore, the general solution of the given equation is

$$y + \theta(z) = \eta(x + y + z).$$

Example 5.28. *Solve* $q^2 r - 2pqs + p^2 t = 0$.

Solution. *Here*

$$R = q^2, S = -2pq, T = p^2, V = 0.$$

Step I (Monge's Subsidiary Equations): The Monge's subsidiary equations are

$$q^2 dy^2 + 2pq\,dx\,dy + p^2 dx^2 = 0 \tag{5.79}$$

$$q^2 dp\,dy + p^2 dq\,dx = 0. \tag{5.80}$$

Equation (5.79) can be factored as

$$(qdy + pdx)^2 = 0$$

which yields that

$$pdx + qdy = 0 \tag{5.81}$$

Step II (Intermediate Integral): Equation (5.81) is equivalent to $dz = 0$ which, on integration, gives rise to

$$z = c_1.$$

From Eq. (5.81), we have $qdy = -pdx$. Using this, Eq. (5.80) reduces to

$$qdp(-pdx) + p^2 dqdx = 0$$

or

$$qdp - pdq = 0$$

or

$$\frac{dp}{p} = \frac{dq}{q}.$$

Integrating, we get $p = c_2 q$. Therefore, we obtain only one intermediate integral

$$p = q\phi(z). \tag{5.82}$$

Step III (General Solution): Rewrite Eq. (5.82) as $p - \phi(z)q = 0$, which is a quasilinear equation of first order. Its Lagrange's equations are

$$\frac{dx}{1} = \frac{dy}{-\phi(z)} = \frac{dz}{0}. \tag{5.83}$$

Taking the last fraction, we have $dz = 0$. Integrating it, we get

$$z = c_3. \tag{5.84}$$

Now, from the first two fractions in (5.83) and (5.84), we have

$$dy + \phi(c_3)dx = 0.$$

On integrating, we obtain

$$y + \phi(c_3)x = c_4$$

which, by using Eq. (5.84), becomes

$$y + x\phi(z) = c_4.$$

Therefore, the general solution of the given equation is

$$y + x\phi(z) = \psi(z).$$

Example 5.29. *Solve $qr - ps = p^3$.*

Solution. *Here*

$$R = q, S = -p, T = 0, V = p^3.$$

Step I (Monge's Subsidiary Equations): The Monge's subsidiary equations are

$$qdy^2 + pdxdy = 0$$
$$qdpdy - p^3dxdy = 0.$$

Since $dy \neq 0$, above equations become, respectively,

$$qdy + pdx = 0 \tag{5.85}$$

$$\text{and} \qquad qdp - p^3dx = 0. \tag{5.86}$$

Step II (Intermediate Integral): Equation (5.85) is equivalent to $dz = 0$ which, on integration, gives rise to

$$z = c_1.$$

From Eq. (5.85), we have $pdx = -qdy$. Using this, Eq. (5.86) reduces to

$$qdp + p^2qdy = 0$$

or

$$dp + p^2dy = 0 \quad (\text{as } q \neq 0)$$

or

$$\frac{dp}{p^2} + dy = 0.$$

Integrating, we get $-\dfrac{1}{p} + y = c_2$. Therefore, we get only one intermediate integral of the form

$$-\frac{1}{p} + y = \phi(z). \tag{5.87}$$

Step III (General Solution): Rewrite Eq. (5.87) as

$$\frac{\partial x}{\partial z} = y - \phi(z)$$

or

$$\partial x = [y - \phi(z)]\partial z.$$

On integrating partially, we obtain

$$x = yz - \int \phi(z)\partial z + \eta(y)$$

or

$$x = yz + \theta(z) + \eta(y)$$

which is the required solution of the given equation.

5.4 Certain Class of Non-linear Partial Differential Equations: Monge–Ampère–Type Equations

Monge's method is not merely applicable to the class of quasilinear equations, but it is also applicable to a wide class of non-linear equations of the form

$$Rr + Ss + Tt + U(rt - s^2) = V. \tag{5.88}$$

Here, the coefficients R, S, T, U, and V are functions of x, y, z, p, q, respectively. A French mathematician *André-Marie Ampère* (1775–1836) extended the idea of Monge's method for Eq. (5.88). Due to this, such types of non-linear equations are called **Monge–Ampère–type equations**. In particular, for $U = 0$, Eq. (5.88) reduces to quasilinear equation, which has already been discussed. In what follows, we shall consider $U \neq 0$. As already discussed in Chapter 2, a PDE of the form (5.88) can be derived by eliminating the arbitrary function ϕ from the relation

$$v = \phi(u) \tag{5.89}$$

where u and v both are known functions of x, y, z, p, q. In order to obtain an intermediate integral of the form (5.89) from Eq. (5.88), we use the following result.

Theorem 5.5. *Let λ_1 and λ_2 be roots of λ-quadratic equation*

$$U^2\lambda^2 + SU\lambda + TR + UV = 0. \tag{5.90}$$

Then, each of the systems

$$\begin{cases} \lambda_1 Udy + Tdx + Udp = 0 \\ Rdy + \lambda_2 Udx + Udq = 0 \end{cases} \quad and \quad \begin{cases} \lambda_2 Udy + Tdx + Udp = 0 \\ Rdy + \lambda_1 Udx + Udq = 0 \end{cases}$$

if integrable, yields an intermediate integral of Eq. (5.88).

Proof. Substituting $r = \dfrac{dp - sdy}{dx}$ and $t = \dfrac{dq - sdx}{dy}$ in Eq. (5.88), we obtain

$$R\left[\frac{dp - sdy}{dx}\right] + Ss + T\left[\frac{dq - sdx}{dy}\right] + U\left[\left(\frac{dp - sdy}{dx}\right)\left(\frac{dq - sdx}{dy}\right) - s^2\right] = V.$$

Multiplying both sides by $dxdy$ and rearranging the terms, we get

$$[Rdydp + Tdxdq + Udpdq - Vdxdy] - s[Rdy^2 - Sdxdy + Tdx^2 + Udxdp + Udydq] = 0.$$

Hence, the Monge's subsidiary equations for Eq. (5.88) are

$$L \equiv Rdy^2 - Sdxdy + Tdx^2 + Udxdp + Udydq = 0$$
$$M \equiv Rdydp + Tdxdq + Udpdq - Vdxdy = 0.$$

Clearly, $L = 0$ cannot be factorised due to the presence of the term $Udxdp + Udydq$. Consider $\lambda L + M = 0$, where λ is so chosen such that it can be resolved into two linear equations. Suppose that

$$\lambda L + M = (A_1 dy + B_1 dx + C_1 dp)(A_2 dy + B_2 dx + C_2 dq)$$

which, using the values of L and M, becomes

$$\lambda(Rdy^2 - Sdxdy + Tdx^2 + Udxdp + Udydq) + Rdydp + Tdxdq + Udpdq$$
$$-Vdxdy = (A_1 dy + B_1 dx + C_1 dp)(A_2 dy + B_2 dx + C_2 dq).$$

Comparing coefficients, we obtain

$$A_1 A_2 = \lambda R, \; A_1 B_2 + B_1 A_2 = -V - S\lambda, \; B_1 B_2 = \lambda T,$$

$$C_1 A_2 = R, \; C_1 B_2 = \lambda U, \; C_1 C_2 = U, \; C_2 A_1 = \lambda U, \; B_1 C_2 = T.$$

The first relation will be satisfied if we take

$$A_1 = \lambda \text{ and } A_2 = R.$$

With this choice, we have

$$C_1 = 1, C_2 = U, B_1 = \frac{T}{U}, B_2 = \lambda U.$$

Putting these values in the second relation, we get

$$\lambda^2 U + \frac{TR}{U} = -V - S\lambda.$$

or

$$U^2\lambda^2 + SU\lambda + TR + UV = 0.$$

Thus, λ satisfies Eq. (5.90). Given that λ_1 and λ_2 are roots of this equation. Therefore, equation $\lambda L + M = 0$ can be factorised as

$$(\lambda_1 Udy + Tdx + Udp)(Rdy + \lambda_1 Udx + Udq) = 0$$
$$(\lambda_2 Udy + Tdx + Udp)(Rdy + \lambda_2 Udx + Udq) = 0.$$

Each factor of the first equation combining with a factor of the second equation makes a system. Therefore, there are four systems.

The system

$$\begin{cases} \lambda_1 Udy + Tdx + Udp = 0 \\ \lambda_2 Udy + Tdx + Udp = 0 \end{cases}$$

implies

$$(\lambda_1 - \lambda_2)Udy = 0$$

and hence $Udy = 0$ if $\lambda_1 \neq \lambda_2$, which will not furnish any solution.

Similarly, the system

$$\begin{cases} Rdy + \lambda_1 Udx + Udq = 0 \\ Rdy + \lambda_2 Udx + Udq = 0 \end{cases}$$

also implies

$$(\lambda_1 - \lambda_2)Udy = 0.$$

Thus, only the following two systems to be considered

$$\begin{cases} \lambda_1 Udy + Tdx + Udp = 0 \\ Rdy + \lambda_2 Udx + Udq = 0 \end{cases}$$

and

$$\begin{cases} \lambda_2 Udy + Tdx + Udp = 0 \\ Rdy + \lambda_1 Udx + Udq = 0 \end{cases}$$

from which we can determine the intermediate integrals of Eq. (5.88). $\qquad\square$

Algorithm for Solving Monge–Ampère–Type Equations: Using Theorem 5.5, we format the intermediate integrals. Now, the process of further integration of these integrals can be distinguished into the following two cases:

Case I: Suppose that Eq. (5.88) has two intermediate integrals

$$v = \phi(u) \ \text{ and } \ v' = \psi(u').$$

Usually, u and u' contain p or q or both, so it is no longer possible to determine the values of p and q directly from intermediate integrals. We consider one of the involved functions, say 'ψ' as a constant function so that $v' = \psi(u') = a$. Thus, from the relations

$$v = \phi(u) \ \text{ and } \ v' = a$$

we can easily determine the values of p and q from these relations. Thereafter, we integrate the relation $dz = pdx + qdy$ to obtain a complete solution of Eq. (5.88). Clearly, this solution consists of one arbitrary function and two constants, namely, 'a' and the constant of integration. Further, from this complete solution, we can obtain the general solution using the technique discussed in Section 4.7.

Alternately, in order to obtain the general solution directly, we introduce two parameters $\alpha = u$ and $\beta = u'$ so that the intermediate integrals take the forms

$$v = \phi(\alpha) \ \text{ and } \ v' = \psi(\beta).$$

Using these four relations, we can determine the values of p and q in terms of x, y as well as the parameters. Then integrating the relation $dz = pdx + qdy$, we obtain a solution, which involves two arbitrary functions and two parameters, namely, α and β. Thus, this solution constitutes the general solution of Eq. (5.88) in parametric form.

Case II: Let we are able to obtain only one intermediate integral, say,

$$v = \phi(u).$$

In this case, with $u = a$ and $v = b$, we can obtain p and q to solve $dz = pdx + qdy$. This solution remains a complete solution containing three arbitrary constants, namely, 'a', 'b', and the constant of integration. On the other hand, to get a general solution, we consider the arbitrary function 'ϕ' to be linear so that the intermediate integral reduces to

$$v = mu + n$$

where m and n are constants. This intermediate integral being a quasilinear equation of order one can be solved by Lagrange's method and gives the general solution of Eq. (5.88).

Let us consider some examples to demonstrate the foregoing procedures of finding the complete as well as general solutions of Monge–Ampère–type equations.

Example 5.30.   *Solve $r + 4s + t + rt - s^2 = 2$.*

Solution.   *Here*

$$R = 1, S = 4, T = 1, U = 1, V = 2.$$

The λ-quadratic equation is

$$\lambda^2 + 4\lambda + 3 = 0.$$

Its roots are

$$\lambda_1 = -1 \ \ and \ \ \lambda_2 = -3.$$

The first system associated with the given equation is

$$\begin{cases} dp + dx - dy = 0 \\ dq + dy - 3dx = 0. \end{cases}$$

Integrating these equations, we get

$$p + x - y = c_1$$
$$q + y - 3x = c_2.$$

Hence, the intermediate integral corresponding to the first system is

$$q + y - 3x = \phi(p + x - y). \tag{5.91}$$

The second system associated with the given equation is

$$\begin{cases} dp + dx - 3dy = 0 \\ dq + dy - dx = 0. \end{cases}$$

Integrating these equations, we get

$$p + x - 3y = c_3$$
$$q + y - x = c_4.$$

Hence, the intermediate integral corresponding to the second system is

$$p + x - 3y = \psi(q + y - x). \tag{5.92}$$

Taking particular integral of intermediate integral (5.91), we have

$$q + y - 3x = a$$

where a is a constant. This yields that

$$q = a - y + 3x. \tag{5.93}$$

Using Eq. (5.93), (5.92) becomes

$$p + x - 3y = \psi(2x + a)$$

which gives rise to

$$p = 3y - x + \psi(2x + a). \tag{5.94}$$

Putting these values of p and q from Eqs (5.93) and (5.94) in dz = pdx + qdy, we obtain

$$\begin{aligned} dz &= [3y - x + \psi(2x + a)]dx + (a - y + 3x)dy \\ &= 3ydx - xdx + \psi(2x + a)dx + ady - ydy + 3xdy \\ &= 3d(xy) - (xdx + ydy) + \psi(2x + a)dx + ady. \end{aligned}$$

Integrating, we have

$$z = 3xy - \frac{1}{2}(x^2 + y^2) + \int \psi(2x + a)dx + ay + b$$

or

$$z = 3xy - \frac{1}{2}(x^2 + y^2) + \theta(2x + a) + ay + b$$

which is the complete solution of the given equation. To obtain the general solution, we take b = η(a), where η is an arbitrary function. The required general solution then obtained by eliminating a between the equations

$$z = 3xy - \frac{1}{2}(x^2 + y^2) + \int \psi(2x + a)dx + ay + \eta(a)$$

and

$$0 = \theta'(2x + a) + y + \eta'(a).$$

Example 5.31. *Find the general solution of the PDE given in Example 5.30 in parametric form.*

Solution.   The intermediate integrals of the given PDE are

$$q + y - 3x = \varphi(p + x - y) \tag{5.95}$$

and

$$p + x - 3y = \psi(q + y - x). \tag{5.96}$$

Eq. (5.95) is equivalent to

$$p + x - y = \varphi(q + y - 3x). \tag{5.97}$$

Define the parameters α and β as

$$\alpha = q + y - 3x \tag{5.98}$$

$$\beta = q + y - x. \tag{5.99}$$

Using the above, Eqs (5.97) and (5.96) reduce to

$$p + x - y = \varphi(\alpha) \tag{5.100}$$

$$p + x - 3y = \psi(\beta). \tag{5.101}$$

Solving (5.98) and (5.99) for x, we have

$$x = \frac{\beta - \alpha}{2} \tag{5.102}$$

so that

$$dx = \frac{1}{2}[d\beta - d\alpha]. \tag{5.103}$$

Solving (5.100) and (5.101) for y, we have

$$y = \frac{\varphi(\alpha) - \psi(\beta)}{2} \tag{5.104}$$

so that

$$dy = \frac{1}{2}\left[\varphi'(\alpha)d\alpha - \psi'(\beta)d\beta\right]. \tag{5.105}$$

From Eqs (5.100) and (5.99), we get, respectively,

$$p = y - x + \varphi(\alpha)$$
$$q = x - y + \beta.$$

Putting these values in $dz = pdx + qdy$, we get

$$dz = [y - x + \varphi(\alpha)]dx + [x - y + \beta]dy$$
$$= d(xy) - xdx - ydy + \varphi(\alpha)dx + \beta dy,$$

which, by using (5.103) and (5.105), becomes

$$dz = d(xy) - xdx - ydy + \varphi(\alpha).\frac{1}{2}(d\beta - d\alpha) + \beta.\frac{1}{2}\left(\varphi'(\alpha)d\alpha - \psi'(\beta)d\beta\right)$$

$$= d(xy) - xdx - ydy + \frac{1}{2}\left[\varphi(\alpha)d\beta + \beta\varphi'(\alpha)d\alpha\right] - \frac{1}{2}\varphi(\alpha)d\alpha - \frac{1}{2}\beta\psi'(\beta)d\beta$$

or

$$2dz = 2d(xy) - 2xdx - 2ydy + d[\beta\varphi(\alpha)] - \varphi(\alpha)d\alpha - \beta\psi'(\beta)d\beta.$$

On integrating, we get

$$2z = 2xy - x^2 - y^2 + \beta\varphi(\alpha) - \int \varphi(\alpha)d\alpha - \int \beta\psi'(\beta)d\beta$$

$$= 2xy - x^2 - y^2 + \beta\varphi(\alpha) - \int \varphi(\alpha)d\alpha - \left[\beta\psi(\beta) - \int 1.\psi(\beta)d\beta\right]$$

or

$$2z = 2xy - x^2 - y^2 + \beta[\varphi(\alpha) - \psi(\beta)] - \int \varphi(\alpha)d\alpha + \int \psi(\beta)d\beta. \qquad (5.106)$$

Let

$$\int \varphi(\alpha)d\alpha = \theta(\alpha) \text{ and } \int \psi(\beta)d\beta = \eta(\beta) \qquad (5.107)$$

so that

$$\varphi(\alpha) = \theta^{'}(\alpha) \text{ and } \psi(\beta) = \eta^{'}(\beta). \qquad (5.108)$$

Using Eqs (5.107) and (5.108), Eqs (5.102), (5.104), and (5.106) become

$$2x = \beta - \alpha$$
$$2y = \theta'(\alpha) - \eta'(\beta)$$
$$2z = 2xy - x^2 - y^2 + \beta[\theta'(\alpha) - \eta'(\beta)] - \theta(\alpha) + \eta(\beta).$$

These three equations constitute the general solution in parametric form, wherein α and β are parameters and θ and η are arbitrary functions.

Example 5.32. *Obtain the complete solution of the PDE*

$$2s + rt - s^2 = 1.$$

Solution. *Here*

$$R = 0, S = 2, T = 0, U = 1, V = 1.$$

The λ-quadratic equation is

$$\lambda^2 + 2\lambda + 1 = 0.$$

Its roots are

$$\lambda_1 = \lambda_2 = -1.$$

The system associated with the given equation is

$$\begin{cases} -dy + dp = 0 \\ -dx + dq = 0. \end{cases}$$

Integrating these equations, we get

$$-y + p = c_1 \Rightarrow p = y + c_1$$
$$-x + q = c_2 \Rightarrow q = x + c_2.$$

Putting these values of p and q in $dz = pdx + qdy$, we obtain

$$dz = (y + c_1)dx + (x + c_2)dy$$

or

$$dz = ydx + xdy + c_1dx + c_2dy.$$

Integrating, we have

$$z = xy + c_1x + c_2y + c_3$$

which is the required complete solution of the given equation.

Example 5.33. *Determine the general solution of the PDE given in Example 5.32.*

Solution. *In the solution of Example 5.32, we have*

$$-y + p = c_1 \quad and \quad -x + q = c_2.$$

Consider the relation

$$-y + p = m(-x + q) + n$$

where m and n are constants. Then the above equation becomes

$$p - mq = n + y - mx$$

which is a quasilinear equation of order one. Hence, the Lagrange's subsidiary equations are

$$\frac{dx}{1} = \frac{dy}{-m} = \frac{dz}{n + y - mx}.$$

From the first two ratios, we have

$$dy + mdx = 0$$

which, on integrating, gives

$$y + mx = a. \tag{5.109}$$

Taking the first and last ratios and using (5.109), we get

$$dz = [n + (a - mx) - mx]dx$$

or

$$dz = (n + a)dx - 2mxdx.$$

Integrating, we get

$$z = (n + a)x - mx^2 + b$$

which by using Eq. (5.109) becomes

$$z = (n + y + mx)x - mx^2 + b$$

or

$$z = nx + xy + b. \tag{5.110}$$

Using (5.109) and (5.110), we get

$$z = nx + xy + \phi(y + mx)$$

which is the general solution of the given PDE.

5.5 Boundary Value Problems in Homogeneous Linear PDEs: Fourier Method

In this section, we discuss the technique for finding the particular solution of boundary value problems (BVPs) in homogeneous linear PDEs. Before this discussion, we study the half range Fourier sine and cosine series, which is crucial to obtain the solution of a BVP.

5.5.1 Half Range Fourier Series

A periodic function satisfying certain conditions defined on an interval of length $2l$ with the origin as its midpoint, *that is,* $(-l, l)$ can be represented as an infinite series containing the trigonometric terms of the form

$$\frac{a_0}{2} + \sum_{n=1}^{\infty} \left(a_n \cos \frac{n\pi x}{l} + b_n \sin \frac{n\pi x}{l} \right).$$

Such a series is called **Fourier series** of the given function. Here, the coefficients a_0, a_n, and b_n are called the **Fourier coefficients**. However, a half range Fourier sine or cosine series is a special form of Fourier series in which only sine terms or only cosine terms are presented, respectively. Half range Fourier series are defined in the interval $(0, l)$, which is half of the interval $(-l, l)$, thus accounting for name 'half range'. These series are applicable to find explicit representations of solutions of the boundary value problems related to homogeneous linear PDEs.

The concept of the Fourier series involves the idea of convergence as well as piecewise continuity, which is beyond the scope of this book. The complete description of Fourier series will hopefully be undertaken later in one of the courses such as real analysis, integral transforms, mathematical methods, or applied mathematics. However, for the detailed study of Fourier series, one can consult the classical book of Churchill [31]. For the sake of being self-contained, we briefly discuss the concept of half range Fourier sine and cosine series.

Fourier Cosine Series: *Half range Fourier cosine series*, or simply, *Fourier cosine series* of $f(x)$ in the interval $(0, l)$ is defined as

$$f(x) = \frac{a_0}{2} + \sum_{n=1}^{\infty} a_n \cos \frac{n\pi x}{l}$$

where Fourier coefficients a_0 and a_n, $n = 1, 2, 3, \ldots$ are determined by the formulae

$$a_0 = \frac{2}{l} \int_0^l f(x) dx$$

$$a_n = \frac{2}{l} \int_0^l f(x) \cos \frac{n\pi x}{l} dx.$$

In particular, taking $l = \pi$, we obtain half range Fourier cosine series of $f(x)$ in the interval $(0, \pi)$ given by

$$f(x) = \frac{a_0}{2} + \sum_{n=1}^{\infty} a_n \cos nx$$

so that

$$a_0 = \frac{2}{\pi} \int_0^{\pi} f(x) dx$$

$$a_n = \frac{2}{\pi} \int_0^{\pi} f(x) \cos nx dx.$$

Fourier Sine Series: *Half range Fourier sine series*, or simply, *Fourier sine series* of $f(x)$ in the interval $(0, l)$ is defined as

$$f(x) = \sum_{n=1}^{\infty} b_n \sin \frac{n\pi x}{l}$$

where Fourier coefficients b_n, $n = 1, 2, 3, \ldots$ are determined by the formula

$$b_n = \frac{2}{l} \int_0^l f(x) \sin \frac{n\pi x}{l} dx.$$

In particular, taking $l = \pi$, we obtain half range Fourier sine series of $f(x)$ in the interval $(0, \pi)$ given by

$$f(x) = \sum_{n=1}^{\infty} b_n \sin nx$$

so that

$$b_n = \frac{2}{\pi} \int_0^{\pi} f(x) \sin nx dx.$$

Example 5.34. *Expand the function $f(x) = x \sin x$, $0 < x < \pi$ in half range cosine series. Also, deduce*

$$\frac{1}{1.3} - \frac{1}{3.5} + \frac{1}{5.7} - \cdots = \frac{\pi - 2}{4}.$$

Solution. Let

$$x \sin x = \frac{a_0}{2} + \sum_{n=1}^{\infty} a_n \cos nx. \tag{5.111}$$

Here, we have

$$a_0 = \frac{2}{\pi} \int_0^{\pi} f(x) dx = \frac{2}{\pi} \int_0^{\pi} x \sin x dx$$

$$= \frac{2}{\pi} \left[x(-\cos x) - 1.(-\sin x) \right]_0^{\pi} = \frac{2}{\pi} [-\pi \cos \pi] = 2$$

$$a_n = \frac{2}{\pi} \int_0^{\pi} f(x) \cos nx dx = \frac{2}{\pi} \int_0^{\pi} x \sin x \cos nx dx$$

$$= \frac{1}{\pi} \int_0^{\pi} x(2 \cos nx \sin x) dx = \frac{1}{\pi} \int_0^{\pi} x[\sin(n+1)x + \sin(n-1)x] dx$$

$$= \frac{1}{\pi} \left[x \left\{ -\frac{\cos(n+1)x}{n+1} + \frac{\cos(n-1)x}{n-1} \right\} - 1. \left\{ -\frac{\sin(n+1)x}{(n+1)^2} - \frac{\sin(n-1)x}{(n-1)^2} \right\} \right]_0^{\pi}$$

$$= \frac{1}{\pi} \left[\pi \left\{ -\frac{\cos(n+1)\pi}{n+1} + \frac{\cos(n-1)\pi}{n-1} \right\} \right] = -\frac{(-1)^{n+1}}{n+1} + \frac{(-1)^{n-1}}{n-1}$$

$$= (-1)^n \left[\frac{1}{n+1} - \frac{1}{n-1} \right] = -\frac{2(-1)^n}{(n-1)(n+1)}, \quad n \neq 1$$

For n = 1, we have

$$a_1 = \frac{2}{\pi} \int_0^\pi f(x) \cos x \, dx = \frac{2}{\pi} \int_0^\pi x \sin x \cos x \, dx = \frac{1}{\pi} \int_0^\pi x \sin 2x \, dx$$

$$= \frac{1}{\pi} \left[x \left\{ -\frac{\cos 2x}{2} \right\} - 1 \cdot \left\{ -\frac{\sin 2x}{2^2} \right\} \right]_0^\pi$$

$$= \frac{1}{\pi} \left[-\frac{\pi \cos 2\pi}{2} \right] = -\frac{1}{2}$$

Putting the values of a_0, a_1, and a_n in Eq. (5.111), we get

$$x \sin x = \frac{2}{2} - \frac{1}{2} \cos x - 2 \sum_{n=2}^{\infty} \frac{(-1)^n}{(n-1)(n+1)} \cos nx$$

or

$$x \sin x = 1 - \frac{1}{2} \cos x - 2 \left(\frac{\cos 2x}{1.3} - \frac{\cos 3x}{2.4} + \frac{\cos 4x}{3.5} - \cdots \right)$$

which is the required Fourier cosine series of $f(x) = x \sin x$. Putting $x = \frac{\pi}{2}$ in the above series, we get

$$\frac{\pi}{2} = 1 - 2 \left(-\frac{1}{1.3} + \frac{1}{3.5} - \frac{1}{5.7} + \cdots \right)$$

which implies that

$$\frac{1}{1.3} - \frac{1}{3.5} + \frac{1}{5.7} - \cdots = \frac{\pi - 2}{4}.$$

Example 5.35. *Find the half range Fourier cosine and sine series of the function $f(x) = x^2$, $0 < x < l$.*

Solution. Let

$$x^2 = \frac{a_0}{2} + \sum_{n=1}^{\infty} a_n \cos \frac{n\pi x}{l}. \tag{5.112}$$

Here, we have

$$a_0 = \frac{2}{l} \int_0^l f(x)dx$$

$$= \frac{2}{l} \int_0^l x^2 dx$$

$$= \frac{2}{3} l^2$$

$$a_n = \frac{2}{l} \int_0^l f(x) \cos \frac{n\pi x}{l} dx$$

$$= \frac{2}{l} \int_0^l x^2 \cos \frac{n\pi x}{l} dx$$

$$= \frac{4l^2(-1)^n}{n^2 \pi^2}.$$

Putting the values of a_0 and a_n in Eq. (5.112), we get

$$x^2 = \frac{l^2}{3} + \frac{4l^2}{\pi^2} \sum_{n=1}^{\infty} \frac{(-1)^n}{n^2} \cos \frac{n\pi x}{l}$$

or

$$x^2 = \frac{l^2}{3} - \frac{4l^2}{\pi^2} \left[\frac{\cos(\pi x/l)}{1^2} - \frac{\cos(2\pi x/l)}{2^2} + \frac{\cos(3\pi x/l)}{3^2} - \cdots \right]$$

which is the Fourier cosine series of $f(x) = x^2$ as desired. Let

$$x^2 = \sum_{n=1}^{\infty} b_n \sin \frac{n\pi x}{l}. \tag{5.113}$$

Here, we have

$$b_n = \frac{2}{l} \int_0^l f(x) \sin \frac{n\pi x}{l} dx$$

$$= \frac{2}{l} \int_0^l x^2 \sin \frac{n\pi x}{l} dx$$

$$= \frac{2l^2(-1)^{n+1}}{n\pi} + \frac{4l^2}{n^3 \pi^3} \left[(-1)^n - 1 \right].$$

Putting the values of b_n in Eq. (5.113), we get

$$x^2 = \sum_{n=1}^{\infty} \left[\frac{2l^2(-1)^{n+1}}{n\pi} + \frac{4l^2}{n^3\pi^3} \left((-1)^n - 1\right) \right] \sin \frac{n\pi x}{l}$$

$$= \frac{2l^2}{\pi} \sum_{n=1}^{\infty} \frac{(-1)^{n+1}}{n} \sin \frac{n\pi x}{l} + \frac{4l^2}{\pi^3} \sum_{n=1}^{\infty} \frac{(-1)^n - 1}{n^3} \sin \frac{n\pi x}{l}$$

or

$$x^2 = \frac{2l^2}{\pi} \left[\frac{\sin(\pi x/l)}{1} - \frac{\sin(2\pi x/l)}{2} + \frac{\sin(3\pi x/l)}{3} - \cdots \right]$$

$$- \frac{8l^2}{\pi^3} \left[\frac{\sin(\pi x/l)}{1^3} + \frac{\sin(3\pi x/l)}{3^3} + \frac{\sin(5\pi x/l)}{5^3} + \cdots \right]$$

which is the Fourier sine series of $f(x) = x^2$ as desired.

5.5.2 Fourier Method

The Fourier method is very effectively used to solve BVP for a second-order homogeneous linear PDE. This method was investigated by the French mathematician and scientist *Jean Baptiste Joseph Fourier* (1768–1830) in order to develop a general theory of heat flow. His research was published in his masterpiece, *Th'eorie Analytique de la Chaleur*, in 1822. This method mainly banks on the technique of separation of variables. Though this method applies to a relatively small class of problems, yet it includes many problems of great physical interest.

Consider a homogeneous linear PDE of order two

$$R(x, y)r + S(x, y)s + T(x, y)t + P(x, y)p + Q(x, y)q + Z(x, y)z = 0. \tag{5.114}$$

Here, the coefficients R, S, and T are not vanishing simultaneously. Recall that Eq. (5.114) possesses a separable solution of the form $z(x, y) = X(x) \cdot Y(y)$ provided

$$\frac{1}{X}f(D)X = \frac{1}{Y}f(D')Y = k \text{ (separation constant)}$$

where $D = \dfrac{d}{dx}$ and $D' = \dfrac{d}{dy}$.

As discussed in Chapter 3, the method of separation of variables requires the ability of presenting the boundary conditions to determine the values of arbitrary constants, wherein using the superposition principle, we obtain a more general solution of the PDE in the form of an infinite series of the product of solutions. As the last step of solving these BVP, we compute the coefficients of this series. In Chapter 3, we have already presented several simple examples, whereas the coefficients of series are determined by easily using given auxiliary conditions, but this is not always possible. In order to solve BVP described by Eq. (5.114), sometimes under a given boundary/initial condition (for which $y = 0$), the general solution reduces to an infinite series containing either cosine or sine terms in the interval $(0, a)$. In such cases, we can determine these coefficients by using the concept of half range Fourier series. This method of solving these specific BVP is often called **Fourier method**.

Such specific BVP, wherein Fourier method is applicable, often arises in physical situations, *for example,* diffusion equation, wave equation, and Laplace equation, which will be discussed a little later. For the limitation of the space, we confine our discussion for the problems, which are similar to these classical equations. Therefore, in what follows, we only consider the BVP for which

- The domain of Eq. (5.114) is $\Omega = \{(x, y) \in \mathbb{R} : 0 < x < a, 0 < y < \infty\}$.
- The prescribed boundary conditions are either of the forms

$$z(0, y) = z(a, y) = 0$$
$$p(0, y) = p(a, y) = 0.$$

Such boundary conditions are called homogeneous BC.

- In view of Eq. (5.115), $f(D) = D^2$ so that separation constant remains $k = \dfrac{X''}{X}$.

An Important Observation: Under the above-mentioned hypotheses, one of the ODEs becomes

$$X''(x) - kX(x) = 0, \quad x \in (0, a), k \in \mathbb{R}$$

satisfying either of the homogeneous boundary conditions

$$X(0) = X(a) = 0 \tag{5.116}$$

or

$$X'(0) = X'(a) = 0. \tag{5.117}$$

Obviously, a homogeneous linear PDE always admits a trivial solution. Here, we are interested to find the criteria for the existence of a non-trivial solution of Eq. (5.114) satisfying the above BC. For this, we distinguish the following three possible cases:

Case I: If $k = 0$, then the given equation becomes

$$X'' = 0. \tag{5.118}$$

On integrating, we get

$$X' = c_1. \tag{5.119}$$

Integrating again, we get the general solution of the Eq. (5.118) given by

$$X(x) = c_1 x + c_2. \tag{5.120}$$

Applying the given BC (5.116), we get

$$c_2 = 0 \quad \text{and} \quad c_1 a + c_2 = 0.$$

On solving, we get $c_1 = c_2 = 0$ and hence the solution of the Eq. (5.118) becomes

$$X(x) = 0.$$

Similarly, applying BC (5.117) and using (5.119), we can get $c_1 = c_2 = 0$ so that

$$X(x) = 0.$$

Hence, under both types of boundary conditions, the solution of Eq. (5.118) is trivial, so it is rejected.

Case II: If $k > 0$, then we can take $k = \lambda^2, \lambda \neq 0$ so that the given equation reduces to

$$X'' - \lambda^2 X = 0. \tag{5.121}$$

whose general solution is

$$X(x) = c_3 e^{\lambda x} + c_4 e^{-\lambda x}. \tag{5.122}$$

Applying the given BC (5.116), we get

$$c_3 + c_4 = 0 \quad \text{and} \quad c_3 e^{\lambda a} + c_4 e^{-\lambda a} = 0.$$

As $a \neq 0$, the above equations imply that $c_3 = c_4 = 0$ and hence the solution of the given equation again becomes

$$X(x) = 0.$$

Now, from Eq. (5.122), we get

$$X'(x) = \lambda(c_3 e^{\lambda x} - c_4 e^{-\lambda x}).$$

Applying BC (5.117) in the above equation, we can get $c_3 = c_4 = 0$ so that

$$X(x) = 0.$$

Thus, under both types of boundary conditions, the solution of Eq. (5.121) being again trivial is rejected.

Case III: If $k < 0$, then we can take $k = -\lambda^2, \lambda \neq 0$ so that the given equation reduces to

$$X'' + \lambda^2 X = 0,$$

whose solution is

$$X(x) = c_5 \cos \lambda x + c_6 \sin \lambda x.$$

Applying the given BC (5.116), we get

$$c_5 = 0 \quad \text{and} \quad c_5 \cos \lambda a + c_6 \sin \lambda a = 0. \tag{5.123}$$

Combining these equations, we obtain $c_6 \sin \lambda a = 0$. But for

$$0 \neq \lambda = \frac{\pi}{a}, \frac{2\pi}{a}, \frac{3\pi}{a}, \cdots$$

we have

$$\sin \lambda a = 0.$$

Therefore, for these infinitely many values of λ, we are able to choose $c_6 \neq 0$. It follows that the system (5.123) has a non-trivial solution for c_5 and c_6, corresponding to which we shall find a non-trivial solution of Eq. (5.121). In the similar manner, we can show that under BC (5.117), this solution of Eq. (5.121) also remains non-trivial.

As $z = XY$, Eq. (5.114) satisfying the given BC has a non-trivial solution iff $k < 0$. Therefore, in practice, we shall use directly $k = -\lambda^2$ in order to obtain a non-trivial solution.

Example 5.36. *Solve the BVP*

$$r = q, \ 0 < x < l, \ 0 < y < \infty,$$

$$z(0, y) = z(l, y) = 0, \ z(x, 0) = x.$$

Solution.   *Let the solution of the given PDE be of the form*

$$z = X(x) \cdot Y(y). \tag{5.124}$$

Now, from (5.124), we have

$$r = X'' Y \text{ and } q = XY'.$$

Using the above, the given equation becomes

$$X'' Y = XY'.$$

On separating the variables, the above equation reduces to

$$\frac{X''}{X} = \frac{Y'}{Y} = k.$$

In order to obtain a non-trivial solution, we choose $k = -\lambda^2$. For this value of k, we get the following two ODEs:

$$X'' + \lambda^2 X = 0 \text{ and } Y' + \lambda^2 Y = 0.$$

The general solutions of the above equations are

$$X = c_1 \cos \lambda x + c_2 \sin \lambda x \text{ and } Y = c_3 e^{-\lambda^2 y}.$$

Putting these values in Eq. (5.124), we get a complete solution given by

$$z = (c_1 \cos \lambda x + c_2 \sin \lambda x) c_3 e^{-\lambda^2 y}$$

or

$$z = (a \cos \lambda x + b \sin \lambda x) e^{-\lambda^2 y}. \tag{5.125}$$

Applying BC $z(0, y) = 0$ in (5.125), we get

$$ae^{-\lambda^2 y} = 0 \Rightarrow a = 0.$$

Thus, the solution (5.125) becomes

$$z = b \sin \lambda x \cdot e^{-\lambda^2 y}. \tag{5.126}$$

Applying BC $z(l, y) = 0$ in (5.126), we get

$$b \sin \lambda l \cdot e^{-\lambda^2 y} = 0 \Rightarrow \sin \lambda l = 0$$

which gives rise to

$$\lambda l = n\pi \Rightarrow \lambda = \frac{n\pi}{l} \quad \forall \, n \in \mathbb{N}.$$

With these values of λ, Eq. (5.126) remains a sequence of solutions of the form

$$z_n = b_n \sin \frac{n\pi x}{l} \cdot e^{-n^2 \pi^2 y / l^2} \quad \forall \, n \in \mathbb{N}.$$

Using the principle of superposition, the most general complete solution is

$$z = \sum_{n=1}^{\infty} b_n \sin \frac{n\pi x}{l} \cdot e^{-n^2 \pi^2 y / l^2}. \tag{5.127}$$

Finally, applying BC $z(x, 0) = x$ in (5.127), we get

$$\sum_{n=1}^{\infty} b_n \sin \frac{n\pi x}{l} = x$$

which is the Fourier sine series for $0 < x < l$. Hence, we have

$$\begin{aligned}
b_n &= \frac{2}{l} \int_0^l x \sin \frac{n\pi x}{l} dx \\
&= \frac{2}{l} \left[-x \frac{l}{n\pi} \cos \frac{n\pi x}{l} + \frac{l^2}{n^2 \pi^2} \sin \frac{n\pi x}{l} \right]_0^l \\
&= \frac{2}{l} \left[-\frac{l^2}{n\pi} \cos n\pi \right] = -\frac{2l}{n\pi} (-1)^n \\
&= \frac{2l}{n\pi} (-1)^{n+1}.
\end{aligned}$$

Finally, Eq. (5.127) becomes

$$z = \sum_{n=1}^{\infty} \frac{2l}{n\pi}(-1)^{n+1} \sin \frac{n\pi x}{l} \cdot e^{-n^2\pi^2 y/l^2}$$

or

$$z = \frac{2l}{\pi} \sum_{n=1}^{\infty} \frac{(-1)^{n+1}}{n} \sin \frac{n\pi x}{l} \cdot e^{-n^2\pi^2 y/l^2}$$

which is the required solution of the given BVP.

Example 5.37. *Solve the BVP*

$$r = q, \ 0 < x < \pi, \ 0 < y < \infty,$$

$$p(0, y) = p(\pi, y) = 0, \ z(x, 0) = \sin x.$$

Solution.   Proceeding on the line similar to the solution of Example 5.36, the complete solution of the given equation is

$$z = (a \cos \lambda x + b \sin \lambda x)e^{-\lambda^2 y}. \tag{5.128}$$

Differentiating Eq. (5.128) partially w.r.t. x, we obtain

$$p = \lambda(-a \sin \lambda x + b \cos \lambda x)e^{-\lambda^2 y}.$$

Applying BC $p(0, y) = 0$ in the above equation, we get

$$\lambda b e^{-\lambda^2 y} = 0 \Rightarrow b = 0.$$

Thus, the solution (5.128) becomes

$$z = a \cos \lambda x \cdot e^{-\lambda^2 y}. \tag{5.129}$$

Differentiating Eq. (5.129) partially w.r.t. x, we obtain

$$p = -\lambda a \sin \lambda x \cdot e^{-\lambda^2 y}.$$

Applying BC $p(\pi, y) = 0$ in the above equation, we get

$$-\lambda a \sin \lambda \pi \cdot e^{-\lambda^2 y} = 0 \Rightarrow \sin \lambda \pi = 0$$

which gives

$$\lambda = 0$$

or

$$\lambda \pi = n\pi \Rightarrow \lambda = n \quad \forall \, n \in \mathbb{N}.$$

With these values of λ, Eq. (5.129) remains a sequence of solutions of the form

$$z_n = a_n \cos nx \cdot e^{-n^2 y} \quad \forall \, n \in \mathbb{N}.$$

Further, for $\lambda = 0$, Eq. (5.129) reduces to $z = a =$constant. Denote this particular solution $z_0 = \dfrac{a_0}{2}$. Therefore, using the principle of superposition, the most general complete solution of the given problem is

$$z = \frac{a_0}{2} + \sum_{n=1}^{\infty} a_n \cos nx \cdot e^{-n^2 y}. \tag{5.130}$$

Finally, applying BC $z(x, 0) = \sin x$ in (5.130), we get

$$\frac{a_0}{2} + \sum_{n=1}^{\infty} a_n \cos nx = \sin x$$

which is the Fourier cosine series for $0 < x < \pi$. Hence, we have

$$a_0 = \frac{2}{\pi} \int_0^{\pi} \sin x dx = \frac{2}{\pi} \left[-\cos x \right]_0^{\pi} = \frac{4}{\pi}$$

and

$$a_n = \frac{2}{\pi} \int_0^{l} \sin x \cos nx dx = \frac{2}{\pi} I$$

where

$$I = \int_0^\pi \sin x \cos nx \, dx$$

$$= \left[\sin x \frac{\sin nx}{n} \right]_0^\pi - \int_0^\pi \cos x \frac{\sin nx}{n} dx$$

$$= 0 - \frac{1}{n} \int_0^\pi \cos x \sin nx \, dx$$

$$= -\frac{1}{n} \left\{ \left[\cos x \left(-\frac{\cos nx}{n} \right) \right]_0^\pi - \int_0^\pi (-\sin x) \left(-\frac{\cos nx}{n} \right) dx \right\}$$

$$= -\frac{1}{n} \left(-\frac{1}{n} \cos \pi \cos n\pi + \frac{1}{n} - \frac{1}{n} I \right)$$

$$= -\frac{1}{n^2} (-1)^n - \frac{1}{n^2} + \frac{1}{n^2} I$$

yielding thereby

$$\left(1 - \frac{1}{n^2} \right) I = -\frac{1}{n^2} \left[(-1)^n + 1 \right]$$

or

$$I = \frac{(-1)^n + 1}{1 - n^2}$$

$$= \begin{cases} 0, & \text{if } n \text{ is odd} \\ \dfrac{2}{1 - n^2}, & \text{if } n \text{ is even} \end{cases}$$

$$= \begin{cases} 0, & \text{if } n = 2m - 1 \\ \dfrac{2}{1 - 4m^2}, & \text{if } n = 2m \quad (\text{where } m = 1, 2, 3, \ldots) \end{cases}$$

Hence, we have

$$a_n = \begin{cases} 0, & \text{if } n = 2m - 1 \\ \dfrac{4}{\pi} \dfrac{1}{1 - 4m^2}, & \text{if } n = 2m \end{cases}$$

Using the values of a_0 and a_n, Eq. (5.130) becomes

$$z = \frac{2}{\pi} + \sum_{n=2m} a_n \cos nx \cdot e^{-n^2 y}.$$

or

$$z = \frac{2}{\pi} + \frac{4}{\pi} \sum_{m=1}^{\infty} \frac{1}{1 - 4m^2} \cos 2mx \cdot e^{-4m^2 y}$$

which is the required solution of the given BVP.

Example 5.38. *Solve the BVP*

$$r = t, \ 0 < x < l, \ 0 < y < \infty,$$

$$z(0, y) = z(l, y) = 0, \ q(x, 0) = 0, \ z(x, 0) = \begin{cases} x, & 0 < x < \dfrac{l}{2} \\ l - x, & \dfrac{l}{2} < x < l. \end{cases}$$

Solution. *Let the solution of the given PDE be of the form*

$$z = X(x) \cdot Y(y). \tag{5.131}$$

Now, from (5.131), we have

$$r = X''Y \text{ and } t = XY''.$$

Using the above, the given equation becomes

$$X''Y = XY''.$$

On separating the variables, the above equation reduces to

$$\frac{X''}{X} = \frac{Y''}{Y} = k.$$

In order to obtain a non-trivial solution, we choose $k = -\lambda^2$. For this value of k, we get the following two ODEs:

$$X'' + \lambda^2 X = 0 \text{ and } Y'' + \lambda^2 Y = 0.$$

The general solutions of the above equations are

$$X = c_1 \cos \lambda x + c_2 \sin \lambda x \text{ and } Y = c_3 \cos \lambda y + c_4 \sin \lambda y.$$

Putting these values in Eq. (5.131), we get a complete solution given by

$$z = (c_1 \cos \lambda x + c_2 \sin \lambda x)(c_3 \cos \lambda y + c_4 \sin \lambda y). \tag{5.132}$$

Applying BC $z(0, y) = 0$ in (5.132), we get

$$c_1(c_3 \cos \lambda y + c_4 \sin \lambda y) = 0 \Rightarrow c_1 = 0.$$

Thus, the solution (5.132) becomes

$$z = c_2 \sin \lambda x (c_3 \cos \lambda y + c_4 \sin \lambda y). \tag{5.133}$$

Differentiating the above equation partially w.r.t. y, we get

$$q = c_2 \sin \lambda x (-\lambda c_3 \sin \lambda y + \lambda c_4 \cos \lambda y).$$

Now, applying BC $q(x, 0) = 0$ in the above equation, we get

$$\lambda . c_4 . c_2 \sin \lambda x = 0 \Rightarrow c_4 = 0 \quad \text{as } \lambda \neq 0 \text{ and } c_2 \neq 0.$$

Thus, the solution (5.133) becomes

$$z = c_2 c_3 \sin \lambda x \cos \lambda y$$

or

$$z = b \sin \lambda x \cos \lambda y \quad (\text{put } c_2 c_3 = b). \tag{5.134}$$

Applying BC $z(l, y) = 0$ in (5.134), we get

$$b \sin \lambda l \cos \lambda y = 0 \Rightarrow \sin \lambda l = 0$$

which gives rise to

$$\lambda l = n\pi \Rightarrow \lambda = \frac{n\pi}{l} \quad \forall n \in \mathbb{N}.$$

With these values of λ, Eq. (5.134) remains a sequence of solutions of the form

$$z_n = b_n \sin \frac{n\pi x}{l} \cdot \cos \frac{n\pi y}{l} \quad \forall n \in \mathbb{N}.$$

Using the principle of superposition, the most general complete solution is

$$z = \sum_{n=1}^{\infty} b_n \sin \frac{n\pi x}{l} \cdot \cos \frac{n\pi y}{l}. \tag{5.135}$$

Finally, applying BC

$$z(x, 0) = f(x) \equiv \begin{cases} x, & 0 < x < \dfrac{l}{2} \\[2mm] l - x, & \dfrac{l}{2} < x < l \end{cases}$$

we get

$$\sum_{n=1}^{\infty} b_n \sin\frac{n\pi x}{l} = f(x). \tag{5.136}$$

Since series (5.136) is a Fourier sine series for $0 < x < l$, we have

$$b_n = \frac{2}{l}\int_0^l f(x)\sin\frac{n\pi x}{l}dx = \frac{2}{l}\left[\int_0^{l/2} f(x)\sin\frac{n\pi x}{l}dx + \int_{l/2}^l f(x)\sin\frac{n\pi x}{l}dx\right]$$

$$= \frac{2}{l}\int_0^{l/2} x\sin\frac{n\pi x}{l}dx + \frac{2}{l}\int_{l/2}^l (l-x)\sin\frac{n\pi x}{l}dx$$

$$= \frac{2}{l}\left[x\left\{-\frac{\cos(n\pi x/l)}{n\pi/l}\right\} - (1)\left\{-\frac{\sin(n\pi x/l)}{n^2\pi^2/l^2}\right\}\right]_0^{l/2} + \frac{2}{l}\left[(l-x)\left\{-\frac{\cos(n\pi x/l)}{n\pi/l}\right\}\right.$$

$$\left. -(-1)\left\{-\frac{\sin(n\pi x/l)}{n^2\pi^2/l^2}\right\}\right]_0^{l/2}$$

$$= \frac{2}{l}\left[-\frac{l^2}{2n\pi}\cos\frac{n\pi}{2} + \frac{l^2}{n^2\pi^2}\sin\frac{n\pi}{2} + \frac{l^2}{2n\pi}\cos\frac{n\pi}{2} + \frac{l^2}{n^2\pi^2}\sin\frac{n\pi}{2}\right]$$

$$= \frac{4l}{n^2\pi^2}\sin\frac{n\pi}{2}$$

so that

$$b_n = \begin{cases} 0, & \text{if } n = 2m \\ \dfrac{4l(-1)^{m+1}}{(2m-1)^2\pi^2}, & \text{if } n = 2m-1 \ (m = 1, 2, 3, \ldots). \end{cases}$$

With these values of b_n, Eq. (5.135) becomes

$$z = \sum_{m=1}^{\infty} \frac{4l(-1)^{m+1}}{(2m-1)^2\pi^2}\sin\frac{(2m-1)\pi x}{l}\cdot\cos\frac{(2m-1)\pi y}{l}$$

or

$$z = \frac{4l}{\pi^2}\sum_{m=1}^{\infty}\frac{(-1)^{m+1}}{(2m-1)^2}\sin\frac{(2m-1)\pi x}{l}\cdot\cos\frac{(2m-1)\pi y}{l}$$

which is the required solution of the given BVP.

Example 5.39. *Solve the BVP*

$$r = t, \ 0 < x < l, \ 0 < y < \infty,$$

$$z(0, y) = z(l, y) = 0, \ z(x, 0) = 0, \ q(x, 0) = kx(l - x).$$

Solution. *Proceeding on the line of the solution of Example 5.38, the solution of the given equation satisfying the BC $z(0, y) = 0$ becomes*

$$z = c_2 \sin \lambda x (c_3 \cos \lambda y + c_4 \sin \lambda y). \tag{5.137}$$

Applying BC $z(x, 0) = 0$ in the above equation, we get

$$c_2 . c_3 \sin \lambda x = 0 \Rightarrow c_3 = 0 \quad \text{as } c_2 \neq 0.$$

Thus, the solution (5.137) becomes

$$z = c_2 c_4 \sin \lambda x \sin \lambda y$$

or

$$z = b \sin \lambda x \sin \lambda y \quad (\text{Put } c_2 c_4 = b). \tag{5.138}$$

Applying BC $z(l, y) = 0$ in (5.138), we get

$$b \sin \lambda l \sin \lambda y = 0 \Rightarrow \sin \lambda l = 0$$

which gives rise to

$$\lambda l = n\pi \Rightarrow \lambda = \frac{n\pi}{l} \quad \forall n \in \mathbb{N}.$$

With these values of λ, Eq. (5.138) remains a sequence of solutions of the form

$$z_n = b_n \sin \frac{n\pi x}{l} \cdot \sin \frac{n\pi y}{l} \quad \forall n \in \mathbb{N}.$$

Using the principle of superposition, the most general complete solution is

$$z = \sum_{n=1}^{\infty} b_n \sin \frac{n\pi x}{l} \cdot \sin \frac{n\pi y}{l}. \tag{5.139}$$

Differentiating partially w.r.t. y, we get

$$q = \frac{\pi}{l} \sum_{n=1}^{\infty} n b_n \sin \frac{n\pi x}{l} \cdot \cos \frac{n\pi y}{l}.$$

Finally, applying BC $q(x, 0) = kx(l - x)$ in the above equation, we get

$$\frac{\pi}{l} \sum_{n=1}^{\infty} nb_n \sin \frac{n\pi x}{l} = kx(l - x)$$

or

$$\sum_{n=1}^{\infty} nb_n \sin \frac{n\pi x}{l} = \frac{kl}{\pi} x(l - x)$$

which is the Fourier cosine series for $0 < x < l$ with Fourier coefficients nb_n. Hence, we have

$$nb_n = \frac{2}{l} \int_0^l \frac{kl}{\pi} x(l - x) \sin \frac{n\pi x}{l} dx$$

$$= \frac{2k}{\pi} \int_0^l (lx - x^2) \sin \frac{n\pi x}{l} dx$$

$$= \frac{2k}{\pi} \left[(lx - x^2) \left\{ \frac{-\cos(n\pi x/l)}{n\pi/l} \right\} - (l - 2x) \left\{ \frac{-\sin(n\pi x/l)}{n^2\pi^2/l^2} \right\} \right.$$

$$\left. + (-2) \left\{ \frac{\cos(n\pi x/l)}{n^3\pi^3/l^3} \right\} \right]_0^l$$

$$= \frac{2k}{\pi} \left[-\frac{2l^3}{n^3\pi^3} (-1)^n + \frac{2l^3}{n^3\pi^3} \right] = \frac{4kl^3}{n^3\pi^4} \left[(-1)^{n+1} + 1 \right]$$

which implies that

$$b_n = \frac{4kl^3}{n^4\pi^4} \left[(-1)^{n+1} + 1 \right]$$

$$= \begin{cases} 0, & \text{if } n = 2m \\ \dfrac{8kl^3}{\pi^4(2m - 1)^4}, & \text{if } n = 2m - 1, \ (m = 1, 2, 3, \dots) \end{cases}$$

Putting the values of b_n, Eq. (5.127) becomes

$$z = \sum_{n=2m-1} b_n \sin \frac{n\pi x}{l} \cdot \sin \frac{n\pi y}{l}$$

or

$$z = \sum_{m=1}^{\infty} \frac{8kl^3}{\pi^4(2m - 1)^4} \sin \frac{(2m - 1)\pi x}{l} \cdot \sin \frac{(2m - 1)\pi y}{l}$$

or

$$z = \frac{8kl^3}{\pi^4} \sum_{n=1}^{\infty} \frac{1}{(2m-1)^4} \sin \frac{(2m-1)\pi x}{l} \cdot \sin \frac{(2m-1)\pi y}{l}$$

which is the required solution of the given BVP.

Example 5.40. *Solve the BVP*

$$r + t = 0,\ 0 < x < a,\ 0 < y < \infty,$$

$$p(0, y) = p(a, y) = 0,\ \lim_{y \to \infty} z(x, y) = 0,\ z(x, 0) = \begin{cases} x, & 0 < x < \dfrac{a}{2} \\ a - x, & \dfrac{a}{2} < x < a. \end{cases}$$

Solution.   Let the solution of the given PDE be of the form

$$z = X(x) \cdot Y(y). \tag{5.140}$$

Now, from (5.140), we have

$$r = X''Y \text{ and } t = XY''.$$

Using the above, the given equation becomes

$$X''Y + XY'' = 0.$$

On separating the variables, the above equation reduces to

$$\frac{X''}{X} = -\frac{Y''}{Y} = k.$$

In order to obtain a non-trivial solution, we choose $k = -\lambda^2$. For this value of k, we get the following two ODEs:

$$X'' + \lambda^2 X = 0 \text{ and } Y'' - \lambda^2 Y = 0.$$

The general solutions of the above equations are

$$X = c_1 \cos \lambda x + c_2 \sin \lambda x \text{ and } Y = c_3 e^{\lambda y} + c_4 e^{-\lambda y}.$$

Putting these values in Eq. (5.131), we get a complete solution given by

$$z = (c_1 \cos \lambda x + c_2 \sin \lambda x)(c_3 e^{\lambda y} + c_4 e^{-\lambda y}). \tag{5.141}$$

Differentiating partially w.r.t. x, we get

$$p = \lambda(-c_1 \sin \lambda x + c_2 \cos \lambda x)(c_3 e^{\lambda y} + c_4 e^{-\lambda y})$$

Applying BC $p(0, y) = 0$ in the above equation, we get

$$\lambda . c_2 . (c_3 e^{\lambda y} + c_4 e^{-\lambda y}) = 0 \Rightarrow c_2 = 0.$$

Thus, the solution (5.141) becomes

$$z = c_1 \cos \lambda x (c_3 e^{\lambda y} + c_4 e^{-\lambda y}). \tag{5.142}$$

Taking $y \to \infty$ in the above equation and using the condition $\lim_{y \to \infty} z(x, y) = 0$, we obtain

$$c_3 = 0.$$

Hence, the solution (5.142) becomes

$$z = c_1 \cos \lambda x \cdot c_4 e^{-\lambda y}$$

or

$$z = A \cos \lambda x \cdot e^{-\lambda y} \quad (\text{put } c_1 c_2 = A) \tag{5.143}$$

Differentiating partially w.r.t. x, we get

$$p = -A\lambda \sin \lambda x . e^{-\lambda y}.$$

Now, applying BC $p(a, y) = 0$ in the above equation, we get

$$-A\lambda \sin \lambda a . e^{-\lambda y} = 0 \Rightarrow \sin \lambda a = 0$$

which gives rise to

$$\lambda = 0 \quad or \quad \lambda a = n\pi \Rightarrow \lambda = \frac{n\pi}{a} \quad \forall n \in \mathbb{N}.$$

With these values of λ, Eq. (5.143) remains a sequence of solutions of the form

$$z_n = A_n \cos \frac{n\pi x}{a} \cdot e^{-n\pi y/a} \quad \forall n \in \mathbb{N}.$$

Further, for $\lambda = 0$, Eq. (5.143) reduces to $z = A$, a constant. Denote this particular solution by $z_0 = A_0/2$. Therefore, using the principle of superposition, the most general complete solution of the given problem is

$$z = \frac{A_0}{2} + \sum_{n=1}^{\infty} A_n \cos \frac{n\pi x}{a} . e^{-n\pi y/a}. \tag{5.144}$$

Finally, applying BC

$$z(x,0) = f(x) \equiv \begin{cases} x, & 0 < x < \dfrac{a}{2} \\[2mm] a - x, & \dfrac{a}{2} < x < a \end{cases}$$

we get

$$\frac{A_0}{2} + \sum_{n=1}^{\infty} A_n \cos \frac{n\pi x}{a} = f(x). \tag{5.145}$$

Since series (5.145) is a Fourier cosine series for $0 < x < a$, we have

$$A_0 = \frac{2}{a} \int_0^a f(x)dx = \frac{2}{a} \left[\int_0^{a/2} f(x)dx + \int_{a/2}^a f(x)dx \right]$$

$$= \frac{2}{a} \left[\int_0^{a/2} x\,dx + \int_{a/2}^a (a-x)dx \right] = \frac{2}{a} \left\{ \left[\frac{x^2}{2} \right]_0^{a/2} + \left[ax - \frac{x^2}{2} \right]_{a/2}^a \right\}$$

$$= \frac{2}{a} \left[\frac{a^2}{8} + \frac{a^2}{2} - \frac{3a^2}{8} \right] = \frac{2}{a} \cdot \frac{a^2}{4} = \frac{a}{2}.$$

$$A_n = \frac{2}{a} \int_0^a f(x) \cos \frac{n\pi x}{a} dx = \frac{2}{a} \left[\int_0^{a/2} f(x) \cos \frac{n\pi x}{a} dx + \int_{a/2}^a f(x) \cos \frac{n\pi x}{a} dx \right]$$

$$= \frac{2}{a} \int_0^{a/2} x \cos \frac{n\pi x}{a} dx + \frac{2}{a} \int_{a/2}^a (a-x) \cos \frac{n\pi x}{a} dx$$

$$= \frac{2}{a} \left[x \frac{\sin(n\pi x/a)}{n\pi/a} + \frac{\cos(n\pi x/a)}{n^2\pi^2/a^2} \right]_0^{a/2} + \frac{2}{a} \left[(a-x) \frac{\sin(n\pi x/a)}{n\pi/a} - \frac{\cos(n\pi x/a)}{n^2\pi^2/a^2} \right]_{a/2}^a$$

$$= \frac{2}{a} \left[\frac{a}{2} \cdot \frac{a}{n\pi} \sin \frac{n\pi}{2} + \frac{a^2}{n^2\pi^2} \cos \frac{n\pi}{2} - \frac{a^2}{n^2\pi^2} \right] + \frac{2}{a} \left[-\frac{a^2}{n^2\pi^2} \cos n\pi - \frac{a}{2} \cdot \frac{a}{n\pi} \sin \frac{n\pi}{2} \right.$$

$$\left. + \frac{a^2}{n^2\pi^2} \cos \frac{n\pi}{2} \right]$$

$$= \frac{2}{a} \left[\frac{2a^2}{n^2\pi^2} \cos \frac{n\pi}{2} - \frac{a^2}{n^2\pi^2} - \frac{a^2}{n^2\pi^2} \cos n\pi \right]$$

$$= \frac{2a}{n^2\pi^2} \left[2 \cos \frac{n\pi}{2} - \cos n\pi - 1 \right]$$

Using the values of A_0 and A_n, Eq. (5.144) becomes

$$z = \frac{a}{4} + \frac{2a}{\pi^2} \sum_{n=1}^{\infty} \frac{1}{n^2} \left(2 \cos \frac{n\pi}{2} - \cos n\pi - 1 \right) \cos \frac{n\pi x}{a} e^{-n^2 \pi^2 y/a^2}.$$

or

$$z = \frac{a}{4} - \frac{8a}{\pi^2} \left[\frac{1}{2^2} \cos \frac{2\pi x}{a} e^{-4\pi^2 y/a^2} + \frac{1}{6^2} \cos \frac{6\pi x}{a} e^{-36\pi^2 y/a^2 + \ldots} \right]$$

which is the required solution of the given BVP.

5.6 Applications of Second-Order PDEs

In applied mathematics, the PDEs generally arise in formulating fundamental laws of nature and in the study of a wide variety of physical models. This section is devoted to discussing some well-known second-order PDEs of mathematical physics, namely, one-dimensional wave equation, one-dimensional heat equation, two- and three-dimensional Laplace equation, Poisson's equation, telegraph equations, and Burgers' equation.

5.6.1 Vibrations of a Stretched String: One-Dimensional Wave Equation

Consider a tightly stretched elastic string of length L along x-axis with its ends fixed at $x = 0$ and $x = L$. If the string is distorted slightly from its equilibrium position and then released, it will begin to vibrate in the plane and hence its points will perform certain motions. We want to discuss wave motion regarding the transverse vibrations of the string. Choose the u-axis perpendicular to the x-axis in the plane of vibration (see Figure 5.1).

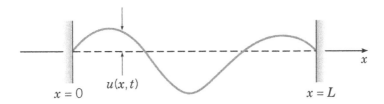

FIGURE 5.1 A vibrating string

Then the vertical displacement or deflection $u(x, t)$ of the string at any point x and at a time t is determined from a second-order PDE given by

$$\frac{\partial^2 u}{\partial t^2} = c^2 \frac{\partial^2 u}{\partial x^2} \tag{5.146}$$

where c^2 is the positive constant depending on the physical nature of the string. Equation (5.146) is called **one-dimensional wave equation**. This is a typical hyperbolic equation. Basically wave equation relates two quantities: $\dfrac{\partial^2 u}{\partial t^2}$, the vertical acceleration of the string, and $\dfrac{\partial^2 u}{\partial x^2}$, the concavity of the position function $u(x,t)$.

Derivation of Wave Equation: Let T be a constant tension in the string, which is considered to be large compared to the weight of the string so that the effect of gravity is negligible. Take the element PQ of the string as shown in Figure 5.2.

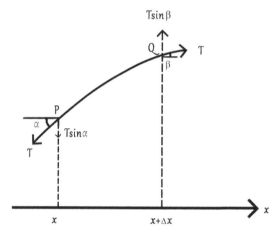

FIGURE 5.2   An element of a vertically displaced string

As the angles α and β are small, we have

$$\sin \alpha = \tan \alpha \ \text{ and } \ \sin \beta = \tan \beta.$$

Thus, the vertical component of the force acting on the element PQ is

$$T \sin \beta - T \sin \alpha = T[\tan \beta - \tan \alpha]$$

$$= T\left[\left(\frac{\partial u}{\partial x}\right)_{x+\triangle x} - \left(\frac{\partial u}{\partial x}\right)_{x}\right].$$

Let m be the mass per unit length of the string. As P and Q are neighbouring points, $PQ \approx \triangle x$. Hence, the mass of the string is $m \triangle x$. Using Newton's second law of motion, we get

$$T\left[\left(\frac{\partial u}{\partial x}\right)_{x+\triangle x} - \left(\frac{\partial u}{\partial x}\right)_{x}\right] = m \triangle x \frac{\partial^2 u}{\partial t^2}$$

or

$$\frac{\partial^2 u}{\partial t^2} = \frac{T}{m}\left[\frac{(\partial u/\partial x)_{x+\triangle x} - (\partial u/\partial x)_{x}}{\triangle x}\right].$$

Taking the limit as $\triangle x \to 0$ (so that $Q \to P$), we obtain

$$\frac{\partial^2 u}{\partial t^2} = c^2 \frac{\partial^2 u}{\partial x^2}$$

where $c^2 = T/m$.

D' Alembert's Solution of Wave Equation: We find the general solution of Eq. (5.146) by reduction to canonical form. As Eq. (5.146) is hyperbolic, the characteristic equation is

$$\frac{dt}{dx} = \frac{\pm 2/c}{2} = \pm \frac{1}{c}$$

or

$$dx \pm c\, dt = 0$$

On integrating, we get the characteristic curves given by

$$x + ct = c_1 \quad \text{and} \quad x - ct = c_2.$$

Hence, the canonical coordinates are

$$\xi = x + ct \quad \text{and} \quad \eta = x - ct.$$

Using these coordinates, the given equation reduces its canonical form given by

$$u_{\xi\eta} = 0.$$

The solution of the above equation $u = \phi(\xi) + \psi(\eta)$, which in terms of original variables, becomes

$$u(x, t) = \phi(x + ct) + \psi(x - ct).$$

This solution of wave equation is often referred to as **D' Alembert's solution**.

Variables Separable Solution of Wave Equation: We find the solution of wave equation using the method of separation of variables. Assume that the solution of Eq. (5.146) is of the form

$$u(x, t) = X(x)T(t). \tag{5.147}$$

From Eq. (5.147), we get

$$\frac{\partial^2 u}{\partial x^2} = X''T \text{ and } \frac{\partial^2 u}{\partial t^2} = XT''.$$

Putting these values in Eq. (5.146), we obtain

$$XT'' = c^2 X''T.$$

On separating the variables, the above equation reduces to

$$\frac{X''}{X} = \frac{T''}{c^2 T} = k \text{ (say)}$$

which gives us two ODEs:

$$X'' - kX = 0 \tag{5.148}$$

and

$$T'' - kc^2 T = 0 \tag{5.149}$$

These are the following three possibilities:

Case I: If $k = 0$, then Eqs (5.148) and (5.149) become

$$X'' = 0 \text{ and } T'' = 0$$

whose solutions are

$$X = c_1 x + c_2 \text{ and } T = c_3 t + c_4.$$

Putting these values in Eq. (5.148), we get a complete solution given by

$$u(x, t) = (c_1 x + c_2)(c_3 t + c_4). \tag{5.150}$$

Case II: If $k > 0$, then we can assume that $k = \lambda^2$ so that Eqs (5.148) and (5.149) become

$$X'' - \lambda^2 X = 0 \text{ and } T'' - \lambda^2 c^2 T = 0$$

whose solutions are

$$X = c_5 e^{\lambda x} + c_6 e^{-\lambda x} \text{ and } T = c_7 e^{\lambda ct} + c_8 e^{-\lambda ct}.$$

Putting these values in Eq. (5.148), we get a complete solution given by

$$u(x, t) = (c_5 e^{\lambda x} + c_6 e^{-\lambda x})(c_7 e^{\lambda ct} + c_8 e^{-\lambda ct}). \tag{5.151}$$

Case III: If $k < 0$, then we can assume that $k = -\lambda^2$ so that Eqs (5.148) and (5.149) become

$$X'' + \lambda^2 X = 0 \text{ and } T'' + \lambda^2 c^2 T = 0$$

whose solutions are

$$X = c_9 \cos \lambda x + c_{10} \sin \lambda x \text{ and } T = c_{11} \cos \lambda ct + c_{12} \sin \lambda ct.$$

Putting these values in Eq. (5.148), we get a complete solution given by

$$u(x,t) = (c_9 \cos \lambda x + c_{10} \sin \lambda x)(c_{11} \cos \lambda ct + c_{12} \sin \lambda ct). \qquad (5.152)$$

Out of the three possible solutions (5.150), (5.151), and (5.152), we have to choose the solution, which is consistent with the physical nature of the problem. As we are working with the problem on vibrations, u must be a periodic function of x and t. Therefore, $u(x,t)$ must involve trigonometric terms. Thus, among these, (5.152) is the only suitable solution of Eq. (5.146), *that is,*

$$u(x,t) = (A \cos \lambda x + B \sin \lambda x)(C \cos \lambda ct + D \sin \lambda ct).$$

Example 5.41. *A tightly stretched string is fixed at two points l apart. Motion is started by displacing the string into the form $u = \mu x(l - x)$, μ is a constant, from which it is released at time $t = 0$. Find the displacement of any point on the string at a distance of x from one end at time $t > 0$.*

Solution. *The motion of the string is governed by one-dimensional wave equation given by*

$$\frac{\partial^2 u}{\partial t^2} = c^2 \frac{\partial^2 u}{\partial x^2}. \qquad (5.153)$$

Given that the string is fixed at two points $x = 0$ and $x = l$, that is,

$$u(0,t) = u(l,t) = 0.$$

Since the initial velocity of any point of the string remains zero, we have

$$\frac{\partial u}{\partial t} \Big|_{t=0} = 0. \qquad (5.154)$$

Also, the initial displacement is given by

$$u(x,0) = \mu x(l - x). \qquad (5.155)$$

The suitable solution of Eq. (5.153) is

$$u(x,t) = (A \cos \lambda x + B \sin \lambda x)(C \cos \lambda ct + D \sin \lambda ct). \qquad (5.156)$$

Applying BC $u(0,t) = 0$ and IC (5.154), we obtain $A = D = 0$ and hence the solution (5.156) reduces to

$$u(x,t) = b \sin \lambda x \cos \lambda ct \quad (where\ b = BC). \qquad (5.157)$$

Now, applying BC $u(l,t) = 0$, we get $\lambda = \dfrac{n\pi}{l}$ and hence the solution (5.157) reduces to

$$u(x,t) = b_n \sin \frac{n\pi x}{l} \cos \frac{n\pi ct}{l}.$$

Using the principle of superposition, we obtain the most general complete solution given by

$$u(x, t) = \sum_{n=1}^{\infty} b_n \sin \frac{n\pi x}{l} \cos \frac{n\pi ct}{l}. \tag{5.158}$$

Finally, applying IC (5.155), we obtain

$$\mu x(l - x) = \sum_{n=1}^{\infty} b_n \sin \frac{n\pi x}{l}$$

which is a Fourier sine series in the interval $0 < x < l$. Hence, we have

$$b_n = \frac{2}{l} \int_0^l \mu x(l - x) \sin \frac{n\pi x}{l}$$

$$= \frac{4\mu l^2}{n^3 \pi^3}(-\cos n\pi + 1) = \frac{4\mu l^2}{n^3 \pi^3}[(-1)^{n+1} + 1] \quad (\text{after simplifying})$$

$$= \begin{cases} 0 & \text{if } n = 2m \\ \dfrac{8\mu l^2}{(2m - 1)^3 \pi^3} & \text{if } n = 2m - 1 \quad (m = 1, 2, 3, \ldots). \end{cases}$$

Putting these values in Eq. (5.158), we get

$$u(x, t) = \frac{8\mu l^2}{\pi^3} \sum_{m=1}^{\infty} \frac{1}{(2m - 1)^3} \sin \frac{(2m - 1)\pi x}{l} \cos \frac{(2m - 1)\pi ct}{l}$$

which is the required displacement of the string.

Example 5.42. *A uniform string of length l held tightly between $x = 0$ and $x = l$ and is initially at rest in its equilibrium position. If it is set vibrating to give each point a velocity v_0, then find its displacement at any instant of time.*

Solution. *The vibration of the string is governed by one-dimensional wave equation given by*

$$\frac{\partial^2 u}{\partial t^2} = c^2 \frac{\partial^2 u}{\partial x^2}. \tag{5.159}$$

As the string held tightly between $x = 0$ and $x = l$, we have

$$u(0, t) = u(l, t) = 0.$$

Also, the string is initially at rest in its equilibrium position. It means that the initial displacement remains zero, that is,

$$u(x, 0) = 0.$$

The initial velocity of the string is

$$\frac{\partial u}{\partial t}\Big|_{t=0} = v_0. \tag{5.160}$$

The suitable solution of Eq. (5.159) is

$$u(x, t) = (A \cos \lambda x + B \sin \lambda x)(C \cos \lambda ct + D \sin \lambda ct). \tag{5.161}$$

Applying BC $u(0, t) = 0$ and IC $u(x, 0) = 0$, we obtain $A = C = 0$ and hence the solution (5.161) reduces to

$$u(x, t) = b \sin \lambda x \sin \lambda ct \quad \text{(where } b = BD\text{)}. \tag{5.162}$$

Now, applying BC: $u(l, t) = 0$, we get $\lambda = \dfrac{n\pi}{l}$ and hence the solution (5.162) reduces to

$$u(x, t) = b_n \sin \frac{n\pi x}{l} \sin \frac{n\pi ct}{l}.$$

Using the principle of superposition, we obtain the most general complete solution given by

$$u(x, t) = \sum_{n=1}^{\infty} b_n \sin \frac{n\pi x}{l} \sin \frac{n\pi ct}{l}. \tag{5.163}$$

Differentiating partially w.r.t. t, we get

$$\frac{\partial u}{\partial t} = \frac{\pi c}{l} \sum_{n=1}^{\infty} n b_n \sin \frac{n\pi x}{l} \cos \frac{n\pi ct}{l}.$$

Finally, applying IC (5.160) in the above equation, we obtain

$$v_0 = \frac{\pi c}{l} \sum_{n=1}^{\infty} n b_n \sin \frac{n\pi x}{l}$$

or

$$\frac{v_0 l}{\pi c} = \sum_{n=1}^{\infty} n b_n \sin \frac{n\pi x}{l}$$

which is a Fourier sine series in the interval $0 < x < l$. Hence, we have

$$nb_n = \frac{2}{l} \int_0^l \frac{v_0 l}{\pi c} \sin \frac{n\pi x}{l} = \frac{2v_0}{\pi c} \left[-\frac{\cos(n\pi x/l)}{n\pi/l} \right]_0^l$$

$$= \frac{2v_0 l}{n\pi^2 c}(-\cos n\pi + 1) = \frac{2v_0 l}{n\pi^2 c}[(-1)^{n+1} + 1]$$

implying thereby

$$b_n = \begin{cases} 0 & \text{if } n = 2m \\[2mm] \dfrac{4v_0 l}{c\pi^2(2m-1)^2} & \text{if } n = 2m-1 \quad (m = 1, 2, 3, \ldots). \end{cases}$$

Putting these values in Eq. (5.163), we get

$$u(x, t) = \frac{4v_0 l}{c\pi^2} \sum_{m=1}^{\infty} \frac{1}{(2m-1)^2} \sin \frac{(2m-1)\pi x}{l} \sin \frac{(2m-1)\pi ct}{l}$$

which is the required displacement of the string.

5.6.2 Conduction of Heat in a Rod: One-Dimensional Heat Equation

Consider a homogeneous rod (or a bar) of length L having a uniform cross-section, which is aligned along the x-axis by taking its one end at the origin. Let us study the process of propagation of heat in the rod such that the direction of heat flow is along the positive x-axis (see Figure 5.3).

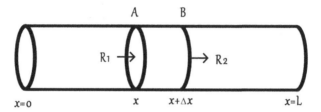

FIGURE 5.3 Heat conduction in a rod

The variation of temperature $u(x, t)$ in the rod at a distance x and at a time t is determined by a second-order PDE given by

$$\frac{\partial u}{\partial t} = \alpha^2 \frac{\partial^2 u}{\partial x^2}. \tag{5.164}$$

Here, α^2 is a constant called thermal diffusivity of the material of the given rod. Equation (5.164) is called one-dimensional heat equation or diffusion equation. This is a typical parabolic equation. Basically, the heat equation relates two quantities: $\dfrac{\partial u}{\partial t}$, the rate of change in temperature with respect to time, and $\dfrac{\partial^2 u}{\partial x^2}$, the concavity of the temperature profile $u(x, t)$.

Derivation of Heat Equation: Let ρ be the density, c the specific heat, k the thermal conductivity, and a the area of cross section of the rod. The rate of increase of heat in a portion AB of thickness $\triangle x$ of the rod is

$$c\rho a \, \triangle x \frac{\partial u}{\partial t} = R_1 - R_2. \tag{5.165}$$

where R_1 is the inflow of heat and R_2 is the outflow of heat. Now

$$R_1 = -ka \left(\frac{\partial u}{\partial x} \right)_x \text{ and } R_2 = -ka \left(\frac{\partial u}{\partial x} \right)_{x+\triangle x}.$$

As u is a decreasing function, its derivative is negative. Putting these values of R_1 and R_2 in Eq. (5.165), we get

$$c\rho a \triangle x \frac{\partial u}{\partial t} = ka \left[\left(\frac{\partial u}{\partial x} \right)_{x+\triangle x} - \left(\frac{\partial u}{\partial x} \right)_x \right]$$

or

$$\frac{\partial u}{\partial t} = \frac{k}{\rho c} \left[\frac{(\partial u / \partial x)_{x+\triangle x} - (\partial u / \partial x)_x}{\triangle x} \right].$$

Taking the limit as $\triangle x \to 0$, we obtain

$$\frac{\partial u}{\partial t} = \alpha^2 \frac{\partial^2 u}{\partial x^2}$$

where $\alpha^2 = k / \rho c$.

Variables Separable Solution of Heat Equation: We find the solution of heat equation using the method of separation of variables. Assume that the solution of Eq. (5.164) is of the form

$$u(x, t) = X(x)T(t). \tag{5.166}$$

From Eq. (5.166), we get

$$\frac{\partial^2 u}{\partial x^2} = X''T \text{ and } \frac{\partial u}{\partial t} = XT'.$$

Putting these values in Eq. (5.164), we obtain

$$XT' = \alpha^2 X''T.$$

On separating the variables, the above equation reduces to

$$\frac{X''}{X} = \frac{T'}{\alpha^2 T} = k \text{ (say)}$$

which gives us two ODEs:

$$X'' - kX = 0 \tag{5.167}$$

$$\text{and } T' - k\alpha^2 T = 0 \tag{5.168}$$

These are the following three possibilities:

Case I: If $k = 0$, then Eqs (5.167) and (5.168) become

$$X'' = 0 \text{ and } T' = 0$$

whose solutions are

$$X = c_1 x + c_2 \text{ and } T = c_3.$$

Putting these values in Eq. (5.167), we get a complete solution given by

$$u(x, t) = (c_1 x + c_2)c_3. \tag{5.169}$$

Case II: If $k > 0$, then we can assume that $k = \lambda^2$ so that Eqs (5.167) and (5.168) become

$$X'' - \lambda^2 X = 0 \text{ and } T' - \lambda^2 \alpha^2 T = 0$$

whose solutions are

$$X = c_4 e^{\lambda x} + c_5 e^{-\lambda x} \text{ and } T = c_6 e^{\lambda^2 \alpha^2 t}.$$

Putting these values in Eq. (5.167), we get a complete solution given by

$$u(x, t) = (c_4 e^{\lambda x} + c_5 e^{-\lambda x})c_6 e^{\lambda^2 \alpha^2 t}. \tag{5.170}$$

Case III: If $k < 0$, then we can assume that $k = -\lambda^2$ so that Eqs (5.167) and (5.168) become

$$X'' + \lambda^2 X = 0 \text{ and } T' + \lambda^2 \alpha^2 T = 0$$

whose solutions are

$$X = c_7 \cos \lambda x + c_8 \sin \lambda x \text{ and } T = c_9 e^{-\lambda^2 \alpha^2 t}.$$

Putting these values in Eq. (5.167), we get a complete solution given by

$$u(x, t) = (c_7 \cos \lambda x + c_8 \sin \lambda x)c_9 e^{-\lambda^2 \alpha^2 t}. \tag{5.171}$$

Out of the three possible solutions (5.169), (5.170), and (5.171), we have to choose the solution, which is consistent with the physical nature of the problem. As we are dealing with the problem on heat conduction, the solution must be transient, *that is, u* must decrease whenever the time *t* increases. Therefore, among these, (5.171) is the only suitable solution of Eq. (5.164), *that is,*

$$u(x, t) = (A \cos \lambda x + B \sin \lambda x)e^{-\lambda^2 \alpha^2 t}.$$

Example 5.43. *An insulated rod 30 cm long has its ends at $20°C$ and $80°C$ until steady state conditions prevail. The temperature at each end then suddenly reduced to $0°C$ and kept so. Find the resulting temperature distribution in the rod.*

Solution.   *The heat conduction in the rod is governed by one-dimensional heat equation given by*

$$\frac{\partial u}{\partial t} = \alpha^2 \frac{\partial^2 u}{\partial x^2}. \tag{5.172}$$

Under steady state conditions, the heat flow is independent of time so that $\dfrac{\partial u}{\partial t} = 0$ *and hence Eq. (5.172) reduces to ODE*

$$\frac{d^2 u_s}{dx^2} = 0$$

whose general solution is

$$u_s(x) = c_1 x + c_2. \tag{5.173}$$

Putting $x = 0$ and $x = 30$ in Eq. (5.173) and using the fact that $u(0) = 20$ and $u(30) = 80$, we obtain $c_2 = 20$ and $c_1 = 2$. Substituting these values in Eq. (5.173), we get the initial condition given by

$$u(x, 0) = u_s(x) = 2x + 20. \tag{5.174}$$

Given BCS are

$$u(0, t) = u(30, t) = 0. \tag{5.175}$$

The suitable solution of Eq. (5.172) is

$$u(x, t) = (A \cos \lambda x + B \sin \lambda x) e^{-\lambda^2 \alpha^2 t}. \tag{5.176}$$

Applying BC (5.175) in Eq. (5.176), we obtain the most general complete solution of (5.172)

$$u(x, t) = \sum_{n=1}^{\infty} b_n \sin \frac{n\pi x}{30} e^{-(n^2 \pi^2 \alpha^2 t)/900}. \tag{5.177}$$

Finally, applying IC (5.174) in Eq. (5.177), we get

$$2x + 20 = \sum_{n=1}^{\infty} b_n \sin \frac{n\pi x}{30}$$

which is a Fourier sine series in the interval $0 < x < 30$. Hence, we have

$$b_n = \frac{2}{30} \int_0^{30} (2x + 20) \sin \frac{n\pi x}{30}$$

$$= \frac{40}{n\pi}[1 + 4(-1)^{n+1}] \quad \text{(after simplifying)}.$$

Putting these values of b_n in Eq. (5.177), we obtain

$$u(x,t) = \frac{40}{\pi} \sum_{n=1}^{\infty} \frac{1 + 4(-1)^{n+1}}{n} \sin \frac{n\pi x}{30} e^{-(n^2 \pi^2 \alpha^2 t)/900}$$

which is the resulting temperature in the rod as desired.

Example 5.44. *The temperatures at the ends A and B of a rod 100 cm in length are maintained at $0°C$ and $100°C$, respectively, until steady state conditions prevail. Thereafter, at $t = 0$, the two ends are suddenly insulated. Find the temperature $u(x,t)$ at a distance x from A at a time t.*

Solution. *The heat conduction in the rod is governed by one-dimensional heat equation given by*

$$\frac{\partial u}{\partial t} = \alpha^2 \frac{\partial^2 u}{\partial x^2}. \tag{5.178}$$

Under steady state conditions, Eq. (5.178) reduces to

$$\frac{d^2 u_s}{dx^2} = 0$$

whose general solution is

$$u_s(x) = c_1 x + c_2.$$

Using $u(0) = 0$ and $u(100) = 100$, the above equation reduces to

$$u(x,0) = u_s(x) = x. \tag{5.179}$$

After steady state condition, the ends are insulated, so boundary conditions are

$$\frac{\partial u}{\partial x}\Big|_{x=0} = \frac{\partial u}{\partial x}\Big|_{x=100} = 0. \tag{5.180}$$

The suitable solution of Eq. (5.178) is

$$u(x,t) = (A \cos \lambda x + B \sin \lambda x)e^{-\lambda^2 \alpha^2 t}.$$

$$\therefore \frac{\partial u}{\partial x} = \lambda(-A \sin \lambda x + B \cos \lambda x)e^{-\lambda^2 \alpha^2 t}.$$

Applying BC (5.180) in the above equation, we obtain $B = 0$ and $\lambda = \dfrac{n\pi}{100}$ and therefore, the most general complete solution of (5.178) is

$$u(x,t) = \frac{a_0}{2} + \sum_{n=1}^{\infty} a_n \cos \frac{n\pi x}{100} e^{-(n^2 \pi^2 \alpha^2 t)/10000}. \tag{5.181}$$

Finally, applying IC (5.179) in Eq. (5.181), we get

$$x = \frac{a_0}{2} + \sum_{n=1}^{\infty} a_n \cos \frac{n\pi x}{100}$$

which is a Fourier cosine series in the interval $0 < x < 100$. Hence, we find

$$a_0 = \frac{2}{100} \int_0^{100} x\,dx = 100$$

$$a_n = \frac{2}{100} \int_0^{100} x \cos \frac{n\pi x}{100}\,dx$$

$$= \begin{cases} 0 & \text{if } n = 2m \\ -\dfrac{400}{\pi^2(2m-1)^2} & \text{if } n = 2m - 1 \quad (m = 1,2,3,\ldots). \end{cases}$$

Putting these values of a_0 and a_n in Eq. (5.181), we obtain

$$u(x,t) = 50 - \frac{400}{\pi^2} \sum_{m=1}^{\infty} \frac{1}{(2m-1)^2} \cos \frac{(2m-1)\pi x}{100} e^{-\frac{(2m-1)^2 \pi^2 \alpha^2 t}{10000}}$$

which is the resulting temperature in the rod as desired.

5.6.3 Steady State Temperature Distribution: Laplace Equation in 2D

Consider the flow of heat in a metal plate of uniform thickness a placed in xy-plane as shown in Figure 5.4. The length of the plate may be finite or infinite. In the steady state, the temperature of the plate is independent of time t and depends only on the spatial coordinates (see Examples 5.43 and 5.44).

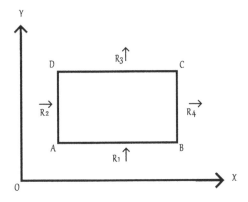

FIGURE 5.4 Heat conduction in a plate

At any point (x, y) of the plate, the steady state temperature $u(x, y)$ is governed by a second-order PDE of the form

$$\frac{\partial^2 u}{\partial x^2} + \frac{\partial^2 u}{\partial y^2} = 0. \tag{5.182}$$

Equation (5.182) is called **two-dimensional Laplace equation** named after *Pierre-Simon Laplace* (1749–1827). This is a typical elliptic equation. Solutions of Laplace equation are called harmonic functions.

Analogously, the steady state temperature $u(x, y, z)$ at a point (x, y, z) of a solid body is governed by three-dimensional Laplace equation:

$$\nabla^2 u = 0$$

where the operator $\nabla^2 := \dfrac{\partial^2}{\partial x^2} + \dfrac{\partial^2}{\partial y^2} + \dfrac{\partial^2}{\partial z^2}$ is called Laplacian.

Derivation of Two-Dimensional Laplace Equation: Consider a rectangular element $ABCD$ of the plate with sides $\triangle x$ and $\triangle y$ as shown in the figure. If $A \equiv (x, y)$, then $B \equiv (x + \triangle x, y)$, $C \equiv (x + \triangle x, y + \triangle y)$, and $D \equiv (x, y + \triangle y)$. Let ρ be the density, c the specific heat, k the thermal conductivity, and a the thickness of the plate. We know that heat amount is proportional to the area and the temperature gradient normal to the area. Therefore, the amount of heat entering from the side AB is

$$R_1 = -ka \triangle x \left(\frac{\partial u}{\partial y}\right)_y.$$

Similarly, the amount of heat entering from the side AD is

$$R_2 = -ka \triangle y \left(\frac{\partial u}{\partial y}\right)_x.$$

Also, the amount of heat leaving through the side CD is

$$R_3 = -ka \triangle x \left(\frac{\partial u}{\partial y}\right)_{y+\triangle y}.$$

Finally, the amount of heat leaving through the side BC is

$$R_4 = -ka \triangle y \left(\frac{\partial u}{\partial y}\right)_{x+\triangle x}.$$

Thus, the total gain of heat by $ABCD$ is

$$(R_1 + R_2) \; - \; (R_3 + R_4) = (R_1 - R_3) + (R_2 - R_4)$$
$$= \; ka \triangle x \triangle y \left[\frac{(\partial u/\partial x)_{x+\triangle x} - (\partial u/\partial x)_x}{\triangle x} + \frac{(\partial u/\partial y)_{y+\triangle y} - (\partial u/\partial y)_y}{\triangle y}\right].$$

The total gain of heat is proportional to the rate $\partial u/\partial t$ of temperature change. But in the steady state, $\partial u/\partial t = 0$. Therefore, we have

$$ka \, \triangle x \, \triangle y \left[\frac{(\partial u/\partial x)_{x+\triangle x} - (\partial u/\partial x)_x}{\triangle x} + \frac{(\partial u/\partial y)_{y+\triangle y} - (\partial u/\partial y)_y}{\triangle y} \right] = 0$$

which yields that

$$\frac{(\partial u/\partial x)_{x+\triangle x} - (\partial u/\partial x)_x}{\triangle x} + \frac{(\partial u/\partial y)_{y+\triangle y} - (\partial u/\partial y)_y}{\triangle y} = 0.$$

Taking the limit as $\triangle x \to 0$ and $\triangle y \to 0$, the above equation reduces to Eq. (5.182).

Variables Separable Solution of Laplace's Equation: By the method of separation of variables, let $u(x, y) = X(x)Y(y)$ be the solution of the Laplace equation, then Eq. (5.182) reduces to

$$X''Y + XY'' = 0.$$

On separating the variables, the above equation reduces to

$$\frac{X''}{X} = -\frac{Y''}{Y} = k \text{ (say)}$$

which gives us two ODEs:

$$X'' - kX = 0 \text{ and } Y'' + kY = 0. \tag{5.183}$$

On solving (5.183), we get

(i) When $k = 0$, then

$$u(x, y) = (c_1 x + c_2)(c_3 y + c_4)$$

(ii) When $k > 0$ (say $k = \lambda^2$), then

$$u(x, y) = (c_5 \cos \lambda x + c_6 \sin \lambda x)(c_7 e^{\lambda y} + c_8 e^{-\lambda y})$$

(iii) When $k < 0$ (say $k = -\lambda^2$), then

$$u(x, y) = (c_9 e^{\lambda x} + c_{10} e^{-\lambda x})(c_{11} \cos \lambda y + c_{12} \sin \lambda y).$$

Out of these three possible solutions, we have to choose the solution, which is consistent with the physical nature of the problem.

Example 5.45. *A square plate is bounded by the lines $x = 0, y = 0, x = 20, y = 20$ and its faces are insulated. The temperature along the upper horizontal edge is given by $u(x, 20) = x(20 - x)$ for $0 < x < 20$, whereas the other three edges are maintained at $0°C$. Find the steady state temperature distribution in the plate.*

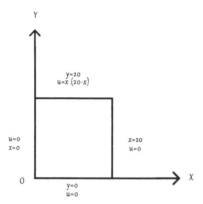

Solution. The steady state temperature distribution in the plate is governed by Laplace equation given by

$$\frac{\partial^2 u}{\partial x^2} + \frac{\partial^2 u}{\partial y^2} = 0. \tag{5.184}$$

Boundary conditions are

$$u(0, y) = u(20, y) = u(x, 0) = 0 \tag{5.185}$$

$$u(x, 20) = x(20 - x) \quad (0 < x < 20). \tag{5.186}$$

In lieu of BC (5.185), the suitable solution of Eq. (5.184) is

$$u(x, y) = (A \cos \lambda x + B \sin \lambda x)(Ce^{\lambda y} + De^{-\lambda y}). \tag{5.187}$$

Applying BC $u(0, y) = 0$, we obtain $A = 0$ and hence the solution (5.187) becomes

$$u(x, y) = B \sin \lambda x (Ce^{\lambda y} + De^{-\lambda y}). \tag{5.188}$$

Applying BC $u(x, 0) = 0$, we obtain $C + D = 0$, *that is,* $D = -C$. Therefore, the solution (5.188) reduces to

$$u(x, y) = BC \sin \lambda x (e^{\lambda y} - e^{-\lambda y})$$

or

$$u(x, y) = b \sin \lambda x \sinh \lambda y, \quad (\text{where } b = 2BC).$$

Now, applying BC $u(20, y) = 0$, we get $\lambda = \dfrac{n\pi}{20}$. Therefore, the most general complete solution of (5.184) is

$$u(x, y) = \sum_{n=1}^{\infty} b_n \sin \frac{n\pi x}{20} \sinh \frac{n\pi y}{20}. \tag{5.189}$$

Finally, applying BC (5.186) in Eq. (5.189), we get

$$20x - x^2 = \sum_{n=1}^{\infty} b_n \sinh n\pi \sin \frac{n\pi x}{20}$$

which is a Fourier sine series in the interval $0 < x < 20$ with Fourier coefficients $b_n \sinh n\pi$. Hence, we find

$$b_n \sinh n\pi = \frac{2}{20} \int_0^{20} (20x - x^2) \sin \frac{n\pi x}{20} dx$$

$$= -\frac{1600}{n^3 \pi^3} \left[(-1)^n - 1 \right] \quad \text{(after simplification)}$$

implying thereby

$$b_n = \frac{1600}{n^3 \pi^3 \sinh n\pi} \left[1 - (-1)^n \right]$$

$$= \begin{cases} 0 & \text{if } n = 2m \\ \dfrac{3200}{(2m-1)^3 \pi^3 \sinh(2m-1)\pi} & \text{if } n = 2m - 1 \quad (m = 1, 2, 3, ...). \end{cases}$$

Putting these values of b_n in Eq. (5.194), we obtain

$$u(x, y) = \frac{3200}{\pi^3} \sum_{m=1}^{\infty} \frac{1}{(2m-1)^3 \sinh(2m-1)\pi} \sin \frac{(2m-1)\pi x}{20} \sinh \frac{(2m-1)\pi y}{20}$$

which is the steady state temperature in the plate as desired.

Example 5.46. *An infinitely long plane uniform plate is bounded by two parallel edges and an end at right angles to them. The breadth is π. This end is maintained at a temperature u_0 at all points and the other edges are at zero temperature. Show that in the steady state, temperature is given by*

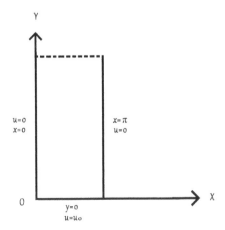

$$u(x, y) = \frac{4u_0}{\pi} \left[e^{-y} \sin x + \frac{1}{3} e^{-3y} \sin 3x + \frac{1}{5} e^{-5y} \sin 3x + \dots \right].$$

Solution. *The steady state temperature distribution in the plate is governed by Laplace equation given by*

$$\frac{\partial^2 u}{\partial x^2} + \frac{\partial^2 u}{\partial y^2} = 0. \tag{5.190}$$

Boundary conditions are

$$u(0, y) = 0 = u(\pi, y) \tag{5.191}$$

$$\lim_{y \to \infty} u(x, y) = 0 \quad (0 < x < \pi) \tag{5.192}$$

$$u(x, 0) = u_0 \quad (0 < x < \pi). \tag{5.193}$$

In lieu of BC (5.191), the suitable solution of Eq. (5.190) is

$$u(x, y) = (A \cos \lambda x + B \sin \lambda x)(Ce^{\lambda y} + De^{-\lambda y}).$$

Applying BC $u(0, y) = 0$, we obtain $A = 0$ and hence the solution (5.6.3) becomes

$$u(x, y) = B \sin \lambda x (Ce^{\lambda y} + De^{-\lambda y}). \tag{5.194}$$

Taking $y \to \infty$ in the above equation and applying BC (5.192), we get

$$B \sin \lambda x \lim_{y \to \infty} (Ce^{\lambda y} + De^{-\lambda y}) = 0 \Rightarrow \lim_{y \to \infty} [Ce^{\lambda y}] = 0,$$

which is possible only when $C = 0$. Thus, the solution (5.194) reduces to

$$u(x, y) = b \sin \lambda x . e^{-\lambda y} \quad (\text{where } b = BD)$$

Now, applying BC (5.192), we get $\lambda = n$. Therefore, the most general complete solution of (5.190) is

$$u(x, y) = \sum_{n=1}^{\infty} b_n \sin nx . e^{-ny}. \tag{5.195}$$

Finally, applying BC (5.193), in Eq. (5.195), we get

$$u_0 = \sum_{n=1}^{\infty} b_n \sin nx$$

which is a Fourier sine series in the interval $0 < x < \pi$. Hence, we find

$$
\begin{aligned}
b_n &= \frac{2}{\pi} \int_0^\pi u_0 \sin nx \, dx \\
&= \frac{2u_0}{n\pi} \left[1 - (-1)^n \right] \\
&= \begin{cases} 0 & \text{if } n = 2m \\ \dfrac{4u_0}{(2m-1)\pi} & \text{if } n = 2m - 1 \quad (m = 1, 2, 3, ...). \end{cases}
\end{aligned}
$$

Putting these values of b_n in Eq. (5.195), we obtain

$$
u(x, y) = \frac{4u_0}{\pi} \sum_{m=1}^\infty \frac{1}{2m - 1} e^{-(2m-1)y} \sin(2m - 1)x
$$

or

$$
u(x, y) = \frac{4u_0}{\pi} \left[e^{-y} \sin x + \frac{1}{3} e^{-3y} \sin 3x + \frac{1}{5} e^{-5y} \sin 3x + ... \right].
$$

5.6.4 Transmission Line: Telegraph Equations

The **telegraph equations** (also known as **telegrapher's equations**) are a pair of second-order hyperbolic linear PDEs, which describe the current and potential on an electrical transmission line (a specialised insulated cable) with distance and time. The equations are investigated by *Oliver Heaviside* (1850–1925), a British mathematician and physicist.

Consider the one-dimensional flow of electricity in a long transmission line placed along x-axis such that one end of the cable is at origin O as shown in Figure 5.5. At a distance x from O and at time t, let $I(x, t)$ and $V(x, t)$ represent the current and the potential on the transmission line, respectively. Suppose R denotes the resistance of transmission line, L denotes the inductance of transmission line, C denotes the capacitance to the ground and G denotes the conductance to the ground. Then by Ohm's law, the voltage drop across the resistor is given by

$$
V = IR.
$$

The voltage drop across the inductor, where an inductor gives a differentiator circuit, is

$$
V = L \frac{dI}{dt}.
$$

The voltage drop across the capacitor, where a capacitor gives an integrator circuit, is

$$
V = \frac{1}{C} \int I \, dt.
$$

On differentiating the above equation w.r.t. t, we find that the current through the capacitor is

$$I = C\frac{dV}{dt}$$

Since the conductance is reciprocal of the resistance, therefore by using Ohm's law, the current through the conductor is

$$I = GV.$$

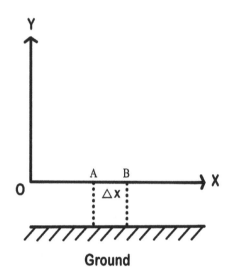

FIGURE 5.5 Current flow in a transmission line

Take a linear element AB of small length $\triangle x$ on the transmission line. Along this element, the voltage drop across the resistor will be $IR \triangle x$, whereas across the inductor will be $L \triangle x\frac{\partial I}{\partial t}$. Hence, the net voltage drop in AB is

$$-\triangle V = IR \triangle x + L \triangle x\left(\frac{\partial I}{\partial t}\right)$$

so that

$$-\frac{\triangle V}{\triangle x} = IR + L\left(\frac{\partial I}{\partial t}\right).$$

Taking the limit as $\triangle x \to 0$, the above equation gives rise to

$$-\frac{\partial V}{\partial x} = IR + L\left(\frac{\partial I}{\partial t}\right). \tag{5.196}$$

Further, along AB, the current loss through the capacitor will be $C \triangle x \dfrac{\partial V}{\partial t}$ whereas through the conductor will be $GV \triangle x$. Hence, the net current loss in AB is

$$- \triangle I = C \triangle x \left(\frac{\partial V}{\partial t} \right) + GV \triangle x$$

so that

$$-\frac{\triangle I}{\triangle x} = C \left(\frac{\partial V}{\partial t} \right) + GV.$$

Taking the limit as $\triangle x \to 0$, the above equation gives rise to

$$- \frac{\partial I}{\partial x} = C \left(\frac{\partial V}{\partial t} \right) + GV. \tag{5.197}$$

Differentiating Eq. (5.196) w.r.t x and Eq. (5.197) w.r.t. t partially, we get respectively

$$\frac{\partial^2 V}{\partial x^2} + R\frac{\partial I}{\partial x} + L\frac{\partial^2 I}{\partial x \partial t} = 0 \tag{5.198}$$

and

$$\frac{\partial^2 I}{\partial x \partial t} + G\frac{\partial V}{\partial t} + C\frac{\partial^2 V}{\partial t^2} = 0. \tag{5.199}$$

Eliminating $\dfrac{\partial I}{\partial x}$ and $\dfrac{\partial^2 I}{\partial x \partial t}$ from (5.197), (5.198), and (5.199), we obtain

$$\frac{\partial^2 V}{\partial x^2} = CL \left(\frac{\partial^2 V}{\partial t^2} \right) + (CR + GL)\frac{\partial V}{\partial t} + RGV. \tag{5.200}$$

Similarly, differentiating (5.196) w.r.t t and Eq. (5.197) w.r.t x partially and eliminating $\dfrac{\partial V}{\partial x}$ and $\dfrac{\partial^2 V}{\partial x \partial t}$, we get

$$\frac{\partial^2 I}{\partial x^2} = CL \left(\frac{\partial^2 I}{\partial t^2} \right) + (CR + GL)\frac{\partial I}{\partial t} + RGI. \tag{5.201}$$

Equations (5.200) and (5.201) are called **telegraph equations**. Both of these are hyperbolic equations. The second and last terms appearing in RHS of telegraph equations are called dissipation terms and dispersion terms, respectively. These equations have dual nature as an interchange of I and V retains the same pair of equations.

Example 5.47. *A transmission line 1000 km long is initially under steady state conditions with potential 1300 volts at the sending end and 1200 volts at receiving end. The terminal end of the line is suddenly grounded, but the potential at the source is kept at 1300 volts. If the inductance and conductance are assumed to be negligible, then find the potential $V(x, t)$.*

Solution. Here $L \approx 0$ and $G \approx 0$. Hence, the telegraph Eq. (5.200) reduces to

$$\frac{\partial V}{\partial t} = \frac{1}{RC} \frac{\partial^2 V}{\partial x^2}. \tag{5.202}$$

Using the method of separation of variables, likewise the heat equation (5.164), a suitable solution of (5.202) is

$$V(x, t) = (A \cos \lambda x + B \sin \lambda x) e^{-\lambda^2 t / RC}. \tag{5.203}$$

At steady state condition, $\dfrac{\partial V}{\partial t} = 0$ *so that* $\dfrac{\partial^2 V_0}{\partial x^2} = 0$, *whose general solution is* $V_0(x) = ax + b$. *Applying* $V_0(0) = 1300$ *and* $V_0(1000) = 1200$, *we get* $b = 0$ *and* $a = -0.1$. *Therefore, the initial steady state potential is*

$$V_0 = 1300 - 0.1x = V(x, 0). \tag{5.204}$$

As the ends suddenly grounded, the BC are

$$V(0, t) = 0 = V(1000, t) \quad \forall\, t > 0.$$

Using these BC, Eq. (5.203) gives the transient part of V by

$$V_t(x, t) = \sum_{n=1}^{\infty} b_n \sin\left(\frac{n\pi x}{1000}\right) \exp\left(-\frac{n^2 x^2 t}{10^6 RC}\right).$$

Further, after grounding the terminal end, the potential at the source is kept at 1300 volts. Thus, the steady voltage, when steady state conditions are ultimately reached, becomes

$$V_s = 1300 - 1.3x. \tag{5.205}$$

Thus, the potential is

$$V(x, t) = V_s + V_t(x, t) = 1300 - 1.3x + \sum_{n=1}^{\infty} b_n \sin\left(\frac{n\pi x}{1000}\right) \exp\left(-\frac{n^2 x^2 t}{10^6 RC}\right).$$

Applying IC (5.204), we get

$$1300 - 0.1x = 1300 - 1.3x + \sum_{n=1}^{\infty} b_n \sin\left(\frac{n\pi x}{1000}\right)$$

so that

$$1.2x = \sum_{n=1}^{\infty} b_n \sin\left(\frac{n\pi x}{1000}\right)$$

which is half range Fourier sine series. We find

$$b_n = \frac{2}{1000} \int_0^{1000} 1.2x \sin\left(\frac{n\pi x}{1000}\right) dx = \frac{2400}{\pi} \frac{(-1)^{n+1}}{n}$$

Therefore, the required electric potential is

$$V(x,t) = 1300 - 1.3x + \frac{2400}{\pi} \sum_{n=1}^{\infty} \frac{(-1)^{n+1}}{n} \sin\left(\frac{n\pi x}{1000}\right) \exp\left(-\frac{n^2\pi^2 t}{10^6 RC}\right).$$

5.6.5 Gravitational Potential: Poisson and Laplace Equations in 3D

Besides determining steady state temperature distributions, Laplace equation plays an important role in potential theory arising in different physical phenomena, *for example,* gravitational potential, electrostatic potential, magnetometric potential, velocity potential occurring in fluid mechanics, harmonic potential, *etc.* Due to this fact, Laplace equation is sometimes referred to as the potential equation and harmonic functions are also termed as potential functions. In this subsection, we derive the Laplace equation and its associated non-homogeneous equation often called Poisson's equation for gravitational potential.

Recall that a gravitational field is indeed a vector field that represents the effect of gravity at a point in space due to the presence of a massive object. Gravitational potential at a point in a gravitational field is defined as the amount of work done in moving unit mass under the attraction of a gravitational field from infinity to that point. Naturally, the gravitational potential is a scalar field. By the conservative property of gravitational fields, it can be easily shown that the gravitational field is the gradient of the gravitational potential.

Consider a closed surface **S** consisting of particles of masses $m_1, m_2, \ldots, m_n$. Let P be any point on surface **S** and $\sum_{i=1}^{n} m_i = M$ be the total mass inside **S**. If ρ is the mass density function and V is the volume in which the masses are distributed throughout, then

$$M = \int\int\int_V \rho dV. \tag{5.206}$$

Let $g_1, g_2, \ldots, g_n$ be the gravitational fields at P due to the masses $m_1, m_2, ..., m_n$ respectively within **S**. Thus, the entire gravitational field at P will be $g = \sum_{i=1}^{n} g_i$. Using Gauss's law, we have

$$\int\int_S \vec{g}.d\vec{S} = -4\pi GM \tag{5.207}$$

where G is the gravitational constant. Applying Gauss divergence theorem, we get

$$\int\int_S \vec{g}.d\vec{S} = \int\int\int_V \nabla.\vec{g}dV. \tag{5.208}$$

Making use of (5.206), (5.207), and (5.208), we obtain

$$\int\int\int_V (\nabla\vec{g} + 4\pi G\rho)dV = 0$$

yielding thereby

$$\nabla.\vec{g} = -4\pi G\rho. \qquad (5.209)$$

As the gravitational field is conservative, we have

$$\vec{g} = \nabla\phi \qquad (5.210)$$

where ϕ is gravitational potential at P. From (5.209) and (5.210), we get

$$\nabla^2\phi = \nabla.\nabla\phi = \nabla.\vec{g} = -4\pi G\rho$$

By a suitable choice of units, we make $G = 1$, then the above equation becomes

$$\nabla^2\phi = -4\pi\rho$$

which is known as **Poisson equation** for gravitational potential. In free space, $\rho = 0$ and hence the above equation reduces to

$$\nabla^2\phi = 0$$

which is **Laplace equation** or **potential equation**. Thus, we conclude that the gravitational potential satisfies Laplace equation in a region containing no masses, whereas it satisfies Poisson's equation in a region containing continuously distributed masses.

In case of a continuous distribution of mass in volume V, if the point $P(x, y, z)$ is outside the body, then we have

$$\phi = \int\int\int_V \frac{\rho}{r}d\xi d\eta d\zeta$$

where ρ is the density of a mass at (ξ, η, ζ) and $r = \sqrt{(x-\xi)^2 + (y-\eta)^2 + (z-\zeta)^2}$. We assume that the resulting derived functions are continuous. Thus, differentiating partially w.r.t. x twice, we get

$$\phi_x = -\int\int\int_V \frac{\rho(x-\xi)}{r^3}d\xi d\eta d\zeta,$$

$$\phi_{xx} = -\int\int\int_V \left[\frac{\rho}{r^3} - \frac{3\rho(x-\xi)^2}{r^5}\right]d\xi d\eta d\zeta.$$

Similarly, we can obtain

$$\phi_{yy} = -\int \int \int_V \left[\frac{\rho}{r^3} - \frac{3\rho(y - \eta)^2}{r^5} \right] d\xi d\eta d\zeta$$

and

$$\phi_{zz} = -\int \int \int_V \left[\frac{\rho}{r^3} - \frac{3\rho(z - \zeta)^2}{r^5} \right] d\xi d\eta d\zeta.$$

Adding the last three equations, we get

$$\nabla^2 \phi = \phi_{xx} + \phi_{yy} + \phi_{zz} = 0.$$

Hence, in this case, we again find Laplace equation.

5.6.6 Burgers' Equation

Burgers' equation is a second-order quasilinear PDE, which occurs in various areas of applied sciences, such as fluid mechanics, non-linear acoustics, gas dynamics, and traffic flow. This equation was first introduced by an English mathematician *Harry Bateman* (1882–1946) in a physical context. Thereafter, a Dutch physicist *Johannes Martinus Burgers* (1895–1981) derived this equation to study the theory of turbulence. Due to this reason, Burgers' equation is sometimes referred to as **Bateman–Burgers equation**. Burgers' equation is given by

$$\frac{\partial u}{\partial t} + u \frac{\partial u}{\partial x} = \mu \frac{\partial^2 u}{\partial x^2} \tag{5.211}$$

where $u(x, t)$ is the **fluid velocity field** and μ is a positive constant called the **kinematic viscosity**. In Eq. (5.211), u_t is the **unsteady term**, uu_x measures **convective term**, whereas μu_{xx} represents **dissipative term**. Burgers' equation is a balance between time evolution, non-linearity, and diffusion.

Around 1950, *Eberhard Hopf* (1902–1983) and *Julian David Cole* (1925–1999) independently obtained an analytic solution of Eq. (5.211) by the two-step transformation

$$u = \psi_x, \quad \psi = -2\mu \log \phi.$$

This transformation is called *Hopf–Cole transformation*. Rewrite the above transformation as

$$u = -2\mu \frac{\phi_x}{\phi}.$$

Thus, we have

$$u_t = -2\mu \left(\frac{\phi_{xt}}{\phi} \right) + 2\mu \left(\frac{\phi_x \phi_t}{\phi^2} \right)$$

$$u_x = -2\mu \left(\frac{\phi_{xx}}{\phi} \right) + 2\mu \left(\frac{\phi_x^2}{\phi^2} \right)$$

$$u_{xx} = -2\mu \left(\frac{\phi_{xxx}}{\phi} \right) + 6\mu \left(\frac{\phi_x \phi_{xx}}{\phi^2} \right) - 4\mu \left(\frac{\phi_x^3}{\phi^3} \right)$$

Substituting these expressions into Eq. (5.211), we obtain

$$\frac{\phi_x}{\phi}(\mu\phi_{xx} - \phi_t) - (\mu\phi_{xx} - \phi_t)_x = 0. \tag{5.212}$$

Therefore, if $\phi(x, t)$ satisfies the diffusion equation

$$\phi_t = \mu\phi_{xx}. \tag{5.213}$$

then it also solves Eq. (5.212) trivially and hence the resulting $u(x, t)$ satisfies (5.211). Therefore, we conclude that under Hopf–Cole transformation, (quasilinear) Burgers' equation reduces to the well-known (linear) diffusion equation, which makes life easy to get the solution of Burgers' equation. Due to this fact, Eq. (5.213) is called linearised Burgers' equation.

▶ Exercises

Find the general solution of the linear PDEs described in problems 1–13.

1. $2yq + y^2t = 1$.   Ans. $z = \log y - \frac{1}{y}\phi(x) + \psi(x)$.

2. $yt - q = xy$.   Ans. $z = \sin y + e^y\theta(x) + \psi(x)$.

3. $ys - p = xy^2 \cos xy$.   Ans. $z = -\cos xy + y\theta(x) + \psi(y)$.

4. $ys + p = \cos(x + y) - y\sin(x + y)$.   Ans. $yz = y\sin(x + y) + \theta(x) + \psi(y)$.

5. $yt + 2q = (9y + 6)e^{2x+3y}$.   Ans. $z = e^{2x+3y} - \frac{1}{y}\phi(x) + \psi(x)$.

6. $xr + 2p = (9x + 6)e^{3x+2y}$.   Ans. $z = e^{3x+2y} - \frac{1}{x}\phi(y) + \psi(y)$.

7. $xs + q = 4x + 2y + 2$.
Ans. $xz = 2x^2y + xy^2 + 2xy + \theta(y) + \phi(x)$.

8. $s + t + q = 0$.   Ans. $ze^x = \theta(x) + \psi(x - y)$.

9. $s + r = x\cos(x + y)$. Ans. $z = -\frac{x}{2}\cos(x + y) + \frac{3}{4}\sin(x + y) + \theta(x - y) + \psi(y)$.

10. $r + \frac{y}{x}s = 15xy^2$.   Ans. $z = \frac{5}{4}x^3y^2 + y\theta\left(\frac{y}{x}\right) + \psi(y)$.

11. $2xr - ys + 2p = 4xy^2$.   Ans. $z = x^2y^2 + \theta(xy^2) + \psi(y)$.

12. $xs + yt + q = 10x^3y$.   Ans. $z = x^3y^2 + \theta\left(\frac{y}{x}\right) + \psi(x)$.

13. $ys - 2xr - 2p = 6xy$.   Ans. $z = -x^2y + \theta(xy^2) + \psi(y)$.

14. Find the surface passing through the parabolas $z = 0, y^2 = 4ax$ and $z = 1, y^2 = -4ax$ and satisfying the equation $xr + 2p = 0$.
Ans. $8axz = 4ax - y^2$.

15. Find a surface satisfying $r + s = 0$ and touching the elliptic paraboloid $z = 4x^2 + y^2$ along its section by the plane $y = 2x + 1$.
Ans. $z + 4x^2 + y^2 - 8xy + 8x - 4y + 2 = 0$.

Reduce the PDEs described in problems 16–31 into canonical form and find their general solutions whenever possible.

16. $4r + 5s + t + p + q = 2$. $\qquad$ Ans. $z_{\xi\eta} = \dfrac{1}{3}z_\eta - \dfrac{8}{9}$.

17. $r + 2s + 5t + p - 2q - 3z = 0$. $\qquad$ Ans. $z_{\xi\xi} + z_{\eta\eta} = \dfrac{1}{4}z_\xi - \dfrac{1}{2}z_\eta + \dfrac{3}{4}z$.

18. $r - 4s + 13t - 9q = 0$. $\qquad$ Ans. $z_{\xi\xi} + z_{\eta\eta} = z_\xi$.

19. $y^2 r - x^2 t = 0$. $\qquad$ Ans. $z_{\xi\eta} = \dfrac{1}{2(\xi^2 - \eta^2)}(\eta z_\xi - \xi z_\eta)$.

20. $t - c^2 r = 0$. Ans. $z_{\xi\eta} = 0$, $z = \phi(x + cy) + \psi(x - cy)$.

21. $\sin^2 x r + \sin 2x s + \cos^2 y t = x$.
Ans. $z_{\eta\eta} = \dfrac{1}{1 - e^{2(\eta - \xi)}} \sin^{-1} e^{\eta - \xi} - z_\xi$.

22. $r - 2\sin x s - \cos^2 y t - \cos x q = 0$.
Ans. $z_{\xi\eta} = 0$, $z = \phi(\cos x + x - y) + \psi(\cos x - x - y)$.

23. $(n - 1)^2 r - y^{2n} t = n y^{2n-1} q$. Ans. $z_{\xi\eta} = 0$, $z = \phi(x + y^{1-n}) + \psi(x - y^{1-n})$.

24. $(y - 1)r - (y^2 - 1)s + y(y - 1)t + p - q = 2ye^{2x}(1 - y)^3$.
Ans. $z_{\xi\eta} = 2\xi$, $z = \theta(x + y) + \psi(ye^x) + y^2(x + y)e^{2x}$.

25. $x^2 r - y^2 t - 2yq = 0$, $x \neq 0, y \neq 0$.
Ans. $z_{\xi\eta} + \dfrac{1}{\xi}z_\eta$, $z = \sqrt{\dfrac{x}{y}}\phi(xy) + \psi\left(\dfrac{y}{x}\right)$.

26. $x^2 r + 2xys + y^2 t = 0$. $\qquad$ Ans. $z_{\eta\eta} = 0$, $z = y\phi\left(\dfrac{y}{x}\right) + \psi\left(\dfrac{y}{x}\right)$.

27. $x^2 r - 2xys + y^2 t - xp + 3yq = \dfrac{8y}{x}$.
Ans. $z_{\eta\eta} = \dfrac{2}{\eta} - \dfrac{2}{\eta}z_\eta$, $z = \dfrac{y}{x} + \phi(xy) + \psi(xy)$.

28. $x^2(y - 1)r - x(y^2 - 1)s + y(y - 1)t + xyp - q = 0$. $\qquad$ Ans. $z_{\xi\eta} = 0$, $z = \phi(xy) + \psi(xe^y)$.

29. $y^2 r + 2xys + 2x^2 t + yq = 0$, $x \neq 0, y \neq 0$. $\qquad$ Ans. $z_{\xi\xi} + z_{\eta\eta} + \dfrac{1}{\xi - \eta}z_\xi + \dfrac{1}{2\eta}z_\eta = 0$.

30. $y^2 r + t = 0$ $\qquad$ Ans. $z_{\xi\xi} + z_{\eta\eta} + \dfrac{1}{2\xi}z_\xi = 0$.

31. $y^2 r + x^2 t = 0$, $x \neq 0, y \neq 0$ $\qquad$ Ans. $z_{\xi\xi} + z_{\eta\eta} + \dfrac{1}{2}\left(\dfrac{1}{\xi}z_\xi + \dfrac{1}{\eta}z_\eta\right) = 0$.

Find the solution of the Cauchy's problems in Exercises 32–34.

32. $t - c^2 r = 0$, $z(x, 0) = \sin x, q(x, 0) = x^2$. $\qquad$ Ans. $z = \sin x \cos cy + x^2 y + \dfrac{1}{3}c^2 y^3$.

33. $r = t$, $z(x, 0) = \sin x, q(x, 0) = \cos x$. $\qquad$ Ans. $z = \sin(x + y)$.

34. $r + t = 0$, $z(x, 0) = 0, q(x, 0) = \dfrac{1}{\lambda}\sin \lambda x$ $\qquad$ Ans. $z = \dfrac{1}{\lambda^2}\sin \lambda x \sinh \lambda y$.

Classify the PDEs described in problems 35–37.

35. $u_{xx} + u_{yy} + u_{zz} = 0$ (Laplace Equation).   Ans. elliptic.

36. $u_{xx} + u_{yy} + u_{zz} = \dfrac{1}{c^2}u_{tt}$, $c \neq 0$(Wave Equation).   Ans. hyperbolic.

37. $u_{xx} + u_{yy} + u_{zz} = \dfrac{1}{\alpha^2}u_t$, $\alpha \neq 0$ (Heat Equation).   Ans. parabolic.

Solve the PDEs described in problems 38–53 by Monge's method.

38. $r = a^2 t$.   Ans. $z = \theta(y + ax) + \eta(y - ax)$.

39. $2x^2 r - 5xys + 2y^2 t + 2(px + qy) = 0$.   Ans. $z = \theta(x^2 y) + \eta(xy^2)$.

40. $r - t\cos^2 x + p \tan x = 0$.   Ans. $z = \theta(y + \sin x) + \eta(y - \sin x)$.

41. $t - r\sec^4 y = 2q \tan y$..   Ans. $z = \theta(x + \tan y) + \eta(x - \tan y)$.

42. $\dfrac{r}{x^2} - \dfrac{t}{y^2} = \dfrac{p}{x^3} - \dfrac{q}{y^3}$.   Ans. $z = \theta(x^2 + y^2) + \eta(x^2 - y^2)$.

43. $y^2 r + 2xys + x^2 t + px + qy = 0$.
Ans. $z = \theta(x^2 - y^2)\log(x + y) + \eta(x^2 - y^2)$.

44. $(1 + q)^2 r - 2(1 + p + q + pq)s + (1 + p)^2 t = 0$.   Ans. $y + x\theta(x + y + z) = \eta(x + y + z)$.

45. $(x - y)(x^2 r - 2xys + y^2 t) = 2xy(p - q)$.   Ans. $z = (x + y)\phi(xy) + \psi(xy)$.

46. $(r - s)x = (t - s)y$.   Ans. $z = (x + y)\theta(x + y) + x\eta(y/x)$.

47. $(q + 1)s = (p + 1)t$.   Ans. $z = \theta(x + y + z) + \eta(x)$.

48. $2yq + y^2 t = 1$.   Ans. $yz = y\log y - \theta(x) + \eta(x)$.

49. $3s + rt - s^2 = 2$. Ans. $x = \beta - \alpha$, $y = \theta'(\alpha) - \eta'(\beta)$, $z = xy + \beta[\theta'(\alpha) - \eta'(\beta)] - \theta(\alpha) + \eta(\beta)$.

50. $rt - s^2 + 1 = 0$. Ans. $2x = \alpha - \beta$, $2y = \eta'(\beta) - \theta'(\alpha)$, $2z = 2xy + \beta[\theta'(\alpha) + \eta'(\beta)] - \theta(\alpha) - \eta(\beta)$.

51. $s^2 - rt = a^2$. Ans. $2ax = \beta - \alpha$, $2ay = \theta'(\alpha) - \eta'(\beta)$, $2az = \beta[\theta'(\alpha) - \eta'(\beta)] - \theta(\alpha) + \eta(\beta) - 2a^2 xy$.

52. $5r + 6s + 3t + 2(rt - s^2) + 3 = 0$.
Ans. $2z = 3xy - \dfrac{3}{2}x^2 - \dfrac{5}{2}y^2 - c_1 x - c_2 y + c_3$.

53. $2pr + 2qt - 4pq(rt - s^2) = 1$.
Ans. $3z = \pm 2(x - a)^{3/2} \pm 2(y - b)^{2/3} + c$.

54. Obtain cosine and sine series of $f(x) = x$ in the interval $0 < x < \pi$. Hence show that $\dfrac{1}{1^2} + \dfrac{1}{3^2} + \dfrac{1}{5^2} + \ldots = \dfrac{\pi^2}{8}$.

Ans. $x = \dfrac{\pi}{2} - \dfrac{4}{\pi}\left[\cos x + \dfrac{\cos 3x}{3^2} + \dfrac{\cos 5x}{5^2} + \ldots\right]$; $x = 2\left[\sin x - \dfrac{\sin 2x}{2} + \dfrac{\sin 3x}{3} - \dfrac{\sin 4x}{4} + \ldots\right]$.

55. Expand the constant function $f(x) = k$, $0 < x < l$ in a sine series.
Ans. $f(x) = \dfrac{4k}{\pi}\left[\dfrac{1}{1}\sin\dfrac{\pi x}{l} + \dfrac{1}{3}\sin\dfrac{3\pi x}{l} + \dfrac{1}{5}\sin\dfrac{5\pi x}{l} + \ldots\right]$.

56. Express $\sin x$ as a cosine series in $0 < x < \pi$.
Ans. $\sin x = \dfrac{2}{\pi} - \dfrac{4}{\pi}\left[\dfrac{\cos 2x}{1.3} + \dfrac{\cos 4x}{3.5} + \dfrac{\cos 6x}{5.7} + \ldots\right]$.

57. Obtain half range sine series of e^x in the interval $0 < x < 1$.
Ans. $e^x = 2\pi\displaystyle\sum_{n=1}^{\infty}\dfrac{n[1 - e(-1)^n]}{1 + n^2\pi^2}\sin n\pi x$.

58. Find the half range Fourier sine series of $f(x) = \begin{cases} x & \text{if } 0 < x \leq \dfrac{\pi}{4} \\ \cos x & \text{if } \dfrac{\pi}{4} \leq x < \dfrac{\pi}{2} \end{cases}.$

Ans. $f(x) = \dfrac{4\sqrt{2}}{\pi}\left[\dfrac{\sin 2x}{1.3} - \dfrac{\sin 6x}{5.7} + \dfrac{\sin 10x}{9.11} - \dots\right].$

59. Find the half range sine series expansion of $f(x) = \dfrac{\pi}{2} - x$ in $(0, \pi)$.

Ans. $f(x) = \sin 2x + \dfrac{1}{2}\sin 4x + \dfrac{1}{3}\sin 6x + \dots.$

60. Find the half range cosine series expansion of $f(x) = \pi - x$ in $(0, \pi)$.

Ans. $f(x) = \dfrac{\pi}{2} + \dfrac{4}{\pi}\left[\cos x + \dfrac{\cos 3x}{3^2} + \dfrac{\cos 5x}{5^2} + \dots\right].$

61. Solve $r = q$, $\quad 0 < x < l$, $0 < y < \infty$ subject to the conditions:

$$z(0, y) = z(l, y) = 0 \; z(x, 0) = 3\sin n\pi x.$$

Ans. $z = 3\displaystyle\sum_{n=1}^{\infty} e^{-n^2\pi^2 y}\sin n\pi x.$

62. Solve $r = q$, $\quad 0 < x < l$, $0 < y < \infty$ subject to the conditions:

$$z(0, y) = z(l, y) = 0 \; z(x, 0) = lx - x^2.$$

Ans. $z = \dfrac{8l^2}{\pi^3}\displaystyle\sum_{m=1}^{\infty}\dfrac{1}{(2m-1)^3}e^{-[(2m-1)^2\pi^2 y]/l^2}\sin\dfrac{(2m-1)\pi x}{l}.$

63. Solve $r = q$, $\quad 0 < x < a$, $0 < y < \infty$ subject to the conditions:

$$p(0, y) = p(a, y) = 0, \; z(x, 0) = x(a - x), \; z \text{ is bounded as } y \to \infty.$$

Ans. $z = \dfrac{a^2}{6} - \dfrac{a^2}{\pi^2}\displaystyle\sum_{m=1}^{\infty}\dfrac{1}{m^2}e^{-4m^2\pi^2 y/a^2}\cos\dfrac{2m\pi x}{a}.$

64. Solve $r = t$, $\quad 0 < x < l$, $0 < y < \infty$ subject to the conditions:

$$z(0, y) = z(l, y) = 0, \; z(x, 0) = \dfrac{4\pi x(l - x)}{l}, \; q(x, 0) = 0.$$

Ans. $z = \dfrac{32}{\pi^2}\displaystyle\sum_{m=1}^{\infty}\dfrac{1}{(2m-1)^3}\sin\dfrac{(2m-1)\pi x}{l}\cos\dfrac{(2m-1)\pi y}{l}.$

65. Solve $r = t$, $\quad 0 \leq x \leq 1$, $y \geq 0$ subject to the conditions:

$$z(0, y) = 0, \; p(1, y) = 0, \; q(x, 0) = 0 \text{ and}$$

$$z(x, 0) = \begin{cases} x & 0 \le x \le \dfrac{l}{4} \\[2mm] \dfrac{1}{2} - x & \dfrac{l}{4} < x \le \dfrac{1}{2} \\[2mm] 0 & \dfrac{l}{2} < x \le 1. \end{cases}$$

Ans. $z = \dfrac{8}{\pi^2} \displaystyle\sum_{m=1}^{\infty} \dfrac{\sin[(2m-1)\pi/4]}{(2m-1)^2} \sin\dfrac{(2m-1)\pi x}{2} \cos\dfrac{(2m-1)\pi y}{2}.$

66. Solve $r = t$, $\quad 0 < x < l, \ 0 < y < \infty$ subject to the conditions:

$$z(0, y) = 0 = z(l, y), \ q(x, 0) = 0 \text{ and}$$

$$z(x, 0) = \begin{cases} \dfrac{3}{l}x & 0 < x \le \dfrac{l}{3} \\[2mm] \dfrac{3}{l}(l - 2x) & \dfrac{l}{3} < x \le \dfrac{2l}{3} \\[2mm] \dfrac{3}{l}(x - l) & \dfrac{2l}{3} < x < 1. \end{cases}$$

Ans. $z = \dfrac{9}{\pi^2} \displaystyle\sum_{m=1}^{\infty} \dfrac{1}{m^2} \sin\dfrac{2m\pi}{3} \sin\dfrac{2m\pi x}{l} \cos\dfrac{2m\pi y}{l}.$

67. Solve $r + t = 0$, $\quad 0 < x < a, \ 0 < y < \infty$ subject to the conditions:

$$z(0, y) = z(a, y) = 0, \ z(x, 0) = kx, \ z(x, \infty) = 0.$$

Ans. $z = \dfrac{2ak}{\pi} \displaystyle\sum_{n=1}^{\infty} \dfrac{(-1)^{n+1}}{n} e^{-n\pi y/a} \sin\dfrac{n\pi x}{a}.$

68. Solve $r + t = 0$, $\quad 0 < x < 1, \ 0 < y < \infty$ subject to the conditions:

$$z(0, y) = z(1, y) = 0, \ z(x, 0) = x(1 - x), \ z = 0 \text{ when } y \text{ is very large.}$$

Ans. $z = \dfrac{8}{\pi^3} \displaystyle\sum_{m=1}^{\infty} \dfrac{1}{(2m-1)^3} e^{-(2m-1)\pi y} \sin(2m - 1)\pi y.$

69. Solve $r + t = 0$, $\quad 0 \le x \le 1, \ 0 \le y \le 1$ subject to the conditions:

$$z(x, 0) = x(1 - x), \ p(0, y) = p(1, y) = p(x, 1) = 0.$$

Ans. $z = \dfrac{1}{6} - \dfrac{1}{\pi^2} \displaystyle\sum_{n=1}^{\infty} \dfrac{1}{n^2 \cosh 2ny} \cos 2n\pi x \cosh 2n\pi y.$

70. A string is stretched and fastened to two points l apart. Motion is started by displacing the string in the form $u = A \sin \frac{\pi x}{l}$ from which it is released at time $t = 0$. Show that the displacement of any point at a distance x from one end at a time t is given by

$$u(x, t) = A \sin \frac{\pi x}{l} \cos \frac{\pi c t}{l}.$$

71. A tightly stretched string with fixed end points $x = 0$ and $x = l$ is initially in a position given by $u(x, 0) = u_0 \sin^3 \frac{\pi x}{l}$. If it is released from rest from this position, find the displacement u at any time and at any distance from the end $x = 0$.

Ans. $u(x, t) = \dfrac{3u_0}{4} \sin \dfrac{\pi x}{l} \cos \dfrac{\pi c t}{l} - \dfrac{u_0}{4} \sin \dfrac{3\pi x}{l} \cos \dfrac{3\pi c t}{l}$.

72. The points of trisection of a tightly stretched string of length l with fixed ends are pulled aside through a distance d on opposite sides of the opposition of equilibrium and the string is released from rest. Obtain an expression for the displacement of the string at any subsequent time and show that the midpoint of the string always remains at rest.

Ans. $u(x, t) = \dfrac{18d}{\pi^2} \displaystyle\sum_{n=1}^{\infty} \dfrac{1}{n^2} \left(\sin \dfrac{n\pi}{3} - \sin \dfrac{2n\pi}{3} \right) \sin \dfrac{\pi x}{l} \cos \dfrac{\pi c t}{l}$.

73. A homogeneous rod of conducting material of length 100 has its ends kept at zero temperature and the initial temperature is

$$u(x, 0) = \begin{cases} x & \text{if } 0 \le x \le 50 \\ 100 - x & \text{if } 50 \le x \le 100. \end{cases}$$

Find the temperature $u(x, t)$ at any time t.

Ans. $u(x, t) = \dfrac{400}{\pi^2} \displaystyle\sum_{n=1}^{\infty} \dfrac{1}{n^2} \sin \dfrac{n\pi}{2} \sin \dfrac{n\pi x}{100} e^{-\alpha^2 n^2 \pi^2 t/10000}$.

74. Find the temperature in a bar of length 2 whose ends are kept at zero and lateral surface insulated if the initial temperature is $\sin \frac{\pi x}{2} + 3 \sin \frac{5\pi x}{2}$.

Ans. $\sin \dfrac{\pi x}{2} e^{-\pi^2 \alpha^2 t/4} + 3 \sin \dfrac{5\pi x}{2} e^{-25\pi^2 \alpha^2 t/4}$.

75. Find the temperature $u(x, t)$ in a slab of length π which is insulated laterally and whose ends $x = 0$ and $x = \pi$ are also insulated. The initial temperature in the slab is $\sin x$.

Ans. $u(x, t) = \dfrac{2}{\pi} - \dfrac{4}{\pi} \displaystyle\sum_{m=1}^{\infty} \dfrac{1}{4m^2 - 1} e^{-4m^2 \alpha^2 t} \cos 2mx$.

76. Determine the steady state temperature distribution in a thin rectangular plate bounded by lines $x = 0, x = a$ and $y = 0, y = b$. The edges $x = 0, x = a$, and $y = 0$ are kept at zero temperature and the edge $y = b$ is kept at 100°C.

Ans. $u(x, y) = \dfrac{400}{\pi} \displaystyle\sum_{m=1}^{\infty} \dfrac{1}{(2m - 1) \sinh[(2m - 1)\pi b/a]} \sin \dfrac{(2m - 1)\pi x}{a} \sinh \dfrac{(2m - 1)\pi y}{a}$.

77. A rectangular plate with insulated surfaces is 10 cm wide and so long compared to its width that it may be considered infinite in length without introducing an appreciable error. If the temperature along the short edge $y = 0$ is given by

$$u(x, 0) = \begin{cases} 20x & \text{if } 0 < x \le 5 \\ 20(10 - x) & \text{if } 5 < x < 10 \end{cases}$$

and the two long edges $x = 0$ and $x = 10$ as well as other short edge are kept at $0°C$. Find the steady state temperature at any point (x, y).

Ans. $u(x, y) = \dfrac{800}{\pi^2} \displaystyle\sum_{n=1}^{\infty} \dfrac{\sin n\pi/2}{n^2} e^{-(n\pi y/10)} \sin \dfrac{n\pi x}{10}$.

78. Find the steady state temperature for a square plate with boundary conditions as given below

$$u(0, y) = u(x, 0) = 0, u(\pi, y) = u(x, \pi) = 1.$$

Ans. $u(x, y) = \dfrac{4}{\pi} \displaystyle\sum_{m=1}^{\infty} \dfrac{1}{(2m - 1) \sinh(2m - 1)\pi} \sin(2m - 1)x \sinh(2m - 1)y+$

$\dfrac{4}{\pi} \displaystyle\sum_{m=1}^{\infty} \dfrac{1}{(2m - 1) \sinh(2m - 1)\pi} \sin(2m - 1)x \sinh(2m - 1)y.$

79. Show that a transmission line with negligible resistance and conductance propagates waves of current and a potential with a velocity equal to $1/\sqrt{LC}$, where L is self-inductance and C the capacitance.

80. In a telephone wire of length l, a steady voltage distribution of 20 volts at the sending end and 12 volts at receiving end is maintained. The receiving end is grounded at $t = 0$. Neglecting the inductance and conductance, find the voltage $V(x, t)$ and $I(x, t)$.

Ans. $V(x, t) = \dfrac{20(l - x)}{l} + \dfrac{24}{\pi} \displaystyle\sum_{n=1}^{\infty} \dfrac{(-1)^{n+1}}{n} \sin \dfrac{n\pi x}{l} e^{-n^2\pi^2 t/RCl^2}$

$I(x, t) = \dfrac{20}{lR} + \dfrac{243}{lR} \displaystyle\sum_{n=1}^{\infty} (-1)^n \sin \dfrac{n\pi x}{l} e^{-n^2\pi^2 t/RCl^2}.$

81. A transmission line has a resistance of 8.8Ω/mile and a conductance of $2\mu\Omega$/mile. The line is broken, leaving it with an open circuit transmission. If the resistance of the line measured at the source is 3000Ω, how far away is the break.

Ans. 200 miles (approximately).

6

Higher Order Linear Partial Differential Equations

In Chapters 4 and 5, we have discussed various methods for finding the solutions of first- and second-order partial differential equations (PDEs). Now, we undertake the solution of those linear PDEs, in which higher order derivatives occur. The theoretical framework and methods of solution developed for first and second-order linear equations are extendable to a very limited class of linear equations of higher (arbitrary) order. Generally, it is very difficult to solve any linear PDE, wherein the coefficients are functions of x and y. Henceforth, we start our journey with the description of certain kinds of linear PDEs, in which the various terms are multiplied by constants only. We devote Sections 6.1 and 6.2 to such linear PDEs. In Sections 6.3–6.5, we discuss some special types of linear equations with variable coefficients, which are reducible easily to equations with constant coefficients and hence become solvable.

6.1 Homogeneous Linear PDEs with Constant Coefficients

The most general linear PDE with constant coefficients of order n is of the form

$$\left[(A_0 D^n + A_1 D^{n-1} D' + \cdots + A_n D'^n) + (B_0 D^{n-1} + \cdots + B_{n-1} D'^{n-1}) \\ + \cdots + (PD + QD') + Z \right] z = f(x, y) \tag{6.1}$$

where the coefficients $A_0, A_1, \ldots, A_n, B_0, \ldots, P, Q, Z$ all are constants and $A_0, A_1, \ldots, A_n$ do not vanish simultaneously. Recall that an equation of the form (6.1) is called homogeneous whenever $f(x, y) = 0$. Otherwise, it is called non-homogeneous.

Conveniently, Eq. (6.1) can be symbolically written as

$$F(D, D')z = f(x, y)$$

where the polynomial

$$F(D, D') = (A_0 D^n + A_1 D^{n-1} D' + \cdots + A_n D'^n) + (B_0 D^{n-1} \\ + \cdots + B_{n-1} D'^{n-1}) + \cdots + (PD + QD') + Z$$

remains a differential operator with constant coefficients.

As mentioned in Chapter 2, the mixed partial derivatives of z involved in a PDE are assumed to be independent of the order of differentiation. Subsequently, for an equation $F(D, D')z = f(x, y)$ with constant coefficients such that $F(D, D') = g(D, D') \cdot h(D, D')$, we have $h(D, D') \cdot g(D, D') = F(D, D')$. This observation can be described in the form of the following result that has an important bearing in the context of linear PDEs.

Proposition 6.1. *Let $F(D, D')z = f(x, y)$ be a linear PDE with constant coefficients. If the polynomial $F(D, D')$ can be decomposed into some factors, then the order in which these factors occur is unimportant.*

In this section, we discuss the techniques for solving the homogeneous linear PDEs with constant coefficients of arbitrary order. For the sake of brevity, we divide this section into four subsections.

6.1.1 Euler Linear PDEs

A homogeneous linear PDE with constant coefficients in which the derivatives involved are all of the same order is called an *Euler linear equation*. In other words, a homogeneous linear equation $F(D, D')z = 0$ with constant coefficients is called an Euler linear equation if the polynomial $F(D, D')$ is homogeneous. Here, it is worth noting that some authors in their respective books used the term 'homogeneous linear PDE' for such equations inspired by the homogeneous property of $F(D, D')$. But this term contradicts with the usual definition of homogeneous linear PDE, which is used by the same authors in the context of first- and second-order linear PDE. Moreover, this term contradicts the same analogy of homogeneous linear ODE. With a view to avoid this confusion, we say such equations as Euler linear equation because this class of PDEs is a natural extension of second-order linear PDE of the form $ar + bs + ct = 0$, which is known as Euler equation following the lines of Myint-U and Debnath [21] (see Chapter 4, Page No. 100). However, to define the terms 'homogeneous' and 'non-homogeneous' linear PDE, we followed the lines of Sneddon [15].

The most general Euler linear equation of order n takes the form

$$(k_0 D^n + k_1 D^{n-1}D' + \cdots + k_n D'^n)z = 0 \tag{6.2}$$

where $k_0, k_1, \ldots, k_n$ are constants and do not all simultaneously vanish. We assume that $k_0 \neq 0$. Due to homogeneous property, the polynomial $F(D, D') = k_0 D^n + k_1 D^{n-1}D' + \cdots + k_n D'^n$ must be decomposed into n linear factors. Analogously as the case of ODE, we define the following notion of 'auxiliary equation' of Eq. (6.2).

Definition 6.1. *Consider an Euler linear PDE represented by Eq. (6.2) with $k_0 \neq 0$. An equation of order n in m (may be real or complex) of the form*

$$k_0 m^n + k_1 m^{n-1} + \cdots + k_n m = 0 \tag{6.3}$$

*is called the **auxiliary equation** (in short, AE) of Eq. (6.2).*

It is clear that the LHS of Eq. (6.3) is obtained by putting $D = m$ and $D' = 1$ in the polynomial $F(D, D')$.

We present the following result concerning the general solution of an Euler linear equation, whose AE has no repeated roots.

Theorem 6.1. *Let the AE of Eq. (6.2) has n distinct roots, namely, $m_1, m_2, \ldots, m_n$, then the general solution of the PDE is*

$$z = \phi_1(y + m_1 x) + \phi_2(y + m_2 x) \cdots + \phi_n(y + m_n x) \tag{6.4}$$

where $\phi_1, \phi_2, \ldots, \phi_n$ are real-valued arbitrary functions of a complex (or real) variable.

Proof. As $m_1, m_2, \ldots, m_n$ are roots of AE (6.3), we have

$$F(D, D') = k_0(D - m_1 D')(D - m_2 D') \cdots (D - m_n D')$$

and hence (6.4) reduces to

$$(D - m_1 D')(D - m_2 D') \cdots (D - m_n D')z = 0. \tag{6.5}$$

This yields that the solution of each of the equations

$$\left. \begin{aligned} (D - m_1 D')z &= 0 \\ (D - m_2 D')z &= 0 \\ &\vdots \\ (D - m_n D')z &= 0 \end{aligned} \right\}$$

is also a solution of (6.2).

First, we consider $(D - m_1 D')z = 0$, that is, $p - m_1 q = 0$, which is a linear PDE of first order and hence it can be solved by Lagrange's method. Its subsidiary equations are

$$\frac{dx}{1} = \frac{dy}{-m_1} = \frac{dz}{0}.$$

Taking the first two fractions, we get $dy + m_1 dx = 0$, which by integrating gives rise to $y + m_1 x = c_1$. Also, from the third fraction, we get $dz = 0$ so that $z = c_2$. Hence, the general solution of the equation $(D - m_1 D')z = 0$ is $z = \phi_1(y + m_1 x)$, where the function ϕ_1 is arbitrary.

Similarly, we can obtain that the general solutions of the differential equations $(D - m_2 D')$ $z = 0, \ldots, (D - m_n D')z = 0$ are $z = \phi_2(y + m_2 x), \ldots, z = \phi_n(y + m_n x)$, respectively. Therefore, (6.2) is satisfied by all these n solutions. Consequently, by principle of superposition, the most general solution of (6.2) must be of the form (6.4). □

Since all the functions $\phi_1, \phi_2, \ldots, \phi_n$ involved in (6.4) are arbitrary, there is no need to add arbitrary constants in each term.

To illustrate Theorem 6.1, we furnish the following examples.

Example 6.1. *Solve $2r + 5s + 2t = 0$.*

Solution.   The given PDE can be written as

$$(2D^2 + 5DD' + 2D'^2)z = 0.$$

The AE is

$$2m^2 + 5m + 2 = 0$$

or

$$(m + 2)(2m + 1) = 0.$$

The roots of AE are

$$m = -2, -1/2.$$

Therefore, the general solution of the given PDE is

$$z = \phi(y - 2x) + \psi^* \left(y - \frac{1}{2}x \right) = \phi(y - 2x) + \psi(2y - x).$$

Example 6.2. *Solve $(D^4 - D^3D' + D^2D'^2 - DD'^3)z = 0$.*

Solution.   The AE of the given PDE is

$$m^4 - m^3 + m^2 - m = 0$$

or

$$m(m - 1)(m^2 + 1) = 0.$$

Its roots are

$$m = 0, 1, \pm i.$$

Hence, the general solution is

$$z = \phi(y) + \psi(y + x) + \theta(y + ix) + \eta(y - ix).$$

A Possible Dual Approach: Consider Euler linear equation of the form (6.2) with $k_n \neq 0$; however, the coefficient k_0 may be equal to zero. Then we may define a dual AE of Eq. (6.2) by putting $D = 1$ and $D' = m'$ so that $F(1, m') = 0$. If this equation has n distinct roots, namely, $m'_1, m'_2, \ldots, m'_n$, then following the lines similar to the proof of Theorem 6.1, the general solution is of the form

$$z = \psi_1(x + m'_1 y) + \psi_2(x + m'_2 y) \cdots + \psi_n(x + m'_n y).$$

In addition, if $k_0 \neq 0$, then we have $m'_i = 1/m_i$, $i = 1, 2, \ldots, n$, wherein $m_1, m_2, \ldots, m_n$ remain the roots of AE (6.3). Since the functions $\psi_1, \psi_2, \ldots, \psi_n$ are arbitrary, therefore in this case, the general solution obtained as above remains the same as determined by Theorem (6.1). For illustration, we consider the following example.

Example 6.3. *Solve* $(D^3 - 6D^2D' + 11DD'^2 - 6D'^3)z = 0$.

Solution. *The AE is*

$$m^3 - 6m^2 + 11m - 6 = 0$$

or

$$(m - 1)(m - 2)(m - 3) = 0.$$

The roots of AE are

$$m = 1, 2, 3.$$

Therefore, the general solution of the given PDE is

$$z = \phi(y + x) + \psi(y + 2x) + \theta(y + 3x).$$

Alternately, we can determine the general solution by dual approach. The dual AE is

$$1 - 6m' + 11m'^2 - 6m'^3 = 0$$

whose roots are

$$m' = 1, \frac{1}{2}, \frac{1}{3}.$$

Hence, the general solution is

$$z = \varphi(x + y) + \eta^* \left(x + \frac{1}{2}y \right) + \mu^* \left(x + \frac{1}{3}y \right)$$

or

$$z = \varphi(y + x) + \eta(y + 2x) + \mu(y + 3x).$$

Therefore, the general solutions determined by AE and dual AE remain the same.

Example 6.4. *Solve* $\dfrac{\partial^3 z}{\partial x^2 \partial y} - \dfrac{\partial^3 z}{\partial y^3} = 0$.

Solution. *The given PDE can be written symbolically as*

$$(D^2D' - D'^3)z = 0.$$

Here, the coefficient of D^3 is zero. Therefore, its AE is of order 2 and hence it cannot be solved by applying Theorem 6.1. We solve it by dual approach.

The dual AE of the given equation is

$$m' - m'^3 = 0$$

whose roots are

$$m' = 0, 1, -1.$$

Thus, the general solution is

$$z = \phi(x) + \psi(x + y) + \theta(x - y).$$

In the following result, we consider the case when some roots of the AE are same.

Theorem 6.2. *Let $m_1, m_2, \ldots, m_n$ be the roots of an Euler linear PDE of order n represented by Eq. (6.2). If r roots are equal (i.e. $m_1 = m_2 = \cdots = m_r = m$ (say)) and rest roots are distinct, then the general solution of the PDE is*

$$z = \phi_1(y + mx) + x\phi_2(y + mx) + x^2\phi_3(y + mx) + \cdots + x^{r-1}\phi_r(y + mx)$$
$$+ \phi_{r+1}(y + m_{r+1}x) + \cdots + \phi_n(y + m_nx) \tag{6.6}$$

or

$$z = \psi_1(y + mx) + y\psi_2(y + mx) + y^2\psi_3(y + mx) + \cdots + y^{r-1}\psi_r(y + mx)$$
$$+ \psi_{r+1}(y + m_{r+1}x) + \cdots + \psi_n(y + m_nx), \text{ provided } m \neq 0 \tag{6.7}$$

where $\phi_1, \phi_2, \ldots, \phi_n$ and $\psi_1, \psi_2, \ldots, \psi_n$ are arbitrary functions.

Proof. First, we prove the result for the case $r = 2$, that is, $m_1 = m_2 = m$. Consider the PDE

$$(D - mD')^2 z = 0. \tag{6.8}$$

Put

$$(D - mD')z = u. \tag{6.9}$$

Then, Eq. (6.8) reduces to

$$(D - mD')u = 0 \tag{6.10}$$

or

$$\frac{\partial u}{\partial x} - m\frac{\partial u}{\partial y} = 0.$$

which is a linear PDE of order one. Its Lagrange's subsidiary equations are

$$\frac{dx}{1} = \frac{dy}{-m} = \frac{du}{0}.$$

From the first two fractions, we get

$$y + mx = c_1.$$

Also, the last fraction yields

$$u = c_2.$$

Hence, by Lagrange's method, the general solution of Eq. (6.10) is

$$u = \phi(y + mx).$$

Putting this value in (6.9), we obtain

$$\frac{\partial z}{\partial x} - m\frac{\partial z}{\partial y} = \phi(y + mx). \qquad (6.11)$$

Its Lagrange's subsidiary equations are

$$\frac{dx}{1} = \frac{dy}{-m} = \frac{dz}{\phi(y + mx)}. \qquad (6.12)$$

The first two fractions give rise to

$$y + mx = c_3.$$

The first and last fractions of (6.12) taken together become

$$dz = \phi(c_3)dx.$$

On integrating, we get

$$z = x\phi(c_3) + c_4$$

implying thereby

$$z = c_4 + x\phi(y + mx).$$

Therefore, the general solution of (6.11) (and hence of (6.8)) is

$$z = \varphi(y + mx) + x\phi(y + mx).$$

If $m \neq 0$, then Eq. (6.12) can be written as

$$\frac{dx}{-1/m} = \frac{dy}{1} = \frac{dz}{\psi(y + mx)}. \qquad (6.13)$$

where $\psi = -\dfrac{1}{m}\phi$. From the first two fractions, we get $y + mx = c_5$. While the last two fractions yields $z = c_6 + y\psi(y + mx)$. Hence, the general solution of (6.8) may be alternately written as

$$z = \eta(y + mx) + y\psi(y + mx).$$

In the similar manner, we can prove that the general solution of the PDE $(D - mD')^3 z = 0$ takes the form

$$z = \theta(y + mx) + x\varphi(y + mx) + x^2\phi(y + mx)$$

or

$$z = \mu(y + mx) + y\eta(y + mx) + y^2\psi(y + mx), \quad \text{provided } m \neq 0.$$

Thus, in all, the general solution of PDE

$$(D - mD')^r z = 0$$

is of the form

$$z = \sum_{i=1}^{r} x^{i-1}\phi_i(y + mx)$$

or

$$z = \sum_{i=1}^{r} y^{i-1}\psi_i(y + mx), \quad \text{provided } m \neq 0.$$

Consequently, (6.6) as well as (6.7) form the general solution of the given Euler PDE. This completes the proof. □

For illustration of the above result, we adopt several examples.

Example 6.5. *Solve* $(4D^2 + 12DD' + 9D'^2)z = 0$.

Solution. *The AE is*

$$4m^2 + 12m + 9 = 0.$$

The roots of the above equation are

$$m = -\frac{3}{2}, -\frac{3}{2}.$$

Thus, both roots of AE are equal and hence the general solution is

$$z = \phi^*\left(y - \frac{3}{2}x\right) + x\psi^*\left(y - \frac{3}{2}x\right) = \phi(2y - 3x) + x\psi(2y - 3x).$$

Example 6.6. *Solve* $\dfrac{\partial^4 z}{\partial x^4} + \dfrac{\partial^4 z}{\partial y^4} = 2\dfrac{\partial^4 z}{\partial x^2 \partial y^2}$.

Solution. *Rewrite the given PDE symbolically as*

$$D^4 - 2D^2 D'^2 + D'^4 = 0.$$

The AE is

$$m^4 - 2m^2 + 1 = 0$$

or

$$(m^2 - 1)^2 = 0$$

whose roots are

$$m = 1, 1, -1, -1.$$

Thus, the general solution is

$$z = \phi(y + x) + x\psi(y + x) + \theta(y - x) + x\eta(y - x)$$

Example 6.7. *Solve* $(D^4 - 2D^3 D' + 2DD'^3 - D'^4)z = 0$.

Solution. *The AE is*

$$m^4 - 2m^3 + 2m - 1 = 0.$$

On factorising the expression of LHS, the above equation reduces to

$$(m + 1)(m - 1)^3 = 0.$$

The roots of AE are

$$m = -1, 1, 1, 1.$$

Hence, the general solution is

$$z = \phi(y - x) + \psi(y + x) + x^2\theta(y + x) + x^3\eta(y + x).$$

Exceptional Case: If $F(D, D')$ does not contain D^n and D'^n (that is, $k_0 = k_n = 0$), then both the corresponding AE and dual AE are not of order n. In this case, we have

$$F(D, D') = D'^k h(D, D')$$

such that $h(D, D')$ remains a homogeneous polynomial of degree $n - k$ containing D^{n-k}. Therefore, the general solution of $F(D, D')z = 0$ is a combination of the solutions of $D'^k z = 0$ and $h(D, D')z = 0$. To illustrate this description, we adopt the following example.

Example 6.8. *Solve* $(D^3D'^2 + D^2D'^3)z = 0$.

Solution. *The given PDE can be written as*

$$D'^2(D^3 + D^2D')z = 0.$$

Thus, the above equation splits into two PDEs:

$$\frac{\partial^2 z}{\partial y^2} = 0 \tag{6.14}$$

and

$$(D^3 + D^2D')z = 0. \tag{6.15}$$

The general solution of Eq. (6.14) is

$$z = y\phi(x) + \psi(x).$$

The AE of Eq. (6.15) is

$$m^3 + m^2 = 0$$

whose roots are

$$m = 0, 0, -1$$

and hence the general solution of (6.15) is

$$z = \theta(y) + x\eta(y) + \mu(y - x).$$

Finally, the general solution of the given PDE is

$$z = y\phi(x) + \psi(x) + \theta(y) + x\eta(y) + \mu(y - x).$$

6.1.2 Reducible Linear PDEs

Now, we shall discuss the solution of the general form of homogeneous linear PDE with constant coefficients, in which all the terms are not necessarily of the same order. The linear PDE with constant coefficients can be divided into two classes:

(i) Reducible Equations: A linear PDE of the form (6.1) is called reducible if the symbolic function $F(D, D')$ can be decomposed into linear factors (that is, each factor is of first degree in D and D'), for example, the PDE $(D^2 - D'^2 + D + D')z = xy + y^2$ is reducible as $D^2 - D'^2 + D + D' = (D + D')(D - D' + 1)$.

In a reducible linear PDE $F(D, D')z = f(x, y)$ of order n, the polynomial $F(D, D')$ can be expressed as

$$F(D, D') = \prod_{i=1}^{n}(\alpha_i D + \beta_i D' + \gamma_i).$$

The general solution of a reducible linear PDE contains as many arbitrary functions as the order of PDE. Obviously, every Euler linear PDE remains reducible.

(ii) Irreducible Equations: A linear PDE of the form (6.1) is called irreducible if it is not reducible (that is, the symbolic function $F(D, D')$ cannot be decomposed into linear factors), for example, the PDE $(D^2 - D')z = \sin(x + 2y)$ is irreducible as the operator $D^2 - D'$ cannot be decomposed into factors of first degree in D and D'.

In this section, we develop the needed ideas to obtain the general solution of reducible homogeneous linear PDE. We begin with the following result, which plays a key role in the computation of general solution.

Theorem 6.3. *Let $F(D, D')z = 0$ be a homogeneous linear PDE with constant coefficients. If $\alpha D + \beta D' + \gamma$ is a linear factor of $F(D, D')$, then*

$$z = e^{-\gamma x/\alpha} \phi(\alpha y - \beta x) \quad \text{if } \alpha \neq 0 \tag{6.16}$$

or

$$z = e^{-\gamma y/\beta} \psi(\alpha y - \beta x) \quad \text{if } \beta \neq 0 \tag{6.17}$$

where ϕ and ψ are arbitrary real-valued functions of a complex (or real) variable, which is a solution of $F(D, D')z = 0$.

Proof. As $\alpha D + \beta D' + \gamma$ is a factor of $F(D, D')$, any solution of

$$(\alpha D + \beta D' + \gamma)z = 0. \tag{6.18}$$

is also a solution of the given PDE. Equation (6.18) can be rewritten as

$$\alpha p + \beta q = -\gamma z.$$

Lagrange's subsidary equations are

$$\frac{dx}{\alpha} = \frac{dy}{\beta} = \frac{dz}{-\gamma z}. \tag{6.19}$$

From the first two ratios, we obtain

$$\beta x - \alpha y = c_1.$$

If $\alpha \neq 0$, then the first and last ratios give

$$\frac{dz}{z} = -\frac{\gamma}{\alpha} dx.$$

On integrating, we get

$$\log z = -\frac{\gamma x}{\alpha} + \log c_2$$

or

$$z = c_2 e^{-\gamma x/\alpha}.$$

Hence, $z = e^{-\gamma x/\alpha} \phi(\beta x - \alpha y)$ is the solution of the given PDE as desired.

If $\beta \neq 0$, then the last two ratios yield

$$\frac{dz}{z} = -\frac{\gamma}{\beta} dy$$

which, on integrating, gives rise to

$$z = c_3 e^{-\gamma x/\beta}.$$

Thus, the required solution may be alternately written as $z = e^{-\gamma x/\beta} \psi(\beta x - \alpha y)$. □

The following result provides us the formulation for the computation of the general solution of a reducible PDE, wherein each linear factor of $F(D, D')$ enjoys multiplicity one.

Theorem 6.4. *Let $F(D, D')z = 0$ be a reducible PDE having n linearly independent factors (that is, no factor is a mere multiple of other) of the form*

$$F(D, D') = \prod_{i=1}^{n} (\alpha_i D + \beta_i D' + \gamma_i).$$

If for each $i(1 \leq i \leq n)$, z_i is a solution of PDE corresponding to the factor $\alpha_i D + \beta_i D' + \gamma_i$ as realised in Theorem 6.3, then $z = \sum_{i=1}^{n} z_i$ forms the general solution of PDE.

Proof. The proof is straightforward in view of Theorem 6.3 together with the principle of superposition. □

From the above two theorems, we derive the following special case:

Corollary 6.1. *Let $F(D, D')z = 0$ be a homogeneous linear equation with constant coefficients. Then*

   (i) *corresponding to a factor $\alpha D + \gamma$ of $F(D, D')$, the term of general solution is $e^{-\gamma x/\alpha} \phi(\alpha y)$,*
   (ii) *corresponding to a factor $\beta D' + \gamma$ of $F(D, D')$, the term of general solution is $e^{-\gamma y/\beta} \psi(\beta x)$.*

With a view to illustrate the foregoing description, we furnish the following examples.

Example 6.9. *Solve $(2D + D' + 1)(D - 3D' + 2)z = 0$.*

Solution. Here, $\alpha_1 = 2, \beta_1 = 1, \gamma_1 = 1; \alpha_2 = 1, \beta_2 = -3, \gamma_2 = 2$. Therefore, the general solution of the given equation is

$$z = e^{-y} \phi(2y - x) + e^{-2x} \psi(y + 3x).$$

Alternately, the required solution may be written as

$$z = e^{-x/2}\theta(2y - x) + e^{2y/3}\eta(y + 3x).$$

Example 6.10. *Solve* $(D^2 - D'^2 - 3D + 3D')z = 0$.

Solution. *The given PDE can be written as*

$$(D - D')(D + D' - 3)z = 0.$$

Here, $\alpha_1 = 1, \beta_1 = -1, \gamma_1 = 0$; $\alpha_2 = 1, \beta_2 = 1, \gamma_2 = -3$. *Therefore, the general solution of the given equation is*

$$z = \phi(y + x) + e^{3x}\psi(y - x).$$

Example 6.11. *Solve* $\dfrac{\partial^3 z}{\partial x^2 \partial y} - 2\dfrac{\partial^3 z}{\partial x \partial y^2} = 3\dfrac{\partial^2 z}{\partial x \partial y}.$

Solution. *The given equation can be expressed symbolically as*

$$(D^2 D' - 2DD'^2 - 3DD')z = 0$$

or,

$$DD'(D - 2D' - 3)z = 0.$$

The solutions corresponding the factors D *and* D' *are* $z = \phi(y)$ *and* $z = \psi(x)$, *respectively (by Corollary 6.1). Further, the solution corresponding to the last factor remains* $e^{3x}\theta(y + 2x)$. *Thus, in all, the general solution of the given PDE becomes*

$$z = \phi(y) + \psi(x) + e^{3x}\theta(y + 2x).$$

Now, we present the following result, whenever $F(D, D')$ admits repeated linear factors.

Theorem 6.5. *Under the hypothesis of Theorem 6.4, if* $F(D, D')$ *contains* r *repeated factors of the form* $(\alpha D + \beta D' + \gamma)^r$ *instead of linearly independent factors, then the terms of general solution corresponding to these repeated factors become*

$$e^{-\gamma x/\alpha} \left[\phi_1(\alpha y - \beta x) + x\phi_2(\alpha y - \beta x) + \cdots + x^{r-2}\phi_{r-1}(\alpha y - \beta x) \right.$$
$$\left. + x^{r-1}\phi_r(\alpha y - \beta x)\right] \quad \text{if } \alpha \neq 0$$

or

$$e^{-\gamma y/\beta} \left[\psi_1(\alpha y - \beta x) + y\psi_2(\alpha y - \beta x) + \cdots + y^{r-2}\psi_{r-1}(\alpha y - \beta x) \right.$$
$$\left. + y^{r-1}\psi_r(\alpha y - \beta x)\right] \quad \text{if } \beta \neq 0$$

where $\phi_1, \phi_2, \dots, \phi_r$ *and* $\psi_1, \psi_2, \dots, \psi_r$ *are arbitrary functions.*

Proof. First, we prove the result for the case of $r = 2$. Thus far, we consider the PDE

$$(\alpha D + \beta D' + \gamma)^2 z = 0. \qquad (6.20)$$

Put

$$(\alpha D + \beta D' + \gamma)z = u. \qquad (6.21)$$

Then Eq. (6.20) reduces to

$$(\alpha D + \beta D' + \gamma)u = 0.$$

By Theorem 6.3, the general solution of the above equation is

$$u = e^{-\gamma x/\alpha} \phi(\beta x - \alpha y) \quad \text{if } \alpha \neq 0$$

or

$$u = e^{-\gamma y/\beta} \phi(\beta x - \alpha y) \quad \text{if } \beta \neq 0. \qquad (6.23)$$

Putting the value of u from (6.22) in Eq. (6.21), we get

$$(\alpha D + \beta D' + \gamma)z = e^{-\gamma x/\alpha} \phi(\beta x - \alpha y)$$

which can be written as

$$\alpha p + \beta q = -\gamma z + e^{-\gamma x/\alpha} \phi(\beta x - \alpha y).$$

Lagrange's subsidiary equations of the above PDE are

$$\frac{dx}{\alpha} = \frac{dy}{\beta} = \frac{dz}{-\gamma z + e^{-\gamma x/\alpha} \phi(\beta x - \alpha y)}. \qquad (6.24)$$

The first two ratios of (6.24) give rise

$$\beta x - \alpha y = c_1. \qquad (6.25)$$

Taking the first and last ratios of (6.24) and using (6.25), we obtain

$$\frac{dx}{\alpha} = \frac{dz}{-\gamma z + e^{-\gamma x/\alpha} \phi(c_1)}$$

or

$$\frac{dz}{dx} + \frac{\gamma z}{\alpha} = \frac{1}{\alpha} e^{-\gamma x/\alpha} \phi(c_1) \qquad (6.26)$$

which is a first-order linear ODE. Its integrating factor (IF) is

$$\text{IF} = e^{\int \frac{\gamma}{\alpha} dx} = e^{\gamma x/\alpha}$$

Hence, the solution of Eq. (6.26) is

$$ze^{\gamma x/\alpha} = \int \frac{1}{\alpha} e^{-\gamma x/\alpha} \phi(c_1).e^{\gamma x/\alpha} dx + c_2 = \frac{1}{\alpha} x\phi(c_1) + c_2$$

implying thereby

$$z = e^{-\gamma x/\alpha} \left[c_2 + \frac{1}{\alpha} x\phi(\beta x - \alpha y) \right]. \tag{6.27}$$

Making use of (6.25) and (6.27) and by Lagrange's method, the general solution of Eq. (6.20) is

$$z = e^{-\gamma x/\alpha} \left[\phi_1(\beta x - \alpha y) + x\phi_2(\beta x - \alpha y) \right]$$

where $\phi_2 = \dfrac{1}{\alpha}\phi$. Similarly, from (6.21) and (6.23), we can conclude that

$$z = e^{-\gamma y/\beta} \left[\psi_1(\beta x - \alpha y) + y\psi_2(\beta x - \alpha y) \right]$$

forms the general solution of Eq. (6.20).

Thus, by induction, it can be easily proved that the general solution of

$$(\alpha D + \beta D' + \gamma)^r z = 0$$

is of the form

$$z = e^{-\gamma x/\alpha} \sum_{i=1}^{r} x^{i-1} \phi_i(\beta x - \alpha y) \text{ if } \alpha \neq 0$$

or

$$z = e^{-\gamma y/\beta} \sum_{i=1}^{r} y^{i-1} \psi_i(\beta x - \alpha y) \text{ if } \beta \neq 0.$$

This completes the proof. $\qquad\square$

We adopt the following examples to illustrate the foregoing description.

Example 6.12. *Solve $r + 2s + t + 2p + 2q + z = 0$.*

Solution. *The given equation can be expressed symbolically as*

$$(D^2 + 2DD' + D'^2 + D + D' + 1)z = 0$$

or

$$(D + D' + 1)^2 z = 0.$$

As LHS occurs a repeated linear factor, the general solution of the PDE is

$$z = e^{-x}[\phi(y-x) + x\psi(y-x)].$$

Example 6.13. *Solve* $(D'-2)(D+D'-1)^3 z = 0.$

Solution. The general solution of the above equation is

$$z = e^{2y}\phi(x) + e^x[\psi(y-x) + x\theta(y-x) + x^2\eta(y-x)].$$

Example 6.14. *Solve* $(DD'-D-D'+1)^2 z = 0.$

Solution. The given equation can be rewritten as

$$(D-1)^2(D'-1)^2 z = 0.$$

Therefore, the general solution is

$$z = e^x\left[\phi(y) + x\psi(y)\right] + e^y\left[\theta(x) + y\eta(x)\right].$$

6.1.3 Exponential-Type Series Solution

Before undertaking irreducible PDEs, we discuss the exponential-type series solution of a homogeneous linear PDE. An important advantage of such types of solutions is that in many physical situations, exponential-type series solutions are useful, whereas a general solution is not all that useful in applications. Interestingly, these solutions always exist for a homogeneous linear PDE (either reducible or irreducible) with constant coefficients. An exponential-type series solution of a PDE may involve as many arbitrary constants as one wishes.

Consider the homogeneous linear PDE of the form

$$F(D, D')z = 0. \tag{6.28}$$

Theorem 6.6. *Corresponding to a factor $h(D, D')$ of the polynomial $F(D, D')$, an exponential-type series solution of Eq. (6.28) is of the form*

$$z = \sum_i A_i e^{a_i x + b_i y}$$

where the constants A_i, a_i, b_i are chosen arbitrarily such that the series is uniformly convergent and for each i, the constants a_i and b_i must be connected by the relation $h(a_i, b_i) = 0$. Moreover, the sum of such solutions corresponding to all possible factors forms the most general exponential-type series solution of Eq. (6.28).

Proof. Due to linearity of Eq. (6.28), the polynomial $h(D, D')$ consists the terms of the form $C_{rs}D^r D'^s$, where $0 \le r + s \le O(h)$. Hence, we have

$$h(D, D') = \sum_r \sum_s C_{rs}D^r D'^s.$$

Suppose that $z = e^{ax+by}$ is a solution of $h(D, D')z = 0$, then we have

$$h(D, D')e^{ax+by} = 0$$

or

$$\sum_r \sum_s C_{rs} D^r D'^2 e^{ax+by} = 0$$

or

$$\sum_r \sum_s C_{rs} a^r b^s e^{ax+by} = 0$$

or

$$h(a, b)e^{ax+by} = 0$$

which will hold if

$$h(a, b) = 0.$$

If follows that $z = e^{ax+by}$ is a solution of the equation $h(D, D')z = 0$ provided that $h(a, b) = 0$. For any arbitrary value of a (or b), the relation $h(a, b) = 0$ gives one or more values of b (or a). Hence, there exists infinitely many pairs (a_i, b_i) satisfying $h(a_i, b_i) = 0$ such that for each i, $z_i = e^{a_i x+b_i y}$ remains a solution of the given equation. By the principle of superposition,

$$z = \sum_i A_i e^{a_i x+b_i y}$$

is the solution of $h(D, D')z = 0$ such that $h(a_i, b_i) = 0$. But $h(D, D')$ is the factor of $F(D, D')$; therefore, this remains also a solution of Eq. (6.28).

If $F(D, D')$ has p possible factors so that

$$F(D, D') = \prod_{k=1}^{p} h_k(D, D')$$

then again by using the principle of superposition, the sum of the exponential-type solutions corresponding to all the factors $h_k(D, D')$ $(k = 1, 2, \ldots, p)$ remains the more general exponential-type solution of Eq. (6.28). $\qquad\square$

With a view to illustrate the foregoing description, we furnish the following examples.

Example 6.15. *Find the exponential-type series solution of $(D^2 - D')z = 0$.*

Solution. Let

$$z = e^{ax+by} \tag{6.29}$$

be a solution of the given PDE. Then, we have

$$a^2 - b = 0$$

which gives rise to $b = a^2$. Using this, Eq.(6.29) reduces to $z = e^{a(x+ay)}$. Hence, the required solution is

$$z = \sum_i A_i e^{a_i(x+a_i y)}.$$

Example 6.16. *Find the exponential-type series solution of $r - s = p - q$.*

Solution. Here, we have

$$F(D, D') = D^2 - DD' - D + D' = (D - 1)(D - D').$$

Let $z = e^{ax+by}$ be a solution corresponding to the factor $D - 1$. Then, we have $a - 1 = 0$ so that $a = 1$. Therefore, $z = e^{x+by}$ is a solution of the given equation.

Now, suppose that corresponding to the factor $D - D'$, $z = e^{cx+dy}$ is a solution of the given equation. Then, we have $c - d = 0$, that is, $d = c$. Thus, $z = e^{c(x+y)}$ is a solution of the given equation. Finally, using the principle of superposition, the required solution is

$$z = e^x \sum_i A_i e^{b_i y} + \sum_i B_i e^{c_i(x+y)}.$$

Example 6.17. *Find the exponential-type series solution of the PDE*

$$(2D^4 - 3D^2 D' + D'^2)z = 0.$$

Solution. Here, we have

$$F(D, D') = (2D^2 - D')(D^2 - D').$$

Let $z = e^{ax+by}$ be a solution corresponding to the factor $2D^2 - D'^2$. Then, we have $2a^2 - b = 0$, which gives rise to $b = 2a^2$. Hence, $z = e^{a(x+2ay)}$ is a solution of the given equation.

Now, suppose that corresponding to the factor $D^2 - D'$, $z = e^{cx+dy}$ is a solution of the given equation. Then, we have $c^2 - d = 0$, that is, $d = c^2$. Thus, $z = e^{c(x+cy)}$ is a solution of the given equation. Finally, using the principle of superposition, the required solution is

$$z = \sum_i A_i e^{a_i(x+2a_i y)} + \sum_i B_i e^{c_i(x+c_i y)}.$$

Example 6.18. *Find the exponential-type series solution of the equation $r + t = m^2 z$.*

Solution.   Here, we have

$$F(D, D') = D^2 + D'^2 - m^2.$$

Let

$$z = e^{ax+by}$$

be a solution of the given PDE. Then, we have

$$a^2 + b^2 = m^2.$$

This relation is satisfied if we take $a = m \cos \alpha$ and $b = m \sin \alpha$, where α is a parameter. Hence, the required solution of the given equation is

$$z = \sum_{\alpha} A_{\alpha} e^{m(x \cos \alpha + y \sin \alpha)}.$$

Example 6.19. *Find the exponential-type series solution and general solution of the equation $(D^2 - DD' - 2D)z = 0$.*

Solution.   The given equation can be written as

$$D(D - D' - 2)z = 0.$$

Let $z = \sum_{i} A_i e^{a_i x + b_i y} + \sum_{i} B_i e^{c_i x + d_i y}$ be the solution of the given equation. Then, we have

$$a_i = 0 \text{ and } c_i = d_i + 2.$$

Therefore, the required exponential-type series solution is

$$z = \sum_{i} A_i e^{b_i y} + e^{2x} \sum_{i} B_i e^{d_i (x+y)}.$$

Also, the general solution is

$$z = \phi(y) + e^{2y} \psi(x + y).$$

This example demonstrates that the exponential-type solution of a reducible PDE remains a particular case of the general solution.

6.1.4 Irreducible Homogeneous Linear PDEs

In order to solve an irreducible homogeneous linear PDE, we decompose $F(D, D')$ in the factors of the least possible degree. Corresponding to a linear factor (if it exists), the part of the solution can be determined by the methods discussed in Subsection 6.1.2. Also, corresponding to a non-linear factor, the part of the solution will be an exponential-type series solution described in Subsection 6.1.3. Thus, in all, the number of arbitrary functions involved in the general solution of an irreducible equation is equal to the total number of possible linear factors in $F(D, D')$. If there is no linear factor in $F(D, D')$, then the solution of PDE will never involve any arbitrary function and in this case, such solution is termed as 'complete solution' instead of 'general solution'.

To illustrate the foregoing description, we furnish several examples of irreducible homogeneous linear PDEs.

Example 6.20. *Solve* $(D - 2D' - 1)(D - 2D'^2 - 1)z = 0$.

Solution. *Part of the solution corresponding to the linear factor $D - 2D' - 1$ is*

$$z_1 = e^x \phi(y + 2x).$$

Part of the solution corresponding to the factor $D - 2D'^2 - 1$ is

$$z_2 = \sum_i A_i e^{a_i x + b_i y}$$

where a_i and b_i are connected by

$$a_i - 2b_i^2 - 1 = 0$$

or

$$a_i = 2b_i^2 + 1.$$

Hence, the general solution of the given equation is

$$z = z_1 + z_2 = e^x \phi(y + 2x) + \sum_i A_i e^{(2b_i^2 + 1)x + b_i y}.$$

Example 6.21. *Solve* $(D^2 + D'^2 - c^2)z = 0$.

Solution. *As $D^2 + D'^2 - c^2$ is irreducible, the solution of the given PDE is*

$$z = \sum_i A_i e^{a_i x + b_i y}$$

where a_i and b_i are connected by

$$a_i^2 + b_i^2 - c^2 = 0$$

or

$$b_i = \pm\sqrt{c^2 - a_i^2}.$$

Thus, we obtain

$$z = \sum_i A_i e^{a_i x \pm \sqrt{c^2 - a_i^2}\, y}$$

which is the required solution of the given PDE.

Example 6.22.   *Solve* $(2D^4 + 3D^2 D' + D'^2)z = 0.$

Solution.   The given PDE can be written as

$$(2D^2 + D')(D^2 + D')z = 0.$$

The PDE has no linear factor. The part of solution corresponding to the first factor is

$$z_1 = \sum_i A_i e^{a_i x + b_i y}$$

where a_i and b_i are connected by

$$2a_i^2 + b_i = 0 \Rightarrow b_i = -2a_i^2.$$

Similarly, the part of solution corresponding to the second factor is

$$z_2 = \sum_i B_i e^{c_i x + d_i y}$$

where c_i and d_i are connected by

$$c_i^2 + d_i = 0 \Rightarrow d_i = -c_i^2.$$

Finally, the complete solution of the given equation is

$$z = z_1 + z_2 = \sum_i A_i e^{a_i(x - 2a_i y)} + \sum_i B_i e^{c_i(x - c_i y)}.$$

6.2 Non-homogeneous Linear PDEs with Constant Coefficients

We now discuss the methods to obtain the general solution of a non-homogeneous linear PDE with constant coefficients. A non-homogeneous linear PDE with constant coefficients of order n can be symbolically represented as

$$F(D, D')z = f(x, y) \tag{6.30}$$

where

$$F(D, D') = \sum_r \sum_s C_{rs} D^r D'^s,$$

$0 \leq r + s \leq n$, the coefficients C_{rs} are constants and the known function f is not identically zero.

Definition 6.2. *A complementary function (CF) of Eq. (6.30) is the general solution of corresponding homogeneous equation $F(D, D')z = 0$.*

The idea involved in determining the general solution of a non-homogeneous linear PDE is contained in the following core result.

Theorem 6.7. *The general solution of a non-homogeneous linear PDE of the form (6.30) is*

$$z = z_c + z_p$$

where z_c is the CF and z_p is a particular integral (PI).

Proof. By the definition of CF, we have

$$F(D, D')z_c = 0. \tag{6.31}$$

As z_p is a PI of the given PDE, we have

$$F(D, D')z_p = f(x, y). \tag{6.32}$$

Adding (6.31) and (6.32), we get

$$F(D, D')(z_c + z_p) = f(x, y).$$

It follows that $z_c + z_p$ forms the general solution. $\qquad\square$

In Section 6.1, we have described various methods for computing CFs. Henceforth, in various subsections, we discuss the methods to compute PIs for different possible cases.

6.2.1 Special Methods for Finding PIs

We attempt to find a PI z_p of Eq. (6.30), which is obtained by setting all arbitrary functions involved in the general solution to identically zero so that $z_c = 0$ as CF contains arbitrary functions, whereas PI contains no arbitrary functions. Such a particular solution is determined by

$$z_p = \frac{1}{F(D, D')} f(x, y)$$

where the operator $\dfrac{1}{F(D, D')}$, being the inverse of the operator $F(D, D')$, is defined by the following identity:

$$F(D, D') \left[\frac{1}{F(D, D')} f(x, y) \right] = f(x, y).$$

In the following lines, we record essential properties of PIs:

(i) The operator $\dfrac{1}{F(D,D')}$ is linear, that is,

$$\frac{1}{F(D,D')}\left[c_1 f_1(x,y) + c_2 f_2(x,y)\right] = c_1 \frac{1}{F(D,D')} f_1(x,y) + c_2 \frac{1}{F(D,D')} f_2(x,y).$$

(ii) $\dfrac{1}{F_1(D,D')F_2(D,D')} f(x,y) = \dfrac{1}{F_1(D,D')}\left[\dfrac{1}{F_2(D,D')} f(x,y)\right]$

$$= \frac{1}{F_2(D,D')}\left[\frac{1}{F_1(D,D')} f(x,y)\right].$$

(iii) $\dfrac{1}{D^k} f(x,y) = \underbrace{\displaystyle\int \int \cdots \int f(x,y)(\partial x)^k}_{k \text{ times}}$ and $\dfrac{1}{D'^k} f(x,y) = \underbrace{\displaystyle\int \int \cdots \int f(x,y)(\partial y)^k}_{k \text{ times}}.$

The methods of finding PI of the non-homogeneous linear PDEs are very similar to the ones carried out in solving the non-homogeneous linear ODEs with constant coefficients. For certain particular forms of the function $f(x,y)$, some special methods for finding PI can be developed. These methods are applicable for both reducible and irreducible PDEs. In the following lines, we explain these methods for various forms of $f(x,y)$ one by one.

(i) Rational Integral Algebraic Functions: The idea of finding PI of a PDE involving a rational algebraic function of integral power is contained in the following theorem.

Theorem 6.8. *If $f(x,y) = x^p y^q$, where p and q are integers, then the PI of Eq. (6.30) is*

$$\frac{1}{F(D,D')} x^p y^q = [F(D,D')]^{-1} x^p y^q$$

where $[F(D,D')]^{-1}$ is expanded as an infinite series in ascending powers of D and D' using binomial theorem and operated on $x^p y^q$ term by term. If $F(D,D')$ does not contain the constant term, then we expand $[F(D,D')]^{-1}$ in power of D/D' or D'/D or another one as the case may be.

To illustrate this case, we adopt the following examples.

Example 6.23. *Solve $(D^3 - D'^3)z = x^3 y^3$.*

Solution. The CF is

$$z_c = \phi_1(y+x) + \phi_2(y+\omega x) + \phi_3(y+\omega^2 x)$$

where ω is the complex cube root of unity. The PI is

$$z_p = \frac{1}{(D^3 - D'^3)} x^3 y^3 = \frac{1}{D^3} \left[1 - \frac{D'^3}{D^3} \right]^{-1} x^3 y^3$$

$$= \frac{1}{D^3} \left[1 + \frac{D'^3}{D^3} + \cdots \right] x^3 y^3$$

$$= \frac{1}{D^3} \left[x^3 y^3 + \frac{1}{D^3} 6x^3 \right] = y^3 \frac{1}{D^3} x^3 + 6 \frac{1}{D^6} x^3$$

$$= \frac{x^6 y^3}{120} + \frac{x^9}{10080}.$$

Hence, the general solution of the given PDE is

$$z = z_c + z_p = \phi_1(y + x) + \phi_2(y + \omega x) + \phi_3(y + \omega^2 x) + \frac{x^6 y^3}{120} + \frac{x^9}{10080}.$$

Example 6.24. *Solve $D(D + D' - 1)(D + 3D' - 2)z = x^2 - 4xy + 2y^2$.*

Solution. *The CF is*

$$z_c = \phi_1(y) + e^x \phi_2(y - x) + e^{2x} \phi_2(y - 3x).$$

The PI is

$$z_p = \frac{1}{D(D + D' - 1)(D + 3D' - 2)}(x^2 - 4xy + 2y^2)$$

$$= \frac{1}{2D} \left[1 - (D + D') \right]^{-1} \left[1 - \frac{(D + 3D')}{2} \right]^{-1} (x^2 - 4xy + 2y^2)$$

$$= \frac{1}{2D} \left[1 + (D + D') + (D^2 + 2DD' + D'^2) + \cdots \right] \left[1 + \frac{(D + 3D')}{2} + \right.$$

$$\left. \frac{(D^2 + 6DD' + 9D'^2)}{4} + \cdots \right] (x^2 - 4xy + 2y^2)$$

$$= \frac{1}{2D} \left[1 + \frac{3D}{2} + \frac{5D'}{2} + \frac{7D^2}{4} + \frac{19D'^2}{4} + \frac{11DD'}{2} + \cdots \right] (x^2 - 4xy + 2y^2)$$

$$= \frac{1}{2D} \left[(x^2 - 4xy + 2y^2) + 3(x - 2y) + 5(2y - 2x) + \frac{7}{2} + 19 - 22 \right]$$

$$= \frac{1}{2} \int \left[x^2 - 4xy + 2y^2 - 7x + 4y + \frac{1}{2} \right] \partial x$$

$$= \frac{1}{6}x^3 - x^2 y + xy^2 - \frac{7}{4}x^2 + 2xy + \frac{1}{4}x.$$

Hence, the general solution of the given PDE is

$$z = z_c + z_p = \phi_1(y) + e^x \phi_2(y - x) + e^{2x} \phi_2(y - 3x) + \frac{1}{6}x^3 - x^2 y + xy^2 - \frac{7}{4}x^2 + 2xy + \frac{1}{4}x.$$

Example 6.25. *Solve* $(D^2 - D')z = 2y - x^2$.

Solution. *The CF is*

$$z_c = \sum_i A_i e^{a_i(x+a_i y)}$$

where A_i and a_i are constants. The PI is

$$z_p = \frac{1}{(D^2 - D')}(2y - x^2)$$

$$= \frac{1}{-D'\left(1 - \dfrac{D^2}{D'}\right)}(2y - x^2) = -\frac{1}{D'}\left[1 - \frac{D^2}{D'}\right]^{-1}(2y - x^2)$$

$$= -\frac{1}{D'}\left[1 + \frac{D^2}{D'} + \frac{D^4}{D'^2} + \cdots\right](2y - x^2)$$

$$= -\frac{1}{D'}\left[(2y - x^2) + \frac{1}{D'}D^2(2y - x^2)\right]$$

$$= -\int (2y - x^2)\partial y - \frac{1}{D'^2}(-2)$$

$$= -y^2 + x^2 y + y^2 = x^2 y.$$

Hence, the general solution of the given PDE is

$$z = z_c + z_p = \sum_i A_i e^{a_i(x+a_i y)} + x^2 y.$$

Example 6.26. *Solve* $\dfrac{\partial^3 z}{\partial x^2 \partial y} - 2\dfrac{\partial^3 z}{\partial x \partial y^2} + \dfrac{\partial^3 z}{\partial y^3} = \dfrac{1}{x^2}$.

Solution. *The given PDE can be written as*

$$(D^2 D' - 2DD'^2 + D'^3)z = x^{-2}$$

or

$$D'(D^2 - 2DD' + D'^2)z = x^{-2}.$$

The CF is

$$z_c = \phi(x) + \psi(y + x) + x\theta(y + x).$$

The PI is

$$z_p = \frac{1}{D'(D^2 - 2DD' + D'^2)} x^{-2} = \frac{1}{(D - D')^2} \cdot \frac{1}{D'} x^{-2}$$

$$= \frac{1}{(D - D')^2} y x^{-2} = \frac{1}{D^2} \left[1 - \frac{D'}{D} \right]^{-2} y x^{-2}$$

$$= \frac{1}{D^2} \left[1 + \frac{2D'}{D} + \frac{3D'^2}{D^2} + \cdots \right] y x^{-2}$$

$$= \frac{1}{D^2} \left[y x^{-2} + \frac{2}{D} x^{-2} \right] = \frac{1}{D} \int (y x^{-2} - 2x^{-1}) \partial x$$

$$= \int (-y x^{-1} - 2 \log x) \partial x = -y \log x - 2 \int \log x \, dx.$$

The second term of z_p being a function of x alone can be assumed to be included in the term $\phi(x)$ of z_c. Hence, the required general solution of the given PDE is

$$z = z_c + z_p = \phi(x) + \psi(y + x) + x\theta(y + x) - y \log x.$$

(ii) Exponential Functions: In the computation of PI of a PDE involving the exponential function, the following theorem is needed.

Theorem 6.9. *If $f(x, y) = e^{ax+by}$, then the PI of Eq. (6.30) is*

$$\frac{1}{F(D, D')} e^{ax+by} = \frac{1}{F(a, b)} e^{ax+by} \text{ provided } F(a, b) \neq 0.$$

Proof. We have

$$D^r D'^s e^{ax+by} = a^r b^s e^{ax+by}.$$

Hence, we get

$$F(D, D') e^{ax+by} = \sum_r \sum_s C_{rs} D^r D'^s e^{ax+by}$$

$$= \sum_r \sum_s C_{rs} a^r b^s e^{ax+by}$$

$$= F(a, b) e^{ax+by}$$

or

$$F(D, D') e^{ax+by} = F(a, b) e^{ax+by}.$$

Operating the inverse operator $\dfrac{1}{F(D, D')}$ on both the sides and using its linearity, we obtain

$$e^{ax+by} = F(a, b) \frac{1}{F(D, D')} e^{ax+by}.$$

As $F(a, b) \neq 0$, dividing both the sides of the above equation by $F(a, b)$, we get

$$\frac{1}{F(D, D')} e^{ax+by} = \frac{1}{F(a, b)} e^{ax+by}.$$

☐

To illustrate this case, we adopt the following examples.

Example 6.27. Solve $\dfrac{\partial^3 z}{\partial x^3} - 3\dfrac{\partial^3 z}{\partial x^2 \partial y} + 4\dfrac{\partial^3 z}{\partial y^3} = e^{x+2y}$.

Solution.   *The given PDE can be written as*

$$(D^3 - 3D^2 D' + 4D'^3) = e^{x+2y}.$$

The CF is

$$z_c = \phi(y - x) + \psi(y + 2x) + x\theta(y + 2x).$$

The PI is

$$z_p = \frac{1}{D^3 - 3D^2 D' + 4D'^3} e^{x+2y}$$

$$= \frac{1}{1^3 - 3 \cdot 1^2 \cdot 2 + 4 \cdot 2^3} e^{x+2y} = \frac{1}{27} e^{x+2y}.$$

Therefore, the general solution of the given equation is

$$z = z_c + z_p = \phi(y - x) + \psi(y + 2x) + x\theta(y + 2x) + \frac{1}{27} e^{x+2y}.$$

Example 6.28. Solve $(D^3 - 3DD' + D' + 4)z = e^{2x+y}$.

Solution.   *The given PDE is irreducible. Therefore, CF is*

$$z_c = \sum_i A_i e^{a_i x + b_i y}$$

where $a_i^3 - 3a_i b_i + b_i + 4 = 0$. Also, PI is

$$z_p = \frac{1}{D^3 - 3DD' + D' + 4} e^{2x+y}$$

$$= \frac{1}{2^3 - 3 \cdot 2 \cdot 1 + 1 + 4} e^{2x+y} = \frac{1}{7} e^{2x+y}.$$

Hence, the complete solution is

$$z = z_c + z_p = \sum_i A_i e^{a_i x + b_i y} + \frac{1}{7} e^{2x+y}$$

where for each i, the pair (a_i, b_i) satisfies the relation $a_i^3 - 3a_i b_i + b_i + 4 = 0$.

(iii) Trigonometric Functions: The following result enables us to compute the PI of a PDE involving sine and cosine functions.

Theorem 6.10. *The PI of the equation $F(D^2, DD', D'^2)z = f(x, y)$ corresponding to the functions $f(x, y) = \sin(ax + by)$ and $f(x, y) = \cos(ax + by)$, are respectively, determined by the following formulae:*

(i) $\dfrac{1}{F(D^2, DD', D'^2)} \sin(ax + by) = \dfrac{1}{F(-a^2, -ab, -b^2)} \sin(ax + by)$

(ii) $\dfrac{1}{F(D^2, DD', D'^2)} \cos(ax + by) = \dfrac{1}{F(-a^2, -ab, -b^2)} \cos(ax + by)$

provided $F(-a^2, -ab, -b^2) \neq 0$.

Proof. We have

$$D^2 \sin(ax + by) = -a^2 \sin(ax + by)$$
$$D'^2 \sin(ax + by) = -b^2 \sin(ax + by)$$
$$DD' \sin(ax + by) = -ab \sin(ax + by)$$

implying thereby

$$F(D^2, DD', D'^2) \sin(ax + by) = F(-a^2, -ab, -b^2) \sin(ax + by).$$

Operating $\dfrac{1}{F(D^2, DD', D'^2)}$ on both the sides, we get

$$\sin(ax + by) = F(-a^2, -ab, -b^2) \frac{1}{F(D^2, DD', D'^2)} \sin(ax + by).$$

As $F(-a^2, -ab, -b^2) \neq 0$, dividing the above equation by $F(-a^2, -ab, -b^2)$, we get

$$\frac{1}{F(D^2, DD', D'^2)} \sin(ax + by) = \frac{1}{F(-a^2, -ab, -b^2)} \sin(ax + by).$$

Hence, (i) is verified. The similar arguments can do the needful for (ii). □

To illustrate these cases, we furnish the following examples.

Example 6.29. *Solve $(D^2 - DD' + D' - 1)z = \cos(x + 2y)$.*

Solution. The CF is

$$z_c = e^x \phi(y) + e^{-x} \psi(y + x).$$

The PI is

$$z_p = \frac{1}{D^2 - DD' + D' - 1} \cos(x + 2y)$$

$$= \frac{1}{-1^2 - (-1 \cdot 2) + D' - 1} \cos(x + 2y) = \frac{1}{D'} \cos(x + 2y)$$

$$= \int \cos(x + 2y) \partial y = \frac{1}{2} \sin(x + 2y).$$

Hence, the general solution is

$$z = z_c + z_p = e^x \phi(y) + e^{-x} \psi(y + x) + \frac{1}{2} \sin(x + 2y).$$

Example 6.30. *Solve $(D^2 - 2DD' + D'^2 - 3D + 3D' + 2)z = \sin(2x + 3y)$.*

Solution. *The CF is*

$$z_c = e^x \phi(y + x) + e^{2x} \psi(y + x).$$

The PI is

$$z_p = \frac{1}{D^2 - 2DD' + D'^2 - 3D + 3D' + 2} \sin(2x + 3y)$$

$$= \frac{1}{-2^2 + 2 \cdot 2 \cdot 3 - 3^2 - 3D + 3D' + 2} \sin(2x + 3y)$$

$$= \frac{1}{-3D + 3D' + 1} \sin(2x + 3y)$$

$$= \frac{(3D' - 3D - 1)}{(3D' - 3D + 1)(3D' - 3D - 1)} \sin(2x + 3y)$$

$$= \frac{(3D' - 3D - 1)}{-81 + 108 - 36 - 1} \sin(2x + 3y) = \frac{1}{10}(3D - 3D' + 1) \sin(2x + 3y)$$

$$= \frac{1}{10} \left[\sin(2x + 3y) - 3 \cos(2x + 3y) \right].$$

Hence, the general solution of the given equation is

$$z = z_c + z_p = e^x \phi(y + x) + e^{2x} \psi(y + x) + \frac{1}{10} \left[\sin(2x + 3y) - 3 \cos(2x + 3y) \right].$$

(iv) Product Functions: We obtain the PI of functions, which are expressible as a product of an exponential function with an arbitrary function. The following theorem paves the way for such computation.

Theorem 6.11. *Let $V(x, y)$ be an arbitrary function. Then*

$$\frac{1}{F(D, D')} e^{ax+by} V(x, y) = e^{ax+by} \frac{1}{F(D + a, D' + b)} V(x, y).$$

Proof. For an arbitrary function $U(x, y)$, we have

$$D[e^{ax+by}U] = aUe^{ax+by} + e^{ax+by}DU$$
$$= e^{ax+by}(D + a)U.$$

Operating D on both the sides, we get

$$D^2[e^{ax+by}U] = D[e^{ax+by}(D + a)U]$$
$$= e^{ax+by}a(D + a)U + e^{ax+by}D(D + a)U$$
$$= e^{ax+by}(D + a)^2U.$$

Continuing so on, we obtain

$$D^r[e^{ax+by}U] = e^{ax+by}(D + a)^rU.$$

Similarly, we can obtain

$$D'^s[e^{ax+by}U] = e^{ax+by}(D' + b)^sU$$

and

$$D^rD'^s[e^{ax+by}U] = e^{ax+by}(D + a)^r(D' + b)^sU.$$

Hence, we have

$$F(D, D')[e^{ax+by}U] = \sum_r \sum_s C_{rs}D^rD'^s e^{ax+by}U$$
$$= e^{ax+by}\sum_r \sum_s C_{rs}(D + a)^r(D' + b)^sU$$
$$= e^{ax+by}F(D + a, D' + b)U.$$

Putting $F(D + a, D' + b)U = V$ so that $U = \dfrac{1}{F(D + a, D' + b)}V$, the above equation reduces to

$$F(D, D')\left[e^{ax+by}\frac{1}{F(D + a, D' + b)}V\right] = e^{ax+by}V$$

which implies that

$$\frac{1}{F(D, D')}e^{ax+by}V(x, y) = e^{ax+by}\frac{1}{F(D + a, D' + b)}V(x, y).$$

$\square$

To illustrate this case, we adopt the following examples.

Example 6.31. *Solve $D(D - 2D')(D + D')z = (x^2 + 4y^2)e^{x+2y}$.*

Solution. The CF is

$$z_c = \phi(y) + \psi(y + 2x) + \theta(y - x).$$

The PI is

$$z_p = \frac{1}{D(D - 2D')(D + D')}(x^2 + 4y^2)e^{x+2y}$$

$$= e^{x+2y} \frac{1}{(D + 1)(D - 2D' - 3)(D + D' + 3)}(x^2 + 4y^2)$$

$$= e^{x+2y} \frac{1}{(D + 1)(D - 2D' - 3)}u \ (say)$$

where

$$u = \frac{1}{D + D' + 3}(x^2 + 4y^2)$$

$$= \frac{1}{3}\left[1 + \frac{1}{3}(D + D')\right]^{-1}(x^2 + 4y^2)$$

$$= \frac{1}{3}\left[1 - \frac{1}{3}(D + D') + \frac{1}{9}(D + D')^2 - \cdots\right](x^2 + 4y^2)$$

$$= \frac{1}{3}\left[x^2 + 4y^2 - \frac{2}{3}(x + 4y) + \frac{10}{9}\right] = \frac{1}{27}(9x^2 + 36y^2 - 6x - 24y + 10).$$

Thus, we have

$$z_p = \frac{1}{27}e^{x+2y} \frac{1}{(D + 1)(D - 2D' - 3)}\left[9x^2 + 36y^2 - 6x - 24y + 10\right]$$

$$= \frac{1}{27}e^{x+2y} \frac{1}{D + 1}v \ (say)$$

where

$$v = \frac{1}{D - 2D' - 3}(9x^2 + 36y^2 - 6x - 24y + 10)$$

$$= -\frac{1}{3}\left[1 + \frac{1}{3}(2D' - D)\right]^{-1}(9x^2 + 36y^2 - 6x - 24y + 10)$$

$$= -\frac{1}{3}\left[1 - \frac{1}{3}(2D' - D) + \frac{1}{9}(2D' - D)^2 - \cdots\right](9x^2 + 36y^2 - 6x - 24y + 10)$$

$$= -\frac{1}{3}(9x^2 + 36y^2 - 72y + 58).$$

Using this, we obtain

$$z_p = -\frac{1}{81}e^{x+2y}\frac{1}{D+1}(9x^2 + 36y^2 - 72y + 58)$$

$$= -\frac{1}{81}e^{x+2y}(1 - D + D^2 - \cdots)(9x^2 + 36y^2 - 72y + 58)$$

$$= -\frac{1}{81}(9x^2 + 36y^2 - 18x - 72y + 76)e^{x+2y}.$$

Therefore, the general solution of the given PDE is

$$z = z_c + z_p = \phi(y) + \psi(y + 2x) + \theta(y - x) - \frac{1}{81}(9x^2 + 36y^2 - 18x - 72y + 76)e^{x+2y}.$$

Example 6.32. *Solve* $(D - 2D' + 5)(D^2 + D' + 3)z = e^{3x+4y}\sin(x - 2y).$

Solution. The CF is

$$z_c = e^{-5x}\phi(y + 2x) + \sum_i A_i e^{a_i x - (a_i^2 + 3)y}.$$

The PI is

$$z_p = \frac{1}{(D - 2D' + 5)(D^2 + D' + 3)}e^{3x+4y}\sin(x - 2y)$$

$$= e^{3x+4y}\frac{1}{(D - 2D')(D^2 + 6D + D' + 16)}\sin(x - 2y)$$

$$= e^{3x+4y}\frac{1}{(D - 2D')(-1 + 6D + D' + 16)}\sin(x - 2y)$$

$$= e^{3x+4y}\frac{1}{6D^2 - 11DD' - 2D'^2 + 15D - 30D'}\sin(x - 2y)$$

$$= e^{3x+4y}\frac{1}{6 \cdot (-1) - 11 \cdot (2) - 2 \cdot (-4) + 15D - 30D'}\sin(x - 2y)$$

$$= \frac{1}{5}e^{3x+4y}\frac{1}{(3D - 6D') - 4}\sin(x - 2y)$$

$$= \frac{1}{5}e^{3x+4y}\frac{(3D - 6D') + 4}{9D^2 - 36DD' + 36D'^2 - 16}\sin(x - 2y)$$

$$= \frac{1}{5}e^{3x+4y}\frac{(3D - 6D') + 4}{9 \cdot (-1) - 36 \cdot (2) + 36 \cdot (-4) - 16}\sin(x - 2y)$$

$$= -\frac{1}{1205}e^{3x+4y}(3D - 6D' + 4)\sin(x - 2y)$$

$$= -\frac{1}{1205}[15\cos(x - 2y) + 4\sin(x - 2y)]e^{3x+4y}.$$

Therefore, the general solution is

$$z = z_c + z_p = e^{-5x}\phi(y + 2x) + \sum_i A_i e^{a_i x - (a_i^2 + 3)y} - \frac{1}{1205}[15\cos(x - 2y) + 4\sin(x - 2y)]e^{3x+4y}.$$

Exceptional Case of Exponential Function: If $F(a, b) = 0$, then the classical method to find the PI of $F(D, D')z = e^{ax+by}$ discussed in Theorem 6.9 fails. Thus far, in such a case, we use the following formula, which can be deducible from Theorem 6.11 by particularising $V(x, y) = 1$.

$$\frac{1}{F(D, D')} e^{ax+by} = e^{ax+by} \frac{1}{F(D + a, D' + b)} (1).$$

Now, we consider an illustrative example for such a case.

Example 6.33. *Solve* $(D^3 + D^2 D' - DD' - D'^2)z = e^{x+y}$

Solution. *The CF is*

$$z_c = \phi(y - x) + \sum_i A_i e^{a_i x + a_i^2 y}.$$

The PI is

$$
\begin{aligned}
z_p &= \frac{1}{(D^2 - D')(D + D')} e^{x+y} = \frac{1}{D^2 - D'} \left[\frac{1}{D + D'} e^{x+y} \right] \\
&= \frac{1}{D^2 - D'} \left[\frac{1}{1 + 1} e^{x+y} \right] = \frac{-1}{2} \cdot \frac{1}{D^2 - D'} e^{x+y} \cdot 1 \\
&= \frac{1}{2} e^{x+y} \cdot \frac{1}{(D + 1)^2 - (D' + 1)} (1), \quad \text{(using Theorem 6.11)} \\
&= \frac{1}{2} e^{x+y} \cdot \frac{1}{D^2 + 2D - D'} (1) = \frac{1}{2} e^{x+y} \cdot \frac{1}{2D} \left[1 + \frac{D^2 - D'}{2D} \right]^{-1} (1) \\
&= \frac{1}{4} e^{x+y} \cdot \frac{1}{D} \left[1 - \left(\frac{D}{2} - \frac{D'}{2D} \right) + \cdots \right] (1) \\
&= \frac{1}{4} e^{x+y} \cdot \frac{1}{D} (1) = \frac{1}{4} x e^{x+y}.
\end{aligned}
$$

Therefore, the general solution of the given equation is

$$z = z_c + z_p = \phi(y - x) + \sum_i A_i e^{a_i x + a_i^2 y} + \frac{1}{4} x e^{x+y}.$$

6.2.2 A Short Method for Finding PIs of a Certain Class of Euler Linear Equations

If the polynomial $F(D, D')$ of a PDE is homogeneous (which amounts to saying that the PDE is Euler linear) and $f(x, y) = \zeta(ax + by)$, whereas ζ is a function of single variable (that is, ζ is a composite function such that the intermediate variable remains a homogeneous linear polynomial in x and y)

then a shorter method can be applied to find the PI. Indeed, this method is based on the following theorem:

Theorem 6.12. *Let $F(D, D')z = \zeta(ax + by)$ be an Euler linear PDE of order n. Then, the PI is*

$$\frac{1}{F(D, D')}\zeta(ax + by) = \frac{1}{F(a, b)}\underbrace{\int \int \cdots \int}_{n \text{ times}} \zeta(u)(du)^n$$

where $u = ax + by$ provided $F(a, b) \neq 0$.

Proof. By direct differentiation, we have

$$D\zeta(ax + by) = a\zeta'(ax + by)$$
$$D^2\zeta(ax + by) = a^2\zeta''(ax + by).$$

On continuing, we get

$$D^r\zeta(ax + by) = a^r\zeta^{(r)}(ax + by).$$

Similarly, we can obtain

$$D'^s\zeta(ax + by) = b^s\zeta^{(s)}(ax + by)$$

and

$$D^rD'^s\zeta(ax + by) = a^rb^s\zeta^{(r+s)}(ax + by).$$

As the given PDE is Euler linear, $F(D, D')$ can be expressed as

$$F(D, D') = k_0D^n + k_1D^{n-1}D' + \cdots + k_nD'^n.$$

Thus, we have

$$F(D, D')\zeta(ax + by) = F(a, b)\zeta^{(n)}(ax + by).$$

Operating $\dfrac{1}{F(D, D')}$ on both the sides, we get

$$\frac{1}{F(D, D')}\zeta^{(n)}(ax + by) = \frac{1}{F(a, b)}\zeta(ax + by) \text{ as } F(a, b) \neq 0.$$

Putting $ax + by = u$ in the above equation, we get

$$\frac{1}{F(D, D')}\zeta^{(n)}(u) = \frac{1}{F(a, b)}\zeta(u).$$

Integrating both the sides n times w.r.t. u, we get

$$\frac{1}{F(D, D')}\zeta(u) = \frac{1}{F(a, b)}\underbrace{\int\int\cdots\int}_{n \text{ times}}\zeta(u)(du)^n$$

where $u = ax + by$. □

Exceptional Case: Whenever $F(a, b) = 0$, the method discussed in Theorem 6.12 cannot be applied. In this case, $(bD - aD')$ must be a factor of $F(D, D')$. Therefore, there exists a natural number k $(1 \le k < n)$ such that

$$F(D, D') = (bD - aD')^k G(D, D')$$

where $G(D, D')$ is a homogeneous function of order $(n - k)$ and $G(a, b) \ne 0$. In such a situation, the following theorem is applicable.

Theorem 6.13. *Given a function $\zeta(ax + by)$, we have*

$$\frac{1}{(bD - aD')^k}\zeta(ax + by) = \frac{1}{k!}\frac{x^k}{b^k}\zeta(ax + by).$$

Proof. Consider the first-order PDE

$$(bD - aD')z = x^k\zeta(ax + by) \tag{6.33}$$

or

$$bp - aq = x^k\zeta(ax + by).$$

Lagrange's AEs are

$$\frac{dx}{b} = \frac{dy}{-a} = \frac{dz}{x^k\zeta(ax + by)}.$$

From the first two fractions, we get

$$adx + bdy = 0$$

which on integrating gives rise to

$$ax + by = c. \tag{6.34}$$

Now, taking the first and last fractions, we get

$$dz = \frac{x^k\zeta(ax + by)}{b}dx = \frac{x^k\zeta(c)}{b}dx. \quad \text{(using (6.34))}$$

Integrating the above, we get

$$z = \frac{x^{k+1}}{b(k+1)}\zeta(c).$$

Replacing 'c' by $ax + by$, the above equation reduces to

$$z = \frac{x^{k+1}}{b(k+1)}\zeta(ax+by).$$

This is a solution of Eq. (6.33) so that

$$(bD - aD')\frac{x^{k+1}}{b(k+1)}\zeta(ax+by) = x^k\zeta(ax+by)$$

or

$$\frac{1}{(bD-aD')}x^k\zeta(ax+by) = \frac{x^{k+1}}{b(k+1)}\zeta(ax+by). \qquad (6.35)$$

Now, we have

$$
\begin{aligned}
\frac{1}{(bD-aD')^k}\zeta(ax+by) &= \frac{1}{(bD-aD')^{k-1}}\left[\frac{1}{(bD-aD')}\zeta(ax+by)\right]\\
&= \frac{1}{(bD-aD')^{k-1}}\frac{x}{b}\zeta(ax+by) \;\; \text{(using (6.35) with } k=0)\\
&= \frac{1}{b}\frac{1}{(bD-aD')^{k-2}}\left[\frac{1}{(bD-aD')}x\zeta(ax+by)\right]\\
&= \frac{1}{b}\frac{1}{(bD-aD')^{k-2}}\frac{x^2}{2b}\zeta(ax+by) \;\; \text{(using (6.35) with } k=1)\\
&= \frac{1}{2!}\frac{1}{b^2}\frac{1}{(bD-aD')^{k-2}}x^2\zeta(ax+by)\\
&= \cdots \cdots \cdots \cdots \cdots\\
&= \frac{1}{k!}\frac{x^k}{b^k}\zeta(ax+by).
\end{aligned}
$$

$\square$

With a view to demonstrate the shorter method as discussed earlier, we furnish several examples.

Example 6.34. *Solve* $(D^2 + 2DD' - 8D'^2)z = \sqrt{2x+3y}$.

Solution. The CF is

$$z_c = \phi(y+2x) + \psi(y-4x).$$

The PI is

$$z_p = \frac{1}{(D^2 + 2DD' - 8D'^2)}(2x + 3y)^{1/2}$$

$$= \frac{1}{2^2 + 2 \cdot 2 \cdot 3 - 8 \cdot 3^2} \int \int (u)^{1/2} du \, du, \quad where \; u = 2x + 3y$$

$$= -\frac{1}{56} \cdot \frac{u^{5/2}}{\frac{3}{2} \cdot \frac{5}{2}} = -\frac{1}{210} u^{5/2}$$

$$= -\frac{1}{210}(2x + 3y)^{5/2}.$$

Hence, the general solution is

$$z = z_c + z_p = \phi(y + 2x) + \psi(y - 4x) - \frac{1}{210}(2x + 3y)^{5/2}.$$

Example 6.35. *Solve* $r + 5s + 6t = \dfrac{1}{y - 2x}.$

Solution. *The given equation can be written symbolically as*

$$(D^2 + 5DD' + 6D'^2)z = (y - 2x)^{-1}.$$

The CF is

$$z_c = \phi(y - 2x) + \psi(y - 3x).$$

The PI is

$$z_p = \frac{1}{(D + 2D')(D + 3D')}(y - 2x)^{-1}$$

$$= \frac{1}{D + 2D'}\left[\frac{1}{D + 3D'}(y - 2x)^{-1}\right]$$

$$= \frac{1}{D + 2D'}\left[\frac{1}{-2 + 3 \cdot 1}\int u^{-1} du\right], \quad where \; u = y - 2x$$

$$= \frac{1}{D + 2D'} \log u = \frac{1}{D + 2D'} \log(y - 2x)$$

$$= \frac{1}{1!} \cdot \frac{x^1}{1^1} \log(y - 2x) = x \log(y - 2x).$$

Therefore, the general solution is

$$z = z_c + z_p = \phi(y - 2x) + \psi(y - 3x) + x \log(y - 2x).$$

Example 6.36.   *Solve $4r - 4s + t = 16 \log(x + 2y)$.*

Solution.   *The given PDE can be written as*

$$(4D^2 - 4DD' + D'^2)z = 16 \log(x + 2y).$$

The CF is

$$z_c = \phi \left(y + \frac{1}{2}x \right) + x\psi \left(y + \frac{1}{2}x \right).$$

The PI of the given equation is

$$z_p = \frac{1}{(2D - D')^2} 16 \log(x + 2y)$$

$$= 16 \frac{1}{2!} \frac{x^2}{2^2} \log(x + 2y) \quad (\text{applying Theorem 6.13})$$

$$= 2x^2 \log(x + 2y).$$

Therefore, the general solution is

$$z = \phi \left(y + \frac{1}{2}x \right) + x\psi \left(y + \frac{1}{2}x \right) + 2x^2 \log(x + 2y).$$

Example 6.37.   *Solve $(D - 3D' - 2)^2 z = 2e^{2x} \tan(3x + y)$.*

Solution.   *The CF is*

$$z_c = e^{2x}\phi(y + 3x) + xe^{2x}\psi(y + 3x).$$

The PI is

$$z_p = \frac{1}{(D - 3D' - 2)^2} 2e^{2x} \tan(3x + y)$$

$$= 2e^{2x} \frac{1}{[(D + 2) - 3(D' + 0) - 2]^2} \cdot \tan(3x + y), \quad (\text{using Theorem 6.11})$$

$$= 2e^{2x} \frac{1}{(D - 3D')^2} \cdot \tan(3x + y)$$

$$= 2e^{2x} \cdot \frac{1}{2!} \cdot \frac{x^2}{1^2} \cdot \tan(3x + y)$$

$$= x^2 e^{2x} \tan(3x + y).$$

Thus, the general solution is

$$z = e^{2x}\phi(y + 3x) + xe^{2x}\psi(y + 3x) + x^2 e^{2x} \tan(3x + y).$$

Example 6.38. *Solve* $(D^2 + DD' - 2D'^2)z = \sinh(x + y)$.

Solution. *The CF is*

$$z_c = \phi(y + x) + \psi(y - 2x).$$

The PI is

$$\begin{aligned}
z_p &= \frac{1}{D^2 + DD' - 2D'^2} \sinh(x + y) = \frac{1}{D - D'} \left[\frac{1}{D + 2D'} \sinh(x + y) \right] \\
&= \frac{1}{D - D'} \left[\frac{1}{1 + 2 \cdot 1} \int \sinh u \, du \right], \quad \text{where } u = x + y \\
&= \frac{1}{3} \cdot \frac{1}{D - D'} \cosh u = \frac{1}{3} \cdot \frac{1}{D - D'} \cosh(x + y) \\
&= \frac{1}{3} \cdot \frac{1}{1!} \cdot \frac{x^1}{1^1} \cosh(x + y) = \frac{1}{3} x \cosh(x + y).
\end{aligned}$$

Hence, the general solution is

$$z = z_c + z_p = \phi(y + x) + \psi(y - 2x) + \frac{1}{3} x \cosh(x + y).$$

In the following lines, we furnish some examples for Euler linear PDEs involving exponential and trigonometric functions, whereas the special methods fail but PI can be determined by applying Theorem 6.13.

Example 6.39. *Solve* $\dfrac{\partial^2 z}{\partial x^2} - 4 \dfrac{\partial^2 z}{\partial x \partial y} + 4 \dfrac{\partial^2 z}{\partial y^2} = e^{2x+y}$.

Solution. *The given PDE can be written as*

$$(D^2 - 4DD' + 4D'^2)z = e^{2x+y}.$$

The CF is

$$z_c = \phi(y + 2x) + x\psi(y + 2x).$$

The PI is

$$z_p = \frac{1}{D^2 - 4DD' + 4D'^2} e^{2x+y} = \frac{1}{(D^2 - 2D')^2} e^{2x+y}.$$

Here, $F(2, 1) = 0$, hence we cannot apply the classical method given by Theorem 6.9. We use Theorem 6.13 to get

$$\begin{aligned}
z_p &= \frac{1}{(D - 2D')^2} e^{2x+y} \\
&= \frac{1}{2!} \cdot \frac{x^2}{1^2} e^{2x+y} = \frac{1}{2} x^2 e^{2x+y}.
\end{aligned}$$

Hence, the general solution is

$$z = \phi(y + 2x) + x\psi(y + 2x) + \frac{1}{2}x^2 e^{2x+y}.$$

Example 6.40. *Solve* $(D^2 - D'^2 - 3D + 3D')z = e^{x+2y}.$

Solution. *The CF is*

$$z_c = \phi(y + x) + e^{3x}\psi(y - x).$$

The PI is

$$
\begin{aligned}
z_p &= \frac{1}{(D + D' - 3)(D - D')}e^{x+2y} = \frac{1}{D + D' - 3}\left[\frac{1}{D - D'}e^{x+2y}\right] \\
&= \frac{1}{D + D' - 3}\left[\frac{1}{1 - 2}e^{x+2y}\right] = -\frac{1}{D + D' - 3}e^{x+2y} \\
&= -e^{3x}\int e^{-3x} \cdot e^{x+2(a+x)}dx, \quad \text{where } a = y - x \\
&= -e^{3x}\int e^{2a}dx = -xe^{3x+2a} = -xe^{x+2y}.
\end{aligned}
$$

Hence, the general solution is

$$z = z_c + z_p = \phi(y + x) + e^{3x}\psi(y - x) - xe^{x+2y}.$$

Example 6.41. *Solve* $(D^3 - 4D^2D' + 4DD'^2)z = 4\sin(2x + y).$

Solution. *The CF is*

$$z_c = \phi(y) + \psi(y + 2x) + x\theta(y + 2x).$$

If we put $D^2 = -4$, $DD' = -2$, *and* $D'^2 = -1$, *then we have*

$$
\begin{aligned}
D^3 - 4D^2D' + 4DD'^2 &= \left[D^2 - 4DD' + 4D'^2\right]D \\
&= \left[-4 - 4 \cdot (-2) + 4 \cdot (-1)\right]D \\
&= 0.
\end{aligned}
$$

Consequently, the classical method of finding PI fails. Applying Theorem 6.13, we get

$$
\begin{aligned}
z_p &= \frac{1}{(D^2 - 4DD' + 4D'^2)D}\left[4\sin(2x + y)\right] \\
&= \frac{4}{D^2 - 4DD' + 4D'^2}\int \sin(2x + y)\partial x \\
&= \frac{4}{D^2 - 4DD' + 4D'^2}\left[\frac{-\cos(2x + y)}{2}\right] \\
&= -2 \cdot \frac{1}{(D - 2D')^2}\cos(2x + y) = -2\left[\frac{1}{2!} \cdot \frac{x^2}{1^2}\cos(2x + y)\right] \\
&= -x^2\cos(2x + y).
\end{aligned}
$$

Hence, the general solution of the given PDE is

$$z = z_c + z_p = \phi(y) + \psi(y + 2x) + x\theta(y + 2x) - x^2 \cos(2x + y).$$

6.2.3 A General Method for Finding PIs of Reducible Linear Equations

We present a general method for finding PI of a reducible linear PDE of the form $F(D, D') = f(x, y)$, whereas $f(x, y)$ is an arbitrary function. As $F(D, D')$ can be factorised into linear factors of the form $\alpha D + \beta D' + \gamma$ (wherein either $\alpha \neq 0$ or $\beta \neq 0$), we can divide such a factor by α or β. Consequently, each factor of $F(D, D')$ takes the form $D + mD' + c$ or $D' + mD + c$ (as the case may be). In a forthcoming theorem, we derive a formula for a single factor of a reducible linear PDE. By repeated application of this formula for all·factors, the PI of the given PDE can be determined easily.

Theorem 6.14. *Let $f(x, y)$ be an arbitrary function. Then*

(i) $\dfrac{1}{(D + mD' + c)} f(x, y) = e^{-cx} \displaystyle\int e^{cx} f(x, a + mx) dx$

(ii) $\dfrac{1}{(D' + mD + c)} f(x, y) = e^{-cy} \displaystyle\int e^{cy} f(b + my, y) dy$

such that after integration, the constant a (and the constant b) must be replaced by $y - mx$ (respectively, by $x - my$).

Proof. To prove (i), consider

$$(D + mD' + c)z = f(x, y)$$

or

$$p + mq = -cz + f(x, y).$$

Lagrange's AEs are

$$\frac{dx}{1} = \frac{dy}{m} = \frac{dz}{-cz + f(x, y)}.$$

Taking the first two fractions, we get

$$dy - mdx = 0.$$

Integrating, we get

$$y - mx = a. \tag{6.36}$$

Now, taking the first and last fractions, we get

$$\frac{dz}{dx} + cz = f(x, y),$$

which on using (6.36) reduces to

$$\frac{dz}{dx} + cz = f(x, a + mx).$$

This is a linear ODE of order one with IF $= e^{cx}$ and hence its solution is

$$z = e^{-cx} \int e^{cx} f(x, a + mx) dx.$$

After integration, the constant 'a' must be appropriately replaced by $y - mx$ as the PI does not contain any arbitrary constant. On the similar lines, we can prove (ii) as well. This concludes the proof. $\square$

Choosing $c = 0$ in Theorem 6.14, we derive the following result corresponding to an Euler linear PDE.

Corollary 6.2. *Let $f(x, y)$ be an arbitrary function. Then*

(i) $\dfrac{1}{(D + mD')} f(x, y) = \displaystyle\int f(x, a - mx) dx$

(ii) $\dfrac{1}{(D' + mD)} f(x, y) = \displaystyle\int f(b - my, y) dy$

such that after integration, the constant a (and the constant b) must be replaced by $y - mx$ (respectively, by $x - my$).

With a view to illustrate the foregoing descriptions, we furnish the following examples.

Example 6.42. *Solve $(D - 3D' - 2)^2 z = 2e^{2x} \tan(y + 3x)$.*

Solution. The CF is

$$z_c = e^{2x}[\phi(y + 3x) + x\psi(y + 3x)].$$

The PI is

$$
\begin{aligned}
z_p &= \frac{1}{(D - 3D' - 2)^2} 2e^{2x} \tan(y + 3x) \\
&= \frac{2}{D - 3D' - 2} \left[\frac{1}{D - 3D' - 2} e^{2x} \tan(y + 3x) \right] \\
&= \frac{2}{D - 3D' - 2} \cdot e^{2x} \int e^{-2x} e^{2x} \tan a\, dx, \quad \text{where } a = y + 3x \\
&= \frac{2}{D - 3D' - 2} \cdot e^{2x} x \tan a = \frac{2}{D - 3D' - 2} \cdot e^{2x} x \tan(y + 3x) \\
&= 2e^{2x} \int e^{-2x} \cdot e^{2x} x \tan b\, dx, \quad \text{where } b = y + 3x \\
&= 2e^{2x} \cdot \frac{x^2}{2} \cdot \tan b = x^2 e^{2x} \tan(y + 3x).
\end{aligned}
$$

Thus, the general solution is

$$z = z_c + z_p = e^{2x}[\phi(y + 3x) + x\psi(y + 3x)] + x^2 e^{2x}\tan(y + 3x).$$

Example 6.43. *Solve* $(D^2 - DD' + D' - 1)z = 1 + xy + e^y + \cos(x + 2y).$

Solution. *The given equation can be written as*

$$(D - 1)(D - D' + 1)z = 1 + xy + e^y + \cos(x + 2y).$$

Thus, the CF is

$$z_c = e^x\phi(y) + e^{-x}\psi(y + x).$$

The PI is

$$
\begin{aligned}
z_p &= \frac{1}{D - 1}\left[\frac{1}{D - D' + 1}\{1 + xy + e^y + \cos(x + 2y)\}\right] \\
&= \frac{1}{D - 1}\left[e^{-x}\int e^x\{1 + x(a - x) + e^{a-x} + \cos(2a - x)\}dx\right], \ (\text{where } a = y + x) \\
&= \frac{1}{D - 1}e^{-x}\left[e^x + (ax - x^2)e^x - (a - 2x)e^x - 2e^x + xe^a + \frac{1}{2}e^x\cos(2a - x)\right. \\
&\hspace{6cm}\left. - \frac{1}{2}e^x\sin(2a - x)\right] \\
&= \frac{1}{D - 1}\left[(x - 1)(y + 1) + xe^y + \frac{1}{2}\cos(x + 2y) - \frac{1}{2}\sin(x + 2y)\right] \\
&= e^x\int e^{-x}\left[(x - 1)(b + 1) + xe^b + \frac{1}{2}\cos(x + 2b) - \frac{1}{2}\sin(x + 2b)\right]dx \\
&= e^x\left[-x(b + 1)e^{-x} - (x + 1)e^{-x}e^b + \frac{1}{2}e^{-x}\sin(x + 2b)\right] \\
&= -x(y + 1) - (x + 1)e^y + \frac{1}{2}\sin(x + 2y).
\end{aligned}
$$

Hence, the general solution is

$$
\begin{aligned}
z = z_c + z_p &= e^x\phi(y) + e^{-x}\psi(y + x) - x(y + 1) - (x + 1)e^y + \frac{1}{2}\sin(x + 2y) \\
&= e^x\phi(y) + e^{y-(x+y)}\psi(x + y) - x(y + 1) - (x + y + 1)e^y + ye^y + \frac{1}{2}\sin(x + 2y) \\
&= e^x\phi(y) + e^y\{e^{-(x+y)}\psi(x + y) - (x + y) - 1\} - x(y + 1) + ye^x + \frac{1}{2}\sin(x + 2y)
\end{aligned}
$$

or

$$z = e^x\phi(y) + e^y\theta(x + y) - x(y + 1) + ye^x + \frac{1}{2}\sin(x + 2y).$$

Example 6.44. *Solve* $(D^2 + DD' - 6D'^2)z = y\cos x$.

Solution. The CF is

$$z_c = \phi(y + 2x) + \psi(y - 3x).$$

The PI is

$$
\begin{aligned}
z_p &= \frac{1}{D^2 + DD' - 6D'^2} y\cos x = \frac{1}{(D - 2D')(D + 3D')} y\cos x \\
&= \frac{1}{D - 2D'}\left[\frac{1}{D + 3D'} y\cos x\right] \\
&= \frac{1}{D - 2D'} \cdot \int (a + 3x)\cos x\,dx, \quad \text{where } y = a + 3x \\
&= \frac{1}{D - 2D'}\left[a\sin x + 3\int x\cos x\,dx\right] = \frac{1}{D - 2D'}\left[a\sin x + 3x\sin x + 3\cos x\right] \\
&= \frac{1}{D - 2D'}\left[y\sin x + 3\cos x\right] \\
&= \int [(b - 2x)\sin x + 3\cos x]dx, \quad \text{where } y = b - 2x \\
&= -b\cos x - 2\int x\sin x\,dx + 3\sin x \\
&= -b\cos x + 2x\cos x - 2\sin x + 3\sin x = -y\cos x + \sin x.
\end{aligned}
$$

Hence, the complete solution is

$$z = z_c + z_p = \phi(y + 2x) + \psi(y - 3x) + \sin x - y\cos x.$$

Example 6.45. *Solve* $(D^2 - 4D'^2)z = \dfrac{4x}{y^2} - \dfrac{y}{x^2}$.

Solution. The CF is

$$z_c = \phi(y + 2x) + \psi(y - 2x).$$

The PI is

$$z_p = \frac{1}{(D + 2D')(D - 2D')}\left[\frac{4x}{y^2} - \frac{y}{x^2}\right] = \frac{1}{D + 2D'}u$$

where

$$u = \frac{1}{D - 2D'} \left[\frac{4x}{y^2} - \frac{y}{x^2} \right]$$

$$= \int \left[\frac{4x}{(a - 2x)^2} - \frac{a - 2x}{x^2} \right] dx, \quad where \ a = y + 2x$$

$$= \log(a - 2x) + \frac{a}{a - 2x} + \frac{a}{x} + 2 \log x, \quad (on \ integration)$$

$$= \log y + \frac{y + 2x}{y} + \frac{y + 2x}{x} + 2 \log x$$

$$= \log y + \frac{2x}{y} + \frac{y}{x} + 3 + 2 \log x.$$

Thus, we have

$$z_p = \frac{1}{D + 2D'} \left[\log y + \frac{2x}{y} + \frac{y}{x} + 3 + 2 \log x \right]$$

$$= \int \left[\log(b + 2x) + \frac{2x}{b + 2x} + \frac{b + 2x}{x} + 3 + 2 \log x \right] dx, \quad where \ b = y - 2x$$

$$= \int \left[\log(b + 2x) + \frac{2x}{b + 2x} + \frac{b}{x} + 5 + 2 \log x \right] dx$$

$$= x \log(b + 2x) + b \log x + 5x + 2x \log x - 2x, \quad (on \ integration)$$

$$= x \log y + y \log x + 3x.$$

Hence, the general solution of the given PDE is

$$z = z_c + z_p = \phi(y + 2x) + \psi(y - 2x) + x \log y + y \log x + 3x.$$

In a reducible linear PDE of the form $F(D, D')z = e^{ax+by}$, if $F(a, b) = 0$, then Theorem 6.9 fails. In this case, the PI can be determined easily by applying Theorem 6.14. For illustration, we present the following example.

Example 6.46. *Solve* $(D^3 - 7DD'^2 - 6D'^3)z = \sin(x + 2y) + e^{3x+y}$.

Solution. *The CF is*

$$z_c = \phi(y - x) + \psi(y - 2x) + \theta(y + 3x).$$

The PI is

$$z_p = \frac{1}{D^3 - 7DD'^2 - 6D'^3} \sin(x + 2y) + \frac{1}{D^3 - 7DD'^2 - 6D'^3} e^{3x+y} = z_1 + z_2$$

where

$$z_1 = \frac{1}{D^3 - 7DD'^2 - 6D'^3} \sin(x + 2y) = \frac{1}{D + D'} \left[\frac{1}{D^2 - DD' - 6D'^2} \sin(x + 2y) \right]$$

$$= \frac{1}{D + D'} \left[\frac{1}{-1 + 2 + 24} \sin(x + 2y) \right] = \frac{1}{25} \cdot \frac{D - D'}{D^2 - D'^2} \sin(x + 2y)$$

$$= \frac{1}{25 \cdot 3} (D - D') \sin(x + 2y) = -\frac{1}{75} \cos(x + 2y)$$

and

$$z_2 = \frac{1}{D^3 - 7DD'^2 - 6D'^3} e^{3x+y} = \frac{1}{D - 3D'} \left[\frac{1}{D^2 + 3DD' + 2D'^2} e^{3x+y} \right]$$

$$= \frac{1}{D - 3D'} \left[\frac{1}{20} e^{3x+y} \right] = \frac{1}{20} \frac{1}{1!} \frac{x^1}{1^1} e^{3x+y}$$

$$= \frac{1}{20} x e^{3x+y}.$$

Thus, we obtain

$$z_p = -\frac{1}{75} \cos(x + 2y) + \frac{1}{20} x e^{3x+y}$$

and hence the general solution of the given equation is

$$z = z_c + z_p = \phi(y - x) + \psi(y - 2x) + \theta(y + 3x) - \frac{1}{75} \cos(x + 2y) + \frac{1}{20} x e^{3x+y}.$$

6.2.4 Method of Undetermined Coefficients

If the known function $f(x, y)$ involved in Eq. (6.30), possesses a finite number of essential different derivatives, then the PI can be easily obtained by the *method of undetermined coefficients*. The basic idea behind this method is to approximate the required PI using undetermined coefficients, which are later determined. The advantage of the method of undetermined coefficients is that it does not require any integration and uses the only algebraic method of solutions.

Given a function $f(x, y)$, the set of $f(x, y)$ and all its linearly independent derivatives, often denoted by Der$[f(x, y)]$, is called the *derivative family* of $f(x, y)$. In order to apply the method of undetermined coefficients, $f(x, y)$ must be either one of the exponential function, the polynomial, the sine/cosine or sum/products of such functions only. The derivative family Der$[f(x, y)]$ thus remains a finite set. If $f(x, y)$ is not included in any term of the CF z_c, then the PI z_p can be undertaken as a linear combination of members of Der$[f(x, y)]$.

If a term of z_c includes $f(x, y)$, then we multiply each member of Der$[f(x, y)]$ by the least power of x or y (the choice of x and y depends on whether terms of z_c corresponding to repeated factors are arranged in powers of x and y) such that the new family thus obtained does not contain any term of z_c. We denote this modified family by $\widetilde{\text{Der}}[f(x, y)]$. In this case, z_p will be a linear combination of members of $\widetilde{\text{Der}}[f(x, y)]$.

This method will be demonstrated by the following examples.

Example 6.47. *Using the method of undetermined coefficients, solve*

$$(D^2 + 5DD' + 5D'^2)z = x\sin(3x - 2y).$$

Solution.   The CF is

$$z_c = \phi\left[y + \frac{1}{2}(-5 + \sqrt{5})x\right] + \psi\left[y + \frac{1}{2}(-5 - \sqrt{5})x\right].$$

The derivative family is

$$\text{Der}[x\sin(3x - 2y)] = \left\{x\sin(3x - 2y), x\cos(3x - 2y), \sin(3x - 2y), \cos(3x - 2y)\right\}.$$

The PI z_p will be the linear combination of members of this family, that is,

$$z_p = Ax\sin(3x - 2y) + Bx\cos(3x - 2y) + C\sin(3x - 2y) + E\cos(3x - 2y).$$

Thus, we have

$$D^2z_p = (6A - 9E)\cos(3x - 2y) - (6B + 9C)\sin(3x - 2y) - 9Ax\sin(3x - 2y)$$
$$-9Bx\cos(3x - 2y),$$

$$DD'z_p = (-2A + 6E)\cos(3x - 2y) + (2B + 6C)\sin(3x - 2y) + 6Ax\sin(3x - 2y)$$
$$+6Bx\cos(3x - 2y),$$

$$D'^2z_p = -4E\cos(3x - 2y) - 4C\sin(3x - 2y) - 4Ax\sin(3x - 2y)$$
$$-4Bx\cos(3x - 2y).$$

As z_p satisfies the given PDE, we have

$$Bx\cos(3x - 2y) + Ax\sin(3x - 2y) + (C + 4B)\sin(3x - 2y)$$
$$+(E - 4A)\cos(3x - 2y) = x\sin(3x - 2y).$$

Comparing the coefficients on both the sides, we get

$$A = 1, \; B = C = 0, \; E = 4.$$

Therefore, the PI is

$$z_p = x\sin(3x - 2y) + 4\cos(3x - 2y).$$

Finally, the general solution is

$$z = z_c + z_p = \phi\left[y + \frac{1}{2}(-5 + \sqrt{5})x\right] + \psi\left[y + \frac{1}{2}(-5 - \sqrt{5})x\right] + x\sin(3x - 2y) + 4\cos(3x - 2y).$$

Example 6.48. *Apply the method of undetermined coefficients, to find the general solution of the following equation*

$$(D^3 + D^2D' - DD'^2 - D'^3)z = e^y \cos 2x.$$

Solution. *The CF is*

$$z_c = \phi(y + x) + \psi(y - x) + x\theta(y - x).$$

The derivative family is

$$\text{Der}[e^y \cos 2x] = \left\{ e^y \cos 2x, e^y \sin 2x \right\}.$$

The PI z_p will be the linear combination of members of this family, that is,

$$z_p = Ae^y \cos 2x + Be^y \sin 2x.$$

Thus, we have

$$D^3 z_p = 8Ae^y \sin 2x - 8Be^y \cos 2x,$$
$$D^2 D' z_p = -4Ae^y \cos 2x - 4Be^y \sin 2x,$$
$$DD'^2 z_p = -2Ae^y \sin 2x + 2Be^y \cos 2x,$$
$$D'^3 z_p = Ae^y \cos 2x + Be^y \cos 2x.$$

As z_p satisfies the given PDE, we have

$$(10A - 5B)e^y \sin 2x + (-5A - 10B)e^y \cos 2x = e^y \cos 2x.$$

Comparing the coefficients on both the sides and simplifying, we get

$$A = -\frac{1}{25}, \ B = -\frac{2}{25}.$$

Therefore, the PI is

$$z_p = -\frac{1}{25} e^y \cos 2x - \frac{2}{25} e^y \sin 2x.$$

Finally, the general solution is

$$z = z_c + z_p = \phi(y + x) + \psi(y - x) + x\theta(y - x) - \frac{1}{25} e^y(\cos 2x + 2\sin 2x).$$

Example 6.49. *Solve $(D^2 - 2DD' + D'^2)z = xe^{x+y}$.*

Solution. *The CF is*

$$z_c = \phi(y + x) + x\psi(y + x).$$

The derivative family is

$$\text{Der}[xe^{x+y}] = \left\{ xe^{x+y}, e^{x+y} \right\}.$$

Both members of the above family are particular cases of the terms of z_c. Therefore, multiplying each member by x^2, we get the modified family given by

$$\widetilde{\text{Der}}[xe^{x+y}] = \left\{ x^3 e^{x+y}, x^2 e^{x+y} \right\}.$$

Therefore, the PI is

$$z_p = (Ax^3 + Bx^2)e^{x+y}.$$

Now, we have

$$D^2 z_p = z_p + (6Ax^2 + 6Ax + 4Bx + 2B)e^{x+y}$$
$$DD' z_p = z_p + (3Ax^2 + 2Bx)e^{x+y}$$
$$D'^2 z_p = z_p.$$

Substituting these values in the given PDE and simplifying, we get

$$(6Ax + 2Bx + 2B)e^{x+y} = xe^{x+y}.$$

Comparing coefficients, we get

$$A = \frac{1}{6}, \ B = 0.$$

Hence, the general solution is

$$z = z_c + z_p = \phi(y+x) + x\psi(y+x) + \frac{1}{6}x^3 e^{x+y}.$$

Here, it can be highlighted that the problems finding PI of $F(D, D')z = e^{ax+by}$ for which $F(a, b)$ vanishes, can also be solved alternately by the method of undetermined coefficients (see problems 100–103 in **Exercises**).

6.3 Cauchy–Euler Linear PDEs

A Cauchy–Euler linear PDE is an equation with variable coefficients of the form

$$\left[(A_0 x^n D^n + A_1 x^{n-1} y D^{n-1} D' + \cdots + A_n y^n D'^n) + (B_0 x^{n-1} D^{n-1} + B_1 x^{n-2} y D^{n-2} D' \right.$$
$$\left. + \cdots + B_{n-1} y^{n-1} D'^{n-1}) + \cdots + (PxD + QyD') + Z\right] z = f(x, y) \tag{6.37}$$

where the coefficients $A_0, A_1, \ldots, A_n, B_0, \ldots, P, Q, Z$ all are constants and $A_0, A_1, \ldots, A_n$ are not vanishing simultaneously.

The problem for finding the general solution of a Cauchy–Euler PDE is embodied in the following result.

Theorem 6.15. *Under the transformation $x = e^u$ and $y = e^v$, Eq. (6.37) reduces to a linear PDE with constant coefficients involving new independent variables u and v, wherein the differential operators $\overline{D} := \dfrac{\partial}{\partial u}$ and $\overline{D}' := \dfrac{\partial}{\partial v}$ involved in reduced PDE are determined by the following relations:*

$$xD = \overline{D}, \ yD' = \overline{D}'$$

$$x^2 D^2 = \overline{D}(\overline{D} - 1), \ y^2 D'^2 = \overline{D}'(\overline{D}' - 1), xyDD' = \overline{D}\,\overline{D}'$$

$$\vdots$$

$$x^n D^n = \overline{D}(\overline{D} - 1)(\overline{D} - 2) \cdots (\overline{D} - n + 1), \ y^n D'^n = \overline{D}'(\overline{D}' - 1)(\overline{D}' - 2) \cdots (\overline{D}' - n + 1), \cdots$$

In general, we have

$$x^r y^s D^r D'^s = \overline{D}(\overline{D} - 1) \cdots (\overline{D} - r + 1)\overline{D}'(\overline{D}' - 1) \cdots (\overline{D}' - s + 1), \quad r, s \geq 0, 0 < r + s \leq n.$$

Proof. Given that $x = e^u$ and $y = e^v$ so that $u = \log x$ and $v = \log y$.

$$\frac{\partial z}{\partial x} = \frac{\partial z}{\partial u}\frac{\partial u}{\partial x} = \frac{1}{x}\frac{\partial z}{\partial u}$$

or

$$x\frac{\partial z}{\partial x} = \frac{\partial z}{\partial u}$$

yielding thereby

$$xD = \overline{D}.$$

Now, we have

$$x\frac{\partial}{\partial x}\left[x^{r-1}\frac{\partial^{r-1} z}{\partial x^{r-1}}\right] = x^r\frac{\partial^r z}{\partial x^r} + (r - 1)x^{r-1}\frac{\partial^{r-1} z}{\partial x^{r-1}}$$

or

$$x^r\frac{\partial^r z}{\partial x^r} = \left[x\frac{\partial}{\partial x} - r + 1\right]x^{r-1}\frac{\partial^{r-1} z}{\partial x^{r-1}}$$

implying thereby

$$x^r D^r = (xD - r + 1)x^{r-1}D^{r-1}.$$

Putting $r = 2, 3, \ldots$, we get

$$x^2 D^2 = (xD - 1)xD = \overline{D}(\overline{D} - 1)$$
$$x^3 D^3 = (xD - 2)x^2 D^2 = \overline{D}(\overline{D} - 1)(\overline{D} - 2)$$
$$\vdots$$
$$x^n D^n = \overline{D}(\overline{D} - 1)(\overline{D} - 2) \cdots (\overline{D} - n + 1).$$

Similarly, we can obtain

$$yD' = \overline{D}'$$
$$y^2 D'^2 = \overline{D}'(\overline{D}' - 1)$$
$$x^3 D'^3 = = \overline{D}'(\overline{D}' - 1)(\overline{D}' - 2)$$
$$\vdots$$
$$y^n D'^n = \overline{D}'(\overline{D}' - 1)(\overline{D}' - 2) \cdots (\overline{D}' - n + 1).$$

Notice that the variables x and y are independent and also the variables u and v are independent. Therefore, by using algebra of operators, we have

$$x^r y^s D^r D'^s = \overline{D}(\overline{D} - 1) \cdots (\overline{D} - r + 1)\overline{D}'(\overline{D}' - 1) \cdots (\overline{D}' - s + 1), \quad r, s \geq 0, 0 < r + s \leq n.$$

On substitution these relations, Eq. (6.37) takes the form

$$G(\overline{D}, \overline{D}')z = f(e^u, e^v) \equiv g(u, v) \tag{6.38}$$

which is a linear PDE with constant coefficients in new variables u and v. This completes the proof. $\qquad\square$

By using the method discussed earlier, we can obtain a solution of (6.38) of the form $z = \phi(u, v)$, which by backward substitution: $u = \log x$, $v = \log y$ gives the solution of Eq. (6.37) of the form $z = \psi(x, y)$. For illustration, we adopt some examples of Cauchy–Euler PDEs.

Example 6.50. *Solve* $x^2 \dfrac{\partial^2 z}{\partial x^2} + 2xy \dfrac{\partial^2 z}{\partial x \partial y} + y^2 \dfrac{\partial^2 z}{\partial y^2} = x^2 y^2.$

Solution. *Rewrite the given PDE in symbolic form*

$$(x^2 D^2 + 2xyDD' + y^2 D'^2) = x^2 y^2.$$

Put $x = e^u$ *and* $y = e^v$ *so that* $u = \log x$ *and* $v = \log y$ *and denote* $\overline{D} := \dfrac{\partial}{\partial u}$ *and* $\overline{D}' := \dfrac{\partial}{\partial v}$. *Under this transformation, the given equation reduces to*

$$\left[\overline{D}(\overline{D} - 1) + 2\overline{DD}' + \overline{D}'(\overline{D}' - 1)\right]z = e^{2u + 2v}.$$

On simplifying, the above equation becomes

$$(\overline{D} + \overline{D}')(\overline{D} + \overline{D}' - 1)z = e^{2u+2v}.$$

Hence, CF is

$$z_c = \phi^*(v - u) + e^u \psi^*(v - u)$$

$$= \phi^*(\log y - \log x) + x\psi^*(\log y - \log x) = \phi^* \left(\log \frac{y}{x} \right) + x\psi^* \left(\log \frac{y}{x} \right)$$

$$= \phi \left(\frac{y}{x} \right) + x\psi \left(\frac{y}{x} \right).$$

Also, PI is

$$z_p = \frac{1}{(\overline{D} + \overline{D}')(\overline{D} + \overline{D}' - 1)} e^{2u+2v}$$

$$= \frac{1}{(2+2)(2+2-1)} e^{2u+2v} = \frac{1}{12} e^{2u+2v}$$

$$= \frac{1}{12} x^2 y^2.$$

Therefore, the general solution of the given PDE is

$$z = z_c + z_p = \phi \left(\frac{y}{x} \right) + x\psi \left(\frac{y}{x} \right) + \frac{1}{12} x^2 y^2.$$

Example 6.51. *Solve $x^2 r - y^2 t + xp - yq = \log x$.*

Solution. *The given equation can be written symbolically as*

$$(x^2 D^2 - y^2 D'^2 + xD - yD')z = \log x.$$

Put $x = e^u$ and $y = e^v$ so that $u = \log x$ and $v = \log y$ and denote $\overline{D} := \dfrac{\partial}{\partial u}$ and $\overline{D}' := \dfrac{\partial}{\partial v}$. Under this transformation, the given equation reduces to

$$\left[\overline{D}(\overline{D} - 1) - \overline{D}'(\overline{D}' - 1) + \overline{D} - \overline{D}' \right] z = u$$

or

$$(\overline{D}^2 - \overline{D}'^2)z = u$$

which is an Euler linear PDE. Hence, its CF is

$$z_c = \phi^*(v + u) + \psi^*(v - u)$$

$$= \phi^*(\log y + \log x) + x\psi^*(\log y - \log x) = \phi^*(\log xy) + x\psi^* \left(\log \frac{y}{x} \right)$$

$$= \phi(xy) + x\psi \left(\frac{y}{x} \right).$$

Also, PI is

$$z_p = \frac{1}{\overline{D}^2 - \overline{D}'^2} u = \frac{1}{\overline{D}^2}\left[1 - \frac{\overline{D}'^2}{\overline{D}^2}\right]^{-1} u$$

$$= \frac{1}{\overline{D}^2}\left[1 + \frac{\overline{D}'^2}{\overline{D}^2} + \cdots\right] u = \frac{1}{\overline{D}^2} u$$

$$= \frac{1}{6} u^3 = \frac{1}{6}(\log x)^3.$$

Therefore, the general solution of the given equation is

$$z = z_c + z_p = \phi(xy) + x\psi\left(\frac{y}{x}\right) + \frac{1}{6}(\log x)^3.$$

6.4 Legendre Linear PDEs

A Legendre linear PDE is an equation with variable coefficients of the form

$$\left[\left\{A_0(ax+b)^n D^n + A_1(ax+b)^{n-1}(cy+d)D^{n-1}D' + \cdots + A_n(cy+d)^n D'^n\right\}\right.$$
$$+ \left\{B_0(ax+b)^{n-1}D^{n-1} + \cdots + B_{n-1}(cy+d)^{n-1}D'^{n-1}\right\} + \cdots \qquad (6.39)$$
$$\left. + \left\{P(ax+b)D + Q(cy+d)D'\right\} + Z\right] z = f(x,y)$$

where the coefficients $A_0, A_1, \ldots, A_n, B_0, \ldots, P, Q, Z$ all are constants and $A_0, A_1, \ldots, A_n$ are not vanishing simultaneously. Particularly, for $a = c = 1$ and $b = d = 0$, Legendre equation reduces to Cauchy–Euler equation.

Proceeding on the lines of the proof of Theorem 6.15, one can prove the following result, which provides the working rule for solving a Legendre linear PDE.

Theorem 6.16. *Under the transformation $ax + b = e^u$ and $cy + d = e^v$ so that $u = \log(ax + b)$ and $v = \log(cx + d)$, Eq. (6.39) reduces to a linear PDE with constant coefficients involving new independent variables u and v, wherein the differential operators $\overline{D} := \dfrac{\partial}{\partial u}$ and $\overline{D}' := \dfrac{\partial}{\partial v}$ involved in reduced PDE are determined by the following relation:*

$$(ax+b)^r(cy+d)^s D^r D'^s = a^r c^s \overline{D}(\overline{D}-1)\cdots(\overline{D}-r+1)\overline{D}'(\overline{D}'-1)\cdots(\overline{D}'-s+1) \qquad (6.40)$$

where $r, s \geq 0,\ 0 < r + s \leq n$.

The proof is left for readers as an exercise.

If we take $s = 0$ and $r = 1, 2, \ldots, n$ in (6.40), then we obtain the following relations:

$$(ax + b)D = a\overline{D}$$
$$(ax + b)^2 D^2 = a^2 \overline{D}(\overline{D} - 1)$$
$$\vdots$$
$$(ax + b)^n D^n = a^n \overline{D}(\overline{D} - 1)(\overline{D} - 2) \cdots (\overline{D} - n + 1).$$

Similarly, by taking $s = 0$ and $r = 1, 2, \ldots, n$ in (6.40), we get the following relations:

$$(cy + d)D' = c\overline{D}'$$
$$(cy + d)^2 D'^2 = c^2 \overline{D}'(\overline{D}' - 1)$$
$$\vdots$$
$$(cy + d)^n D'^n = c^n \overline{D}'(\overline{D}' - 1)(\overline{D}' - 2) \cdots (\overline{D}' - n + 1).$$

Furthermore, from Eq. (6.40), we can obtain similar relations for mixed derivatives.

For illustration, we adopt the following example of Legendre linear PDE.

Example 6.52. *Solve* $[(x - 1)^2 D^2 - (y + 2)^2 D'^2]z = xy + 2.$

Solution. *Put* $x - 1 = e^u$ *and* $y + 2 = e^v$ *so that* $u = \log(x - 1)$ *and* $v = \log(y + 2)$ *and denote* $\overline{D} :=$ $\dfrac{\partial}{\partial u}$ *and* $\overline{D}' := \dfrac{\partial}{\partial v}$. *Under this transformation, the given equation reduces to*

$$\left[1^2 \cdot \overline{D}(\overline{D} - 1) - 1^2 \cdot \overline{D}'(\overline{D}' - 1) \right] z = (e^u + 1)(e^v - 2) + 2$$

which becomes

$$(\overline{D} - \overline{D}')(\overline{D} + \overline{D}' - 1)z = e^{u+v} + e^v - 2e^u.$$

The CF is

$$z_c = \phi^*(v + u) + e^u \psi^*(v - u)$$

$$= \phi^*(\log(x - 1)(y + 2)) + (x - 1)\psi^* \left(\log \frac{y + 2}{x - 1} \right)$$

$$= \phi((x - 1)(y + 2)) + (x - 1)\psi \left(\frac{y + 2}{x - 1} \right).$$

The PI is

$$z_p = \frac{1}{(\overline{D} - \overline{D}')(\overline{D} + \overline{D}' - 1)}(e^{u+v} + e^v - 2e^u)$$

$$= \frac{1}{(\overline{D} - \overline{D}')(1 + 1 - 1)}e^{u+v} + \frac{1}{(0-1)(\overline{D} + \overline{D}' - 1)}e^v - \frac{2}{(1-0)(\overline{D} + \overline{D}' - 1)}e^u$$

$$= \int e^{u+(a-u)}du - e^v \int e^{-v}e^v dv - 2e^u \int e^{-u}e^u du, \quad where \ a = v + u$$

$$= ue^{u+v} - ve^v - 2ue^u = (e^v - 2)e^u u - e^v v$$

$$= (xy - y)\log(x - 1) - (y + 2)\log(y + 2).$$

Therefore, the general solution is

$$z = z_c + z_p = \phi((x - 1)(y + 2)) + (x - 1)\psi\left(\frac{y + 2}{x - 1}\right) + (xy - y)\log(x - 1) - (y + 2)\log(y + 2).$$

6.5 Equations Reducible in Linear Canonical Forms with Constant Coefficients

In Chapter 5, we have discussed the method for finding the general solution of second-order semilinear and linear PDE by reduction of canonical form. Sometimes, a second-order linear PDE with variable coefficients reduces to a linear equation with constant coefficients. Consequently, one of the earlier discussed methods can be applied to solve this equation. Here, we present several examples of such PDEs.

Example 6.53. *Find the general solution of the PDE*

$$y^2 r - 2xys + x^2 t + xy^2 p + x^2 yq = 4x^2 y^2 e^{2y^2}.$$

Solution. Here,

$$R = y^2, \quad S = -2xy, \quad T = x^2.$$

Step I (Classification): The discriminant of the given equation is

$$\triangle = S^2 - 4RT = 4x^2 y^2 - 4x^2 y^2 = 0.$$

Therefore, the given equation is parabolic.

Step II (Characteristics): The characteristic equation is

$$\frac{dy}{dx} = \frac{S}{2R} = -\frac{x}{y}$$

or

$$xdx + ydy = 0.$$

On integrating, we get

$$x^2 + y^2 = c$$

which is the characteristic curve of the given PDE.

Step III (Canonical Form): The canonical coordinates are

$$\xi = x^2 + y^2 \quad \text{and} \quad \eta = x^2 - y^2.$$

Using these coordinates, the given equation reduces to

$$4z_{\eta\eta} + z_\xi = e^{\xi - \eta} \tag{6.41}$$

which is the canonical form of the given PDE.

Step IV (General Solution): Equation (6.41) is an irreducible linear PDE with constant coefficients. Denoting $D := \dfrac{\partial}{\partial \xi}$ and $D' := \dfrac{\partial}{\partial \eta}$, Eq. (6.41) can be written as

$$(4D'^2 + D)z = e^{\xi - \eta}.$$

Hence, CF is

$$z_c = \sum_i A_i e^{-4b_i^2 \xi + b_i \eta} = \sum_i A_i \exp\left[b_i\{(1 - 4b_i)x^2 - (1 + 4b_i)y^2\} \right].$$

Also, PI is

$$z_p = \frac{1}{4D'^2 + D} e^{\xi - \eta} = \frac{1}{5} e^{\xi - \eta} = \frac{1}{5} e^{2y^2}.$$

Therefore, the complete solution of the given equation is

$$z = z_c + z_p = \sum_i A_i \exp\left[b_i\{(1 - 4b_i)x^2 - (1 + 4b_i)y^2\} \right] + \frac{1}{5} \exp(2y^2).$$

Example 6.54. *Find the general solution of the equation*

$$y(x + y)(r - s) - xp - yq - z = 0$$

Solution. Here, we have

$$R = y(x + y), \quad S = -y(x + y), \quad T = 0.$$

Step I (Classification): The discriminant of the given equation is

$$\triangle = S^2 - 4RT = y^2(x + y)^2 > 0$$

which yields that the given PDE is hyperbolic.

Step II (Characteristics): The characteristic equation is

$$\frac{dy}{dx} = \frac{S \pm \sqrt{\triangle}}{2R} = \frac{-y(x+y) \pm y(x+y)}{2y(x+y)} = 0, -1$$

which becomes

$$dy = 0 \quad \text{and} \quad dx + dy = 0.$$

On integrating these equations, we get

$$y = c_1 \quad \text{and} \quad x + y = c_2$$

which are the characteristic curves of the given PDE.

Step III (Canonical Form): The canonical coordinates are

$$\xi = y \quad \text{and} \quad \eta = x + y.$$

Using these coordinates, the given equation reduces to

$$\xi \eta z_{\xi\eta} + \xi z_{\xi} + \eta z_{\eta} + z = 0 \tag{6.42}$$

which is the canonical form of the given PDE.

Step IV (General Solution): Equation (6.42) is a Cauchy–Euler homogeneous linear PDE. Put $\xi = e^u$ and $\eta = e^v$ so that $u = \log \xi$ and $v = \log \eta$ and denote $D := \dfrac{\partial}{\partial u}$ and $D' := \dfrac{\partial}{\partial v}$. Under this transformation, Eq. (6.42) reduces to

$$(D + 1)(D' + 1)z = 0.$$

Its solution is

$$z = e^{-u}\phi^*(v) + e^{-v}\psi^*(u) = \frac{1}{\xi}\phi^*(\log \eta) + \frac{1}{\eta}\psi^*(\log \xi) = \frac{1}{\xi}\phi(\eta) + \frac{1}{\eta}\psi(\xi).$$

Therefore, the general solution of the given PDE is

$$z = \frac{1}{y}\phi(x+y) + \frac{1}{(x+y)}\psi(y).$$

Example 6.55. *Find the general solution of the equation*

$$(1+x^2)r + (1+y^2)t + xp + yq = 0.$$

Solution. *Here,*

$$R = 1 + x^2, \quad S = 0, \quad T = 1 + y^2.$$

Step I (Classification): The discriminant of the given equation is

$$\triangle = S^2 - 4RT = -4(1+x^2)(1+y^2) < 0.$$

Hence, the given PDE is elliptic.

Step II (Characteristics): The characteristic equation is

$$\frac{dy}{dx} = \frac{S \pm \sqrt{\triangle}}{2R} = \pm i \frac{\sqrt{1+y^2}}{\sqrt{1+x^2}}$$

or

$$\frac{dx}{\sqrt{1+x^2}} \pm i \frac{dy}{\sqrt{1+y^2}} = 0.$$

On integrating these equations, we get

$$\log\left(x + \sqrt{1+x^2}\right) + i\log\left(y + \sqrt{1+y^2}\right) = c_1$$

and

$$\log\left(x + \sqrt{1+x^2}\right) - i\log\left(y + \sqrt{1+y^2}\right) = c_2$$

which are the characteristic curves of the given PDE.

Step III (Canonical Form): The canonical coordinates are

$$\xi = \log\left(x + \sqrt{1+x^2}\right) \quad \text{and} \quad \eta = \log\left(y + \sqrt{1+y^2}\right).$$

Using these coordinates, the given equation reduces to

$$z_{\xi\xi} + z_{\eta\eta} = 0 \tag{6.43}$$

which is the canonical form of the given PDE.

Step IV (General Solution): Equation (6.43) is an Euler homogeneous linear PDE with constant coefficients. Hence, its general solution is

$$z = \phi(\eta + i\xi) + \psi(\eta - i\xi).$$

Putting the values of ξ and η, we get

$$z = \phi \left[\log \left(y + \sqrt{1 + y^2} \right) + i \log \left(x + \sqrt{1 + x^2} \right) \right]$$
$$+ \psi \left[\log \left(y + \sqrt{1 + y^2} \right) - i \log \left(x + \sqrt{1 + x^2} \right) \right]$$

which is the general solution of the given PDE as desired.

▶ Exercises

Find the general solutions of the homogeneous linear PDEs described in problems 1–21.

1. $r - t = 0.$ Ans $z = \phi(y + x) + \psi(y - x).$

2. $r + t = 0.$ Ans $z = \phi(y + ix) + \psi(y - ix).$

3. $r - s - 6t = 0.$ Ans $z = \phi(y - 2x) + \psi(y + 3x).$

4. $(D^3 - 3D^2D' + 2DD'^2)z = 0.$ Ans $z = \phi(y) + \psi(y + x) + \theta(y + 2x).$

5. $(D^3 - 3D^2D' + 3DD'^2 - D'^3)z = 0.$ Ans
   $z = \phi(y + x) + x\psi(y + x) + x^2\theta(y + x).$

6. $(D^3 - D^2D' - 8DD'^2 + 12D'^3)z = 0.$ Ans $z = x\phi_1(y + 2x) + \phi_2(y + 2x) + \phi_3(y - 3x).$

7. $(D^3 + DD'^2 - 10D'^3)z = 0.$ Ans $z = \phi_1(y + 2x) + \phi_2(y - x + 2ix) + \phi_3(y - x - 2ix).$

8. $25r - 40s + 16t = 0.$ Ans $z = \phi(5y + 4x) + x\psi(5y + 4x).$

9. $(D^3 - 4D^2D' + 4DD'^2)z = 0.$ Ans
   $z = x\phi_1(y + 2x) + \phi_2(y + 2x) + \phi_3(y).$

10. $(D^3 - 3D^2D' + 3DD'^2 - D'^3)z = 0.$ Ans $z = x^2\phi_1(y + x) + x\phi_2(y + x) + \phi_3(y + x).$

11. $\dfrac{\partial^4 z}{\partial x^4} - \dfrac{\partial^4 z}{\partial y^4} = 0.$ Ans $z = \phi_1(y + x) + \phi_2(y - x) + \phi_3(y + ix) + \phi_4(y - ix).$

12. $r - s - 2t + p - 2q = 0.$ Ans $z = \phi(y + 2x) + e^{-x}\psi(y + x).$

13. $(D + D' - 1)(D + 2D' - 2)z = 0.$ Ans $z = e^x\phi(y - x) + e^{2x}\psi(y - 2x).$

14. $(D^2 - D'^2 + D - D')z = 0.$ Ans $z = \phi(y + x) + e^{-x}\psi(y - x).$

15. $(3D^2 + 7DD' + 2D'^2 + 7D + 4D' + 2)z = 0.$

Ans
$z = e^{-x/3}\phi(x - 3y) + e^{-2x}\psi(2x - y).$

16. $(2D^2 - DD' - D'^2 - D + D')z = 0.$

Ans $z = \phi(x + y) + e^{x/2}\psi(x - 2y).$

17. $(D^4 - 2D^2D'^2 + D'^4)z = 0.$

Ans $z = \phi(x + y) + x\psi(y + x) + \theta(y - x) + x\eta(y - x).$

18. $(2D + D' - 1)^2(D - 2D' + 2)^3z = 0.$

Ans $z = e^y[\phi(2y - x) + y\psi(2y - x) + \theta(y + 2x) + y\eta(y + 2x) + y^2\mu(y + 2x)].$

19. $(D - 2D' + 5)^2z = 0.$

Ans
$z = e^{-5x}\phi(y + 2x) + xe^{-5x}\psi(y + 2x).$

20. $(D^2 - a^2D'^2 + 2abD + 2abD')z = 0.$

Ans $z = \phi(y - ax) + e^{-2abx}\psi(y + ax).$

21. $(D^3 + DD'^2 - 10D'^3)z = 0.$

Ans $z = \phi(y + 2x) + \psi(y - x + 2ix) + \theta(y - x - 2ix).$

Find the exponential-type series solutions of the homogeneous linear PDEs described in problems 22–26.

22. $(D^2 + D + D')z = 0.$

Ans. $z = \sum_i A_i e^{a_i(x - y - a_i y)}.$

23. $(D^2 + DD' + D + D' + 1)z = 0.$

Ans. $z = \sum_i A_i e^{a_i x + \phi(a_i)y}$, where

$$\phi(a_i) = -\frac{a_i^2 + a_i + 1}{a_i + 1}.$$

24. $r - t = z,$

Ans $z = \sum_\alpha A_\alpha e^{x \sec\alpha + y \tan\alpha}.$

25. $\dfrac{\partial^2 z}{\partial x^2} = \dfrac{1}{\lambda}\dfrac{\partial z}{\partial y}.$

Ans $z = \sum_n A_n \cos(nx + \varepsilon_n)e^{-kn^2 y}.$

26. $(2D^4 + 3D^2D' + D'^2)z = 0.$

Ans.
$$z = \sum_i A_i e^{a_i(x - 2a_i y)} + \sum_i C_i e^{c_i(x - c_i y)}.$$

Find the general solutions of the homogeneous linear PDEs described in problems 27–31.

27. $(3D^2 - 2D'^2 + D - 1)z = 0.$

Ans $z = \sum_i A_i e^{a_i x + b_i y}$ where $3a_i^2 - 2b_i^2 + a_i - 1 = 0.$

28. $(D + 2D' - 3)(D^2 + D')z = 0.$

Ans $z = e^{3x}\phi(y - 2x) + \sum_i A_i e^{a_i x + a_i^2 y}.$

29. $(D - 2D')(D - 2D' + 1)(D - D'^2)z = 0.$

Ans $z = \phi(y + 2x) + e^{-x}\psi(y + 2x) + \sum_i A_i e^{b_i(b_i x + y)}.$

30. $(D - 2D' + 5)(D^2 + D' + 3)z = 0.$

Ans
$z = e^{-5x}\phi(y + 2x) + \sum_i A_i e^{a_i x - (a_i^2 + 3)y}.$

31. $(D - D' + 1)(D^2 + 2DD'^2 - 2D' + 3)z = 0.$

Ans $z = e^y\phi(x + y) + \sum_i A_i e^{a_i x + b_i y}$

where $a_i^2 + 2a_i b_i^2 - 2b_i + 3 = 0.$

Solve the non-homogeneous linear PDEs described in problems 32–84.

32. $(D-1)(D-D'+1)z = 1 + xy.$

Ans.
$z = e^x\phi(y) + e^{-x}\psi(y+x) - xy - x.$

33. $(D^2 - D'^2)z = x - y.$

Ans.
$z = \phi(y+x) + \psi(y-x) + \dfrac{1}{6}x^3 - \dfrac{1}{2}x^2 y.$

34. $(D + D' - 1)(D + 2D' - 3)z = 4 + 3x + 6y.$

Ans. $z = e^x\phi(y-x) + e^{2x}\psi(y-2x) + 6 + x + 2y.$

35. $(D^2 - 6DD' + 9D'^2)z = 12x^2 + 36xy.$

Ans. $z = \phi(y+3x) + x\psi(y+3x) + 10x^4 + 6x^3 y.$

36. $\dfrac{d^3 z}{dx^3} - \dfrac{d^3 z}{dy^3} = x^3 y^3.$

Ans
$z = \phi(y+x) + \psi(y+wx)$
$\quad + \theta(y+w^2 x) + \dfrac{x^6 y^3}{120} + \dfrac{x^9}{10080},$
where w is a cube root of unity.

37. $(D^3 - DD'^2 - D^2 - DD')z = \dfrac{x+2}{x^3}.$

Ans $z =$
$\phi(y) + \psi(y+2) + e^x\theta(y-x) + \log x.$

38. $(D^2 - D'^2)z = x^{-2}.$

Ans $z = \phi(y+x) + \psi(y-x) - \log x.$

39. $(D^2 D' - 2DD' + D'^3)z = \dfrac{1}{x^3}.$

Ans.
$z = \phi(x) + \psi(y+x) + x\theta(y+x) + \dfrac{y}{2x}.$

40. $(2D^2 - DD' - D'^2 + D - D')z = e^{2x+y}.$

Ans. $z =$
$\phi(x+y) + e^{-x/2}\psi(2y-x) - \dfrac{1}{8}e^{2x+y}.$

41. $(D^3 - 3DD' + D' + 4)z = e^{2x+y}.$

Ans. $z = \sum_i A_i e^{a_i x + b_i y} + \dfrac{1}{7}e^{2x+y}$ where $a_i^3 - 3a_i b_i + b_i + 4 = 0.$

42. $(D + D' - 1)(D + 2D' + 2)z = e^{3x+4y} + y(1 - 2x).$

Ans. $z = e^x\phi(y-x) + e^{-y}\psi(y-2x) + \dfrac{1}{78}e^{3x+4y} + xy + \dfrac{3}{2}.$

43. $(D^2 - D'^2)z = 1 + e^{x+2y}.$

Ans $z =$
$\phi(x+y) + \psi(y-x) + \dfrac{1}{2}x^2 - \dfrac{1}{3}e^{x+2y}.$

44. $2\dfrac{\partial^2 z}{\partial x^2} - 3\dfrac{\partial^2 z}{\partial x\partial y} + \dfrac{\partial^2 z}{\partial y^2} = \sin(x - 2y).$

Ans. $z =$
$\phi(x+y) + \psi(2y+x) - \dfrac{1}{12}\sin(x - 2y).$

45. $(D^2 - 2DD')z = \sin x \cos 2y.$

Ans. $z = \phi(y) + \psi(y+2x) + \dfrac{1}{6}\sin(x + 2y) - \dfrac{1}{10}\sin(x - 2y).$

46. $(D^2 - DD')z = \cos x \cos 2y.$

Ans. $z = \phi(y) + \psi(y+x) + \dfrac{1}{2}\cos(x + 2y) - \dfrac{1}{6}\cos(x - 2y).$

47. $(D^3 - 4D^2 D' + 4DD'^2)z = \cos(2x + 3y).$

Ans. $z = \phi(y) + \psi(y+2x) + x\psi(y + 2x) - \dfrac{1}{32}\sin(2x + 3y).$

48. $(D^2 + DD' + D' - 1)z = \sin(x + 2y)$.

Ans. $z = e^{-x}\phi(y) + e^x\psi(y - x) - \dfrac{1}{10}\cos(x + 2y) - \dfrac{1}{5}\sin(x + 2y)$.

49. $(D - D' - 1)(D - D' - 2)z = \sin(2x + 3y)$.

Ans. $z = e^x\phi(y + x) + e^{2x}\psi(y + x) + \dfrac{1}{10}\sin(2x + 3y) - \dfrac{3}{10}\cos(2x + 3y)$.

50. $(D^2 - DD' - 2D'^2)z = 16xe^{2y}$.

Ans.
$z = \phi(y + 2x) + \psi(y - x) + 2xye^{2y}$.

51. $(3D^2 - 2D'^2 + D - 1)z = 4e^{x+y}\cos(x + y)$.

Ans.
$z = \displaystyle\sum_i A_i e^{a_i x + b_i y} + \dfrac{4}{3}e^{x+y}\sin(x + y)$,

where $3a_i^2 - 2b_i^2 + a_i - 1 = 0$.

52. $(D^2 - D')z = xe^{\alpha x + \alpha^2 y}$.

Ans. $z =$
$\displaystyle\sum_i A_i e^{a_i x + a_i^2 y} + \left[\dfrac{x^2}{4\alpha} - \dfrac{x}{4\alpha^2}\right]e^{\alpha x + \alpha^2 y}$.

53. $(D - 3D' - 2)^3 z = 6e^{2x}\sin(3x + y)$.

Ans.
$z = e^{2x}\phi(y + 3x) + xe^{2x}\psi(y + 3x) + x^2 e^{2x}\theta(y + 3x) + x^3 e^{2x}\sin(3x + y)$.

54. $(2D^2 - D')z = 4e^{x+2y}$.

Ans. $z = \displaystyle\sum_i A_i e^{a_i x + 2a_i^2 y} + xe^{x+2y}$.

55. $(3D^2 - 2D'^2 + D - 1)z = 4e^{x+y}\cos(x + y)$.

Ans.
$z = \displaystyle\sum_i A_i e^{a_i x + b_i y} + \dfrac{4}{3}e^{x+y}\sin(x + y)$,

where $3a_i^2 - 2b_i^2 + a_i - 1 = 0$.

56. $(D^3 - 4D^2 D' + 4DD'^2)z = \cos(y + 2x)$.

Ans. $z = \phi(y) + \psi(y + 2x) + x\theta(y + 2x) + \dfrac{1}{4}x^2\sin(y + x)$.

57. $(2D^2 - 5DD' + 2D'^2)z = 5\sin(2x + y)$.

Ans. $z = \phi(2y + x) + \psi(y + 2x) - \dfrac{5}{3}x\cos(2x + y)$.

58. $(D^2 + 3DD' + 2D'^2) = e^{x-y}$.

Ans. $z = \phi(y - x) + \psi(y - 2x) - xe^{x-y}$.

59. $(4D^2 + 12DD' + 9D'^2)z = e^{3x-2y}$.

Ans. $z =$
$\phi(2y - 3x) + x\psi(2y - 3x) + \dfrac{1}{8}x^2 e^{3x-2y}$.

60. $(D^3 - 3DD'^2 + 2D'^3)z = \sqrt{x + 2y}$.

Ans. $z = \phi(y + x) + x\psi(y + x) + \theta(y - 2x) + \dfrac{8}{525}(x + 2y)^{7/2}$.

61. $(D^2 - 2DD' + D'^2)z = \tan(x + y)$.

Ans. $z =$
$\phi(x + y) + x\psi(x + y) + \dfrac{1}{2}\tan(x + y)$.

62. $r + s - 2t = \sqrt{2x + y}$.

Ans. $z =$
$\phi(y + x) + \psi(y - 2x) + \dfrac{1}{15}(2x + y)^{5/2}$.

63. $\dfrac{\partial^2 z}{\partial x^2} - 4\dfrac{\partial^2 z}{\partial x\partial y} + 3\dfrac{\partial^2 z}{\partial y^2} = (x + 3y)^{-1/2}$.

Ans. $z =$
$\phi(y + x) + \psi(y + 3x) + \dfrac{1}{12}(x + 3y)^{3/2}$.

64. $(D^2 - 4DD' + 3D'^2)z = \sqrt{x + 3y}$.

Ans. $z =$
$\phi(y + x) + \psi(y + 3x) + \dfrac{1}{60}(x + 3y)^{5/2}$.

65. $(D^2 - 6DD' + 5D'^2)z = e^x \sinh y + xy$.

Ans. $z = \phi(y + x) + \psi(y + 5x) -$
$\dfrac{1}{8}xe^{x+y} - \dfrac{1}{24}e^{x-y} + \dfrac{1}{6}x^3y + \dfrac{1}{4}x^4$.

66. $(D^2 + DD' - 2D'^2)z = 8\log(x + 5y)$.

Ans. $z = \phi(y + x) + \psi(y - 2x) - 4x^2 +$
$\dfrac{1}{22}(x + 5y)^2[2\log(x + 5y) - 1]$.

67. $r - t = \tan^3 x \tan y - \tan x \tan^3 y$.

Ans. $z =$
$\phi(y + x) + \psi(y - x) + \dfrac{1}{2}\tan x \, \tan y$.

68. $r - s - 2t = (2x^2 + xy - y^2)\sin xy - \cos xy$.

Ans. $z = \phi(y - x) + \psi(y + 2x) + \sin xy$.

69. $(D^2 - DD' + 2D'^2)z = (y - 1)e^x$.

Ans. $z = \phi(y + 2x) + \psi(y - x) + ye^x$.

70. $(D^2 + 2DD' + 6D'^2)z = 2\cos y - x\sin y$.

Ans. $z = \phi(y - x) + x\psi(y - x) + x\sin y$.

71. $(D + D')(D + D' - 2)z = \sin(x + 2y)$.

Ans. $z = \phi(y - x) + e^{2x}\psi(y - x) +$
$\dfrac{2}{39}\cos(x + 2y) - \dfrac{1}{13}\sin(x + 2y)$.

72. $(D^3 - DD'^2 - D^2 + DD')z = \dfrac{x + 2}{x^3}$.

Ans. $z =$
$\phi(y) + \psi(y + x) + e^x\theta(y - x) + \log x$.

73. $(D^2 + DD' + D' - 1)z = 4\sinh x$.

Ans.
$z = e^{-x}\phi(y) + e^x\psi(x - y) + xe^x + xe^{-x}$.

74. $(D^2 - 3DD' + 2D'^2 - D + 2D')z = (2 + 4x)e^{-y}$.

Ans. $z = \phi(y + 2x) + e^{-y}\psi(y + 2x) -$
$(y^2 + 2xy)e^{-y}$.

75. $(D^2 - DD' + D' - 1)z = \cos(x + 2y) + e^y$.

Ans. $z = e^x\phi(y) + e^{-x}\psi(y + x) +$
$\dfrac{1}{2}\sin(x + 2y) - xe^y$.

76. $(D^2 - DD' + D' - 1)z = e^x + e^{-x}$.

Ans. $z =$
$e^{-x}\phi(y) + e^x\psi(y - x) + \dfrac{1}{2}xe^x - \dfrac{1}{2}xe^{-x}$.

77. $x^2\dfrac{\partial^2 z}{\partial x^2} - 4xy\dfrac{\partial^2 z}{\partial x\partial y} + 4y^2\dfrac{\partial^2 z}{\partial y^2} + 6y\dfrac{\partial z}{\partial y} = x^3y^4$.

Ans. $z = \phi(yx^2) + x\psi(yx^2) + \dfrac{1}{30}x^3y^4$.

78. $x^2\dfrac{\partial^2 z}{\partial x^2} - y^2\dfrac{\partial^2 z}{\partial y^2} = xy$.

Ans. $z = \phi(xy) + x\psi\left(\dfrac{y}{x}\right) + xy\log x$.

79. $x^2r + 2y^2t + px - 3xys + 2qy = x + 2y$.

Ans. $z = \phi(xy) + \psi(x^2y) + x + y$.

80. $(x^2D^2 - 2xyDD' - 3y^2D'^2 + xD - 3yD')z = x^2y\cos(\log x^2)$.

Ans. $z = \phi(x^3y) + \psi\left(\dfrac{y}{x}\right) - \dfrac{1}{65}[7\cos(\log x^2) - 4\sin(\log x^2)]$.

81. $(x^2D^2 - xyDD' - 2y^2D'^2 + xD - 2yD')z = \log\left(\dfrac{y}{x}\right) - \dfrac{1}{2}$.

Ans. $z = \phi(x^2y) + \psi\left(\dfrac{y}{x}\right) + \dfrac{1}{2}(\log x)^2\log y + \dfrac{1}{2}\log x \log y$.

82. $(x - 2)^2\dfrac{\partial^2 z}{\partial x^2} - 2(x - 2)(2y + 1)\dfrac{\partial^2 z}{\partial x\partial y} + (2y + 1)^2\dfrac{\partial^2 z}{\partial y^2} + 3(2y + 1)\dfrac{\partial z}{\partial y} = 2xy + x - 4y - 2$.

Ans. $(x + 1)^2r - 3(x + 1)(y - 1)s + 2(y - 1)^2t + (x + 1)p + 2(y - 1)q = x + 2y - 1$

83. $(x + 1)^2r - 3(x + 1)(y - 1)s + 2(y - 1)^2t + (x + 1)p + 2(y - 1)q = x + 2y - 1$.

Ans. $z = \phi\left((x + 1)(y - 1)\right) + \psi\left((x + 1)^2(y - 1)\right) + x + y$.

84. $\left[(3x+4)^2 D^2 - 2(3x+4)(2y-1)DD' + (2y-1)^2 D'^2 + 3(3x+4)D + 2(2y-1)D' - 1\right]z =$ $\log\left[\dfrac{2y-1}{3x+4}\right].$

Ans. $z = \sqrt[3]{3x+4}\,\phi\left((3x+4)^2(2y-1)^3\right) + \dfrac{\psi\left((3x+4)^2(2y-1)^3\right)}{\sqrt[3]{3x+4}} + \log\left[\dfrac{3x+4}{2y-1}\right].$

85. Show that under the transformation $u = \dfrac{1}{2}x^2$ and $v = \dfrac{1}{2}y^2$, the PDE

$$\frac{1}{x^2}\frac{\partial^2 z}{\partial x^2} - \frac{1}{x^3}\frac{\partial z}{\partial x} = \frac{1}{y^2}\frac{\partial^2 z}{\partial y^2} - \frac{1}{y^3}\frac{\partial z}{\partial y}$$

reduces to a homogeneous linear PDE with constant coefficient and then solve it.

Ans. $z = \phi(x^2 + y^2) + \psi(y^2 - x^2).$

86. Find a real function $u = (x, y)$, which reduces to zero function when $y = 0$ and satisfy the differential equation $\dfrac{\partial^2 u}{\partial x^2} + \dfrac{\partial^2 u}{\partial y^2} = -4\pi(x^2 + y^2).$ Ans. $u(x, y) = -2\pi x^2 y^2.$

87. Find the surface satisfying $r - 4s + 4t = 0$ and passing through the lines $z = x = 0$ and $z - 1 = x - y = 0$. Ans. $z(y + 2x) = 3x.$

88. Find the surface satisfying $r = 6x + 2$ and touching $z = x^3 + y^3$ along its section by the plane $x + y + 1 = 0$. Ans. $z = x^3 + y^3 + (x + y + 1)^2.$

89. A surface is drawn satisfying $r + t = 0$ and touching the circle $x^2 + y^2 = 1$ along its section by $y = 0$. Obtain its equation in the form $x^2(x^2 + z^2 - 1) = y^2(x^2 + z^2)$. Ans. $2z = (2 - x)e^{-x}\cos y.$

90. Show that a surface passing through the circle

$$z = 0,\ x^2 + y^2 = 1$$

and satisfying the differential equation $s = 8xy$ is

$$z = (x^2 + y^2)^2 - 1.$$

91. Show that the surface $z = y(1 - 18x^2) + 2x^2 y^3$ satisfies the differential equation $\dfrac{\partial^2 z}{\partial y^2} = 12x^2 y$ and contain the lines $y = 0 = z$ and $y = 3 = z$.

92. Show the surface satisfying the PDE

$$r - 2s + t = 0$$

and conditions $bz = y^2$ when $x = 0$ and $az = x^2$ when $y = 0$, is

$$z = (x + y)\left(\frac{x}{a} + \frac{y}{b}\right).$$

93. Show that a surface satisfying $r - 2s + t = 6$ and touching the hyperbolic paraboloid $z = xy$ along its section by the plane $y = x$ is

$$z = x^2 - xy + y^2.$$

Using the method of undetermined coefficients, solve the PDEs described in problems 94–99.

94. $(D^3 + D^2D' - DD'^2 - D'^3)z = e^x \cos 2y$.

Ans. $z = \phi(y + x) + \psi(y - x) + x\theta(y - x) - \dfrac{1}{25}e^x(\cos 2y + 2\sin 2y)$.

95. $(D^2 + DD' - 6D'^2)z = y \cos x$.

Ans. $z = \phi(y + 2x) + \psi(y - 3x) - y\cos x + \sin x$.

96. $(D^2 + 2DD' + D'^2)z = 2\cos y - x\sin y$.

Ans. $z = \phi(y - x) + x\psi(y - x) + x\sin y$.

97. $(D^2 + DD' - 6D'^2)z = x^2 \sin(x + y)$.

Ans. $z = \phi(y + 2x) + \psi(y - 3x) + \dfrac{1}{4}\left(x^2 - \dfrac{13}{8}\right)\sin(x + y) - \dfrac{3}{8}x\cos(x + y)$.

98. $r + s - 6t = x^2 \cos(x + y)$.

Ans. $z = \phi(x + y) + \psi(y - 3x) + \dfrac{1}{4}\left(x^2 - \dfrac{13}{8}\right)\cos(x + y) + \dfrac{3}{2}x\sin(x + y)$.

99. $(3D^2 - 2D'^2 + D - 1)z = 4e^{x+y}\cos(x + y)$.

Ans.
$$z = \sum_i A_i e^{a_i x + b_i y} + \frac{4}{3}e^{x+y}\sin(x + y),$$
where $2b_i^2 = 2a_i^2 + a_i - 1$.

100. Use the method of undetermined coefficients to solve the PDE given in Example 6.33.

101. Use the method of undetermined coefficients to solve the PDE given in Example 6.39.

102. Use the method of undetermined coefficients to solve the PDE given in Example 6.40.

103. Use the method of undetermined coefficients to solve the PDE given in Example 6.46.

Reduce the linear PDEs described in problems 104–108 to canonical form and then solve them.

104. $r + y^2t = y$.

Ans. $z_{\xi\xi} + z_{\eta\eta} - z_\xi = e^\xi$; $z = \sum_i A_i \exp\left[a_i \log y \pm x\sqrt{a_i - a_i^2}\right] + y \log y$.

105. $r - 2s + t + p - q = e^x(2y - 3) - e^y$.

Ans. $z_{\eta\eta} - z_\eta = (2\eta - 3)e^{\xi - \eta} - e^\eta$; $z = \phi(x + y) + e^y \psi(x + y)$.

106. $yr + (x + y)s + xt = 0$.

Ans $\xi\eta z_{\xi\eta} + \eta z_\eta = 0$; $z = \dfrac{1}{y - x}\phi(y^2 - x^2) + \psi(y - x)$.

107. $x^2r - y^2t = 0$.

Ans. $2\xi\eta z_{\xi\eta} - \eta z_\eta = 0$; $z = \phi(xy) + x\psi\left(\dfrac{x}{y}\right)$.

108. $x^2r - 2xys + y^2t - xp + 3yq = \dfrac{8y}{x}$.

Ans. $\eta^2 z_{\eta\eta} - \eta z_\eta = \dfrac{8\xi}{\eta^2}$; $z = \phi(xy) + x^2\psi(xy) + \dfrac{y}{x}$.

Bibliography

[1] A. R. Forsyth, *A Treatise on Differential Equations*, 6th ed. (London: Macmillan & Co Ltd., 1929).

[2] H. T. Herbert Piaggio, *An Elementary Treatise on Differential Equations and Their Applications* (London: G. Bell & Sons Ltd., 1929).

[3] D. A. Murray, *Introductory Course in Differential Equations* (Hyderabad: Orient Longman, 1967).

[4] A. Ayres Jr., *Theory and Problems of Differential Equations* (McGraw-Hill Book Co, 1972).

[5] R. Bronson and G. B. Costa, *Differential Equations* (McGraw-Hill, Inc., 1973).

[6] S. L. Ross, *Differential Equations*, 3rd ed. (New York: John Wiley & Sons, 1984).

[7] P. W. Berg and J. L. McGregor, *Elementary Partial Differential Equations* (San Francisko: Holden-Day Inc., 1964).

[8] H. F. Weinberger, *A First Course in Partial Differential Equations* (New York: Blaisdell Publishing Co., 1965).

[9] R. Dennemeyer, *Introduction to Partial Differential Equations and Boundary Value Problems* (New York: McGraw-Hill, 1968).

[10] M. M. Smirnov, In S. Chomet, ed., and translated by Scripta Technica Ltd., *Second-order Partial Differential Equations* (Netherlands: Noordhoff, 1966).

[11] E. T. Coption, *Partial Differential Equations* (Cambridge University Press, 1975).

[12] F. Treves, *Basic Linear Partial Differential Equations* (London: Academic Press, Inc., 1975).

[13] E. C. Zachmanoglou and D. W. Thoe, *Introduction to Partial Differential Equations with Applications* (New York: Dover Publications, Inc., 1976).

[14] P. Prasad and R. Ravindran, *Partial Differential Equations* (New York: John Wiley & Sons, 1984).

[15] I. N. Sneddon, *Elements of Partial Differential Equations* (Singapore: McGraw Hill, 1988).

[16] F. John, *Partial Differential Equations,* 4th ed. (New York: Springer-Verlag, 1991).

[17] S. J. Farlow, *Partial Differential Equations for Scientists and Engineers* (New York: Dover Publications, Inc., 1993).

[18] I. P. Stavroulakis and S. A. Tersian, *Partial Differential Equations: An Introduction with Mathematica and Maple*, 2nd ed. (Singapore: World Scientific Publishing Co. Re. Ltd., 2004).

[19] T. Amaranath, *An Elementary Course in Partial Differential Equations*, 2nd ed. (New Delhi: Narosa Publishing House, 2003).

[20] Y. Pinchover, J. Rubinstein, *An Introduction to Partial Differential Equations* (Cambridge University Press, 2005).

[21] T. Myint-U and L. Debnath, *Linear Partial Differential Equation for Scientists and Engineers*, 4th ed. (Springer, Indian reprint, 2006).

[22] L. C. Evans, *Partial Differential Equations*, 2nd ed. (Providence, Rhode Island: American Mathematical Society, 2010).

[23] G. L. Velazquez, *Partial Differential Equations of First Order and Their Applications to Physics*, 2nd ed. (Singapore: World Scientific Publishing Co., 2012).

[24] M. P. Coleman, *An Introduction to Partial Differential Equations with MATLAB*, 2nd ed. (CRC Press, 2013).

[25] P. V. O'Neil, *Beginning Partial Differential Equations*, 3rd ed. (Wiley-Interscience, 2014).

[26] M. Shearer and R. Levy, *Partial Differential Equations: An Introduction to Theory and Applications* (Princeton University Press, 2015).

[27] K. Shankara Rao, *Introduction to Partial Differential Equations*, 3rd ed. New Delhi: (Prentice Hall of India, 2016).

[28] D. V. Widder, *Advanced Calculus*, 2nd ed. (New York: Dover Publications Inc., 1989).

[29] R. J. T. Bell, *An Elementary Treatise on Coordinate Geometry* (London: MacMillon & Co Ltd, 1960).

[30] B. O'Neill, *Elementary Differential Geometry*, 2nd ed. (Elsevier, 2006).

[31] R. Churchill, *Fourier Series and Boundary Value Problems* (New York: McGraw-Hill, 1962).

[32] P. Allan and S. Zafrany, *Fourier Series and Integral Transforms* (Cambridge University Press, 1997).

[33] F. B. Hilderbrand, *Advanced Calculus for Applications* (Prentice-Hall, 1976).

[34] R. Courant and D. Hilbert, *Methods of Mathematical Physics*, vol. I & II (Willey-InterScience, 1962).

[35] G. Arfken, *Mathematical Methods of Physics*, 2nd ed. (New York: Academic Press, 1970).

[36] E. Kreyszig, *Advanced Engineering Mathematics* (New York: John Wiley & Sons, Inc., 1999).

[37] D. G. Zill and M. R. Cullen, *Advanced Engineering Mathematics*, 3rd ed. (Burlington: Jones & Bartlett Publishers, 2006).

[38] R. K. Jain and S. R. K. Iyenger, *Advanced Engineering Mathematics* (New Delhi: Narosa Publishing House, 2009).

Index

Printed in the United States
by Baker & Taylor Publisher Services